Brass Button Broadcasters

By Trent Christman

TURNER PUBLISHING COMPANY
Paducah, Kentucky

DEDICATION

*With Love and Affection
to
Pat
who waited and watched patiently for a
quarter of a century while my attention was
divided between her and AFRTS and who,
bless her, continued to provide unlimited
support during the many lonely hours she
spent alone while this book was being written.*

Colonel Tom Lewis, the man who organized and was the first Commander of AFRTS also was responsible for helping develop the vinyl transcription in order to supply programming to his stations scattered around the globe. He is seen here accepting the 1 millionth pressing back in 1943. Since then, untold millions more have been supplied to AFRTS stations in far-away places.

Turner Publishing Company
P.O. Box 3101
Paducah, KY 42002-3101
(502) 443-0121

Created and designed by
David A. Hurst
Publishing Consultant

Copyright © 1992
Trent Christman

This book or any part thereof may not be reproduced without the written consent of the Author and Publisher.

The materials were compiled using available and submitted materials; the publisher regrets they cannot assume liability for errors or omissions.

Library of Congress Catalog No. 91-67895

ISBN: 1-56311-086-5

Printed in the USA
LIMITED EDITION

Additional copies may be available directly from the publisher.

THE PROGRAM AND ORDER OF EVENTS

The BRASS BUTTON BROADCASTERS
The Armed Forces Radio and Television Service Story

Author's Warmup	Page 4
Chapter 1. **On the Air...Over There**	Page 7
Chapter 2. **In the Beginning**	Page 11
Chapter 3. **Getting Off the Ground...and On the Air**	Page 15
Chapter 4. **Meanwhile, Back at the Flagpole**	Page 21
Chapter 5. **Ready...Aim...Talk**	Page 27

STATION BREAK Number 1

Chapter 6. **Getting the Tunes to the Dunes**	Page 35
Chapter 7. **The European Vocation**	Page 39
Chapter 8. **Terrific in the Pacific**	Page 43
Chapter 9. **The Camp Followers**	Page 51
Chapter 10. **When Johnny Came Marching Home**	Page 59
Chapter 11. **While Johnny <u>Waited</u> to Come Marching Home**	Page 65

STATION BREAK Number 2

Chapter 12. **Here We Go Again**	Page 75
Chapter 13. **Well, Picture That! AFRTS Gets TV**	Page 81
Chapter 14. **Television Goes Overseas**	Page 87
Chapter 15. **Television Goes Even Farther Overseas**	Page 95
Chapter 16. **The Golden Days of Radio**	Page 105
Chapter 17. **Big Things from Small Packages**	Page 111

STATION BREAK Number 3

Chapter 18. **What Light? What Tunnel?**	Page 127
Chapter 19. **Extra-Censory Perceptions**	Page 135
Chapter 20. **Elsewhere in the News**	Page 143
Chapter 21. **Hooray for Hollywood!**	Page 151
Chapter 22. **Around the World in Many Ways**	Page 161
Chapter 23. **We're Beaming on a Slight Isthmus**	Page 171

STATION BREAK Number 4

Chapter 24. **Dropping in on Noriega**	Page 181
Chapter 25. **Saddam Shame, But It's Back to the Dunes**	Page 187

CLOSING CREDITS

Chronology **When it all happened 1942-1992**	Page 197

AUTHOR'S WARMUP

The author, Trent Christman

It is important to note that this volume is the <u>story</u> of the Armed Forces Radio and Television Service and is in no sense a dry-bones **history,** crammed into an academic dustbin filled with footnotes, endnotes and citations. Anyone looking for tiny numbers after every other sentence [1] or strings of *"ibids"* and *"cited hereafter ases"* had best look elsewhere. Although the facts are as accurate as fading documents and fading memories can make them, there will inevitably be those who say, "Hey, I was there and here is what actually happened..." In truth, there will never be a complete history of this unique organization. A full account of the activities of hundreds of stations and tens of thousands of former and present staff members over a fifty year period would require a shelf-full of volumes. The number of readers requiring that much information and willing to plow through that shelf would not fill a phone booth.

The purpose of this volume is to tell the story of AFRTS using the recollections of many hundreds of former and present military broadcasters who generously contributed their time, memories and anecdotes. Much of the "history" portion has been supplied by the American Forces Information Service (AFIS) and is based on a three-year research project contracted for by that organization. The result was a rough draft which seemed to prove the observation that a complete history was both unwieldy and unreadable. This draft was later rewritten and edited by Lieutenant Colonel Michael Lee Laning, Manager of Internal Information Plan, AFIS. The rewritten material was made available to the author of this volume and certain portions of it were used in greatly altered form to avoid replowing the same fallow ground.

To avoid inflicting readers with terminal boredom, it became apparent that the AFRTS story should be told from the point of view of the men and women who made it a daily companion to the millions of service personnel whom it served for some fifty years. Former AFRTS people were contacted through broadcast trade magazines, old address books, old and new personnel lists and membership lists of broadcast associations including the Armed Forces Broadcasters Association. The response was overwhelming. Their stories ranged from deadly serious through delightfully delirious to wildly hilarious. Almost without exception, the respondents noted that their assignment to one or more of the various stations or networks was the best time of their life. One noted that it was the best time he had ever had with his clothes on.

The author sincerely hopes the reader will feel the same way.

This book is a sincere salute to every man and woman who, during the past 50-plus years, has contributed to the group of broadcast stations and networks around the world which, along with their various Headquarters, form AFRTS, the Armed Forces Radio and Television Service.

Since its shakey beginnings in World War II, with tiny transmitters beaming out programs to troops listening in tanks, foxholes and barracks, AFRTS

[1] Like This.

has grown into a world-girdling giant, beaming news, entertainment, information, sports and the sounds of home from satellites or local studios to transmitters located wherever the troops are. And it has done so, in war and peace, since 1942.

Many present day broadcasters who are successful across the U.S. got their first taste of broadcasting with AFRTS while in the military service. Others, if they were experienced, were able to hone and maintain their communications skills while assigned to the organization. For some talented military broadcast personnel, AFRTS became almost a permanent assignment. Many spent, and are still spending, the majority of their military career in varied and ever-changing assignments; sometimes on tiny Pacific islands, sometimes in world capitals. Also contributing to the continuity and continuing organizational memory of AFRTS are the hundreds of dedicated civilian broadcasters, some of whom have spent a working lifetime with the organization.

Poll after poll has shown that the soldier, sailor, marine and airman considers AFRTS, along with mail from home, to be the most important single morale factor while serving overseas. They are the beneficiaries of the incredibly small group -- compared to normal commercial broadcasting groups -- which make up the staffs at the AFRTS headquarters and the broadcast outlets around the world.

They and the American taxpayer are indebted to America's commercial broadcast industry -- the production companies, the syndicators, the talent, the craft guilds and unions -- who make very special and extraordinary efforts to provide the results of their labors at near cost, or less, in order that the troops may enjoy the same programming as their contemporaries at home.

In recent years, AFRTS audiences have seen more than 90 percent of the top-rated U.S. television programs. The best estimate is that if standard industry rates had been paid to buy this programming, the taxpayer would have been handed a bill for $120-million each year. Instead, because normal profits and many other fees and costs are waived, the tab is a mere eight percent of that; barely enough to cover the supplier's administrative costs.

The Pentagon cliche is to "get the most bang for the buck." It is doubtful that anywhere in the military procurement system does the buck bang louder than it does at AFRTS.

Is the system perfect? No. Does it ever make mistakes or do silly things? You bet your bippy. Could it be improved? Sure.

How do I know?

Because I contributed to a number of the mistakes. I also added an improvement or two here and there. Or tried.

Please excuse a personal word of explanation; after this I'll keep myself out of the book as much as possible and let others, who contributed far more than I, do the talking.

My first contact with AFRS (the **T** for television didn't come along until the 50's) was as an NBC page boy, standing in the wings of an NBC studio in 1942 watching a radio program being recorded. On stage were the likes of Bob Hope, Bing Crosby, Judy Garland, Mel Blanc and a gaggle of other well known talents. Meredith Willson fronted a big orchestra and the script was hysterically funny. It turned out the show was called **Command Performance**, and it was being done for service personnel stationed overseas. I recall thinking it was too bad we poor civilians would never get to hear it.

A couple of months later, the Army solved that problem. It sent me overseas, still poor but no longer a civilian. Now I could hear **Command Performance** every week and as a kindness they allowed me to volunteer to help out at the AFRS station. First it was at a station in New Guinea. First task: climb a palm tree and attach an antenna wire. Then, on to a microscopic speck of coral called Biak. First task: climb a palm tree and attach an antenna wire. And some people think broadcasting isn't glamorous!

Then came real broadcasting work. I got to sweep the wooden floor of the tent housing the station. Then I got to write copy reminding troops to take their anti-malaria Atabrine pills. The copy must have worked because everyone on the island turned yellow, a side effect of Atabrine.

Later came a short stint in Manila where the studio moved three-quarters of the way indoors, thanks to one wall having been blown away by a bomb. Then a ride in a terrifyingly small boat to Japan and helping set up the Tokyo station there. So impressed were the bosses by my skill at turning people yellow, they assigned me to write copy telling the sex-starved occupation forces about the joys and delights of celibacy. It's difficult to judge the effectiveness of this campaign. Personal observations at the time lead me to believe it was overwhelmingly ineffective. At least nobody turned yellow, although a number did turn sullen.

Twenty years of civilian broadcasting followed, at NBC and a major advertising agency. While on a lengthy European vacation with my wife, the AFRTS network there offered me a job. Europe was a joyous experience in those days and, because they promised I wouldn't have to sell celibacy, a 21-year stay in Europe began, followed by four years at the AFRTS headquarters in Hollywood.

But this book will not be my story. It's the story of the hundreds of talented characters who made up, and make up, the AFRTS group around the world. They have been contacted by mail or phone and have been extremely generous in sharing their experiences, memorabilia and personal stories which form the core of this book. They all have my gratitude for their friendship and their help.

I'll intrude my personal experiences as little as possible. The triumphs are theirs. The inadvertent errors and the sometimes crochety opinions are mine. All mine.

Trent Christman
Las Vegas, Nevada
1990-1991

Buzz Rizer, (L), Director of the American Forces Information Service which manages AFRTS, discusses AFRTS plans and policies with Colonel Bruce Eaton, Director of the Army Broadcasting Service, (center). Their dinner partner is Mrs. Nancy Cole, wife of Colonel David Cole, former AFRTS Los Angeles commander. Colonel Eaton has held a number of prestigious positions with the AFRTS system including that of Deputy Commander, AFRTS Los Angeles, Commander of AFN Europe and Director of ABS. Highly decorated and well respected, he has only one slight ecentricity. It is so slight, in fact, it has never even been noted in his efficiency reports. He thinks he's a rabbit. How this began, no one knows, but today he is the frequent recipient of rabbit suits, rabbit statues, live rabbits and other bunny paraphernalia. Easter is his favorite holiday.

Right, Not easily recognizable without his ears and twitchy nose, Eaton is seen here in his role as an army colonel; at this time he is commander of AFN where his favorite shows were "Starsky and **Hutch**", "America's **Bunniest** Home Videos" and "Bugs **Bunny**" cartoons. (Photos courtesy Mary Carnes)

CHAPTER ONE

"On the Air...Over There"

To the uninitiated, military broadcasting is weird.

To the initiated, such as professional commercial broadcasters, it is totally weird.

Such a perception is easy to explain. Military Broadcasting *is* weird, for reasons which will be made clear as this story progresses.

It was called AFRS when it was formed in 1942. Now, with globe-girdling television added to radio it is AFRTS -- the Armed Forces Radio and Television System. It has evolved through the years into a broadcast entity unique in its mission, techniques and operation. In most ways it is totally different from its civilian counterparts.

Charles DeGaulle asked how anyone could possibly administer a country populated by a diverse people who made 465 varieties of cheese. Perhaps he should have looked into the AFRTS operation if he wanted to learn how to deal with complicated situations. Consider just a few of these differences.

Size for example. By AFRTS standards, even the largest commercial broadcast group or network has relatively few stations under direct control. AFRTS and the military services operate radio or television or both in 130 different countries, many of which have strict regulations which apply to AFRTS as well as to local broadcasters. And, depending on the size of the vessel, there is an AFRTS presence on every sizeable ship in the Navy, varying from a fairly simple playback unit to multi-channel radio-television with live capability on capital ships such as carriers.

Broadcast regulations are generally standard for U. S. broadcasters. AFRTS not only follows applicable Federal Communications Commission rules but the regulations of the country in which it operates. In the present world in which words are a weapon and in which many governments are highly sensitive to what words go out over their airwaves, many of the AFRTS stations are forced to operate under strictures unthinkable to their U.S. counterparts.

Communications are often difficult. Networks and groups in the States can phone or fax with relative ease. It becomes more difficult when an AFRTS executive or operator needs to quickly get in touch with someone in the Far East, twelve hours out of sync, or with a station in Antarctica where the telephone service leaves much to be desired. Nor is it always quick and easy to get word to a Navy ship underway or to a land-based station operating under wartime conditions.

The CEO of a domestic station or net need only please the stockholders, the FCC, the advertisers and the audience. AFRTS is alike only in that it must please its audience. The others are not a consideration. Congress is, though. It, through the military budget, controls the money faucet. The Department of Defense is another prime consideration. It determines not only how much of the money Congress dribbles out will go to buy transmitters, tubes, transistors and thumb tacks, it has the authority to shrink or increase the size of the broadcast staffs. It decides where stations can open, which will close, which will move and which should go on about their business. Every station is under the day-to-day control of one of the three services -- Army, Navy or Air Force -- which, like Portugal and Spain centuries ago, have divided the world between them. Each service has a broadcast headquarters and each can also add layers of control to the budgets, personnel levels and policy of stations and networks under its banner.

A commercial station or network can determine the number of personnel it requires to operate efficiently. An AFRTS station or network is told by a headquarters half a world away what the personnel level *should* be according to a complicated set of equations developed over a number of years. Depending on the number of available people in the total manpower pool, divided by the number needed by all the outlets, personnel are rationed out according to experience, availability, qualifications and, though seldom admitted, the aggressiveness with which the local operator screams, moans and cries. Although commercial broadcast personnel are traditionally mobile and constantly looking for better jobs in larger communities, military broadcasters make their commercial opposite numbers look positively immovable. Like all military personnel, the Brass Button Broadcasters all serve a standard-length tour of duty before being reassigned to another job in another area of the world. These tours can be a year in a so-called hardship or unaccompanied area, or three years or more in locations providing facilities for families. The net result is a constant turnover of personnel and the almost certain knowledge that management will have a staff which will be about one-third entry level broadcasters learning on the job, one third who are moderately well trained and one third experienced broadcasters, most of whom will have previously served in other AFRTS situations. Some of the larger operations also have a number of civilian

7

employees, generally at the managerial level, who provide continuity and corporate memory to the operation. This frequently helps ease the constant personnel turmoil. In any case, the manager must learn to live with the inevitable fact that he or she will more often than not be anywhere from ten to fifty percent understaffed and will be continually losing people due to the military policy requiring regular rotation of assignments.

The only solution is to work the staff longer and harder, a difficult solution made possible both by the absence of unions or guilds in the overseas operations and the presence of the military system of discipline. It should be noted here that the vast majority of those who contributed their experiences to this book were unanimous that, while they generally worked exceptionally long hours while with AFRTS, the ordeal was made more palatable by preparing them for future employment which, by comparison, was a piece of cake.

A significant difference between the grey flannel and khaki broadcasters is the procurement of equipment. If a station in Kansas City blows a thingamajig or burns out a whatchamacallit, the engineers can call their local supplier and get a new one within hours. When the same essential piece of equipment dies in Eastern Turkey or McMurdo Sound, Antarctica, chances are the nearest supplier is also in Kansas City. Unless the AFRTS engineers are ingenious enough to patch up the problem with camel dung or penguin droppings, chances are they are going to have to commit themselves to hours of paperwork justifying why a replacement part is needed; show that funds exist to purchase it; fill out endless forms required by the military procurement system; send a wireless message to their headquarters asking procurement action be started; get the paperwork into the sometimes exasperatingly slow postal system and begin answering listener and viewer complaints. Because all three services maintain a joint procurement and supply facility in Sacramento which buys, maintains and ships all the non-standard equipment used by AFRTS, chances are excellent that the part will be on its way back very quickly once the necessary documentation is received.

Going down to the corner thingamajig store is easier and quicker, though.

And then there is the matter of programming.

First of all, there's Radio

Local U. S. radio stations can take aim at a specific segment of their potential audience and program directly to and for it. They know the demographic makeup of their target -- young males, teens, 36 plus, upscale, downscale, heavy-metal fans, country-western fans, adult contemporary or nostalgia. In any metropolitan market, should an audience niche become unfilled, bet on a station to begin filling it. Not so with an AFRTS station. Invariably, it is the only English language station available, with the possible exception of shortwave propaganda stations which direct somewhat heavy-handed or thinly disguised propaganda to foreign listeners. The AFRTS station is the only place to find, in English, the local weather, local happenings, local news, local interviews, local movie schedules, local specials at the commissary, uncensored and unbiased news direct from the major networks. Listeners can also find a warm, fuzzy feeling that someone-out-there-likes-me. The AFRTS manager knows the audience, too. It's the area commander who wants to know how "his" station is serving "his" troops. It's the local public affairs officer who is concerned that the station might say something to embarrass the boss, the area commander. It's the local chaplain, or his wife, who is concerned that some of the music might be offensive and cause rioting, rapine and ribaldry. It is also the 45-year old Colonel's wife who wants to hear Mozart and the 18-year old Private's wife who want to hear AC-DC. It's the young soldier from Tennessee who wants Reba McIntyre and the guy in the adjoining barracks bay who digs Madonna. It's an audience, heavily laced with minorities; urban and country; white, black, brown and red; male and female, of all educational levels and economic backgrounds. They come from 50 different states and, understandably, they want *what they want, when they want it.*

What they get is an eclectic mix of a little of everything. AFRTS furnishes a number of radio packages for use at the local station level. Local disc jockeys are provided the latest hits in a weekly shipment of both vinyl and CD recordings. The local manager determines the music mix depending on his or her audience make up. Many stations have both AM and FM channels and are able to provide alternative services. AFRTS also sends out a weekly unit of complete 25-, 30- or 55-minute programs done by top rated U. S. disk jockeys which generally are designed to appeal to one type of music fan. Frequent shipments of library material on both tape and CD are also made for use as the station desires.

The stations receive newscasts by satellite from the Broadcast Center in Los Angeles which offers all major network news broadcasts on a continuous basis. Overseas listeners are generally given a five-minute newscast at the top of each hour and they hear it, uncensored except in the rarest of circumstances, at the same time as their friends at home. Several expanded newscasts are also normally scheduled by overseas stations. The same satellite brings a wealth of sports events to the predominately male audience. At posts and bases ashore and on ships at sea, Super Bowl, the NBA playoffs and the World Series -- plus all other major events from the Kentucky Derby to the Indy 500 -- keep the military awake and cheering for their favorites even though it might be 4 a.m. local time where they are.

Because the Broadcast Center has practically unlimited access to news sources, it is able to provide the overseas stations all major special events on both radio and television. Troops around the world see or hear not only a continuous menu of sports and news but receive extended coverage of everything from State of the Union speeches, major network news specials, bulletins concerning fast breaking stories, presidential inaugurations and such running stories as the Ollie North Show and the Watergate hearings. As they choose, the military person overseas can be just as well informed, or as dismally ignorant, as are the civilian neighbors back home. AFRTS offers both options, but provides somewhat heavier doses of news and information than the average radio station in the States.

If there is a doubt about this, consider that many of the troops participating in Desert Storm knew that the Patriot missiles were knocking down the incoming Scuds overhead. They were watching CNN live television coverage, or listening to news on radio, while tuned to the local AFRTS stations as they waited in their bunkers for the all-clear. And more than a few of them learned from AFRTS radio or television that they were scheduled to return home. You've got to love a station that tells you things like that.

For many years, Navy ships were dependent on shortwave service from AFRTS for news and sports. Today they receive an infinitely improved signal via Inmarsat, the International Maritime Satellites which cover the world. Although Inmarsat is primarily for navigation and communication between ship and shore, dedicated transponders on the satellite transmit quality audio 24 hours a day to ships everywhere in the world from the Broadcast Center in Los Angeles.

...And then there's Television

While radio is ubiquitous and the overseas listener can easily carry a receiver in his or her fatigue jacket pocket or, unless caught, can listen in tanks, planes or foxholes, this is still a television generation.

The AFRTS television operation is unique in the world, and it is immense. As of the current writing, there are 34 full-service land-based television outlets. They have the capability to originate local programming, field news crews in mobile vans, produce local entertainment shows, newscasts and weather shows, create sophisticated graphics and act very much like any major network affiliate. Three of them -- in Frankfurt, Germany, Vicenza, Italy and Incerlik, Turkey -- are classified as "Super Stations." Each creates a program schedule which is sent up to a satellite and back down to as many as 300 receive-only ground sites. Germany feeds American installations in that country as well as the rest of northern Europe. Italy feeds southern Europe, including Spain, and most of the middle-East. Turkey covers that country and is also seen in parts of the middle-East. Each has manned affiliates capable of providing live or videotaped reports for transmission to the entire network. Currently under consideration is a plan to consolidate Germany and Italy.

The Super-Station signals are encrypted to protect the copyrighted programming from piracy and in some cases to comply with host country regulations. By so doing, the signal is restricted to the American audience.

The other ground-based stations serve specific locations in Japan, Korea, Panama, Iceland, Diego Garcia, the Philippines, Kwajalein in the Marshall Islands, New Zealand, Australia, Honduras and elsewhere.

Programming for these, and other types of outlets, is supplied by the Broadcast Center in Los Angeles in a number of ways. Major stations and networks receive around-the-clock satellite service uplinked and encrypted at the Center. This consists of time-sensitive programming such as news, sports and specials with a time element (such as the Emmys, the Academy Awards or a Presidential address.) Piggy-backed on the satellite signal are a number of other channels including two stereo music channels, a mono music and voice channel and a communications channel for internal use such as last minute information on program changes.

Programs with no particular time sensitivities such as network and syndicated entertainment series are decommercialized and dubbed to videotape at the Los Angeles AFRTS Broadcast Center. They are then duplicated and sent by mail to 160 land based stations and 120 ships. These programs are circuited between groups of stations or ships. As one outlet completes running the weekly shipment, it sends it on to the next in line. Each weekly shipment is approximately 80 hours in length for stations with family members in their audience. Ships and areas without children in their audience get about 72 hours on the reasonable theory that these audiences aren't particularly enthusiastic about *Sesame Street* or other children's programming.

Although the entertainment programming is circuited among a group of stations, each individual station receives a weekly shipment of videotaped priority programming so it will be reasonably current when seen.

Getting programming to the land based stations does not present the problems that supplying programming to ships at sea does. Land stations tend to remain where put and their satellite dishes are easily aimed. The Navy, however, refuses to build ships that stay level in heavy seas or remain in one place on the earth's surface. Also, the Bureau of Ships has some sort of problem with finding a place on deck for a 35 foot diameter satellite receiving dish. Even if they find space, until they stop building ships that roll, it will be difficult to keep the dish aimed precisely at the satellite.

For this reason, AFRTS runs so-called DUPFACs or Duplication Facilities. They are located near major fleet areas in Christchurch, New Zealand, Sigonella, Sicily, Bahrain in the Persian Gulf, Rota, Spain, Diego Garcia Island in the Indian Ocean, Subic Bay in the Philippines, Roosevelt Roads, Puerto Rico and, currently, in Jedda, Saudi Arabia. They duplicate time-sensitive programming and get it out immediately to ships in their area, often by helicopter or to carriers by fixed-wing aircraft.

Finally, it must be mentioned that there are hundreds of so-called Mini-TV stations. These are essentially nothing more than a videocassette playback unit and a monitor. AFRTS supplies each of these about 54 hours of programming weekly. Each is located in an isolated area which cannot receive a satellite downlink. The units are often set up in a day-room or a mess hall and the audience consists of people spending a tour of duty at a communications site on a mountain top in Upper Slobovia, at a Marine guard detachment in the North Framistan embassy or a group of weather researchers on an arctic ice-flow. Some of these people are so isolated they still think *Leave it to Beaver* is number one in the ratings. They *need* AFRTS.

At any given time, there are in excess of 80,000 videotape cassettes arriving, in use or being returned around the world. This would make a stack more than a mile and a quarter high should anyone be stupid enough to try. Of more interest is the fact that AFRTS knows where each of them is at any time.

Later, what is recorded on to all those videotapes will be covered in more detail. Until then, just like on your local television station, this chapter ends with a promo: **AFRTS Television...bringing you 93 percent of the nation's top 75 shows. No other station or network can make that claim. ...coming up next, a story of pirates, privates and procrastination -- how AFRTS got to be what it is. Stay tuned.**

AFN Headquarters in the castle at Hoechst

CHAPTER TWO

In which the War Department slowly begins to realize that getting the message out by broadcasting it is faster than doing it by carrier pigeon. Cleaner, too.

IN THE BEGINNING...

...the War Department said, "Let there **NOT** be radio."

The troops said, "We want radio."

The War Department said, "Baseballs and doughnuts were good enough for the troops in the last war."

The troops said, "Bob Hope and Jack Benny and Dinah Shore and the music of the big bands and the latest news and sports scores are what we want." And they mumbled something about where the brass could put their lousy doughnuts.

The War Department said, "Let us think about this for awhile."

And the troops understood very well that "awhile" comes just moments before "never."

So, let us return now to what that announcer on the *Lone Ranger* used to call "those exciting days of yesteryear." It's now 1939, a time of a pitifully small American peacetime army. Adolf Hitler invaded Poland and World War II began. Franklin D. Roosevelt was president. America was slowly and tortuously recovering from the great depression. Although polls showed most of the U.S. population of 131-million was worried about being drawn into the rapidly escalating European conflict, the U.S. military was far from ready. The Air Corps consisted of 400 fighter planes. Twenty-one dollars a month was starting pay for Army privates. The Navy could adequately patrol only one ocean with its few over-aged ships and overseas garrisons were for the most part limited to detatchments in the Philippines, Panama and Hawaii.

The troops in Hawaii, where English was the primary language, had civilian radio stations available to them. Troops in the Philippines and Panama had to make do with stations broadcasting in Spanish or Tagalog which couldn't care less about those things of vital concern to American troops -- like, who won the World Series? (The Yankees did this year of 1939, for the fourth year in a row.) None of this did much to boost the morale of the poorly paid troops.

Too frequently, the troop's desire for radio was considered of minor importance by the top brass in Washington, most of whom had served in France during World War I, and who knew that the doughboys in the trenches got along pretty well without Eddie Cantor or Charlie McCarthy.

Active sports, they said. That's what the men need. That'll toughen 'em up. West Point football stars were routinely assigned to coach Army unit football teams and supervise Army sports programs. It toughened up the men, the commanders felt, while the officers got their exercise on the tennis courts or golf course.

Military recreational and morale activities were recognized by some officers as essential to the efficiency of a fully functional force. Others considered such fripperies unnecessary. There was no single military section responsible for the supervision of such activities.

During World War I, General John J. Pershing had noted the complete and total confusion caused by the mish-mash of private organizations, each offering help to the soldiers and sailors on active duty. Among them were the YMCA, YWCA, the National Catholic War Council, the Jewish Welfare Board, the Salvation Army, the American Library Association and the Red Cross. Pershing and his advisors couldn't help but note that in many cases their activities were completely uncoordinated and the groups were spending more time duplicating effort, and justifying their existence, than they were in providing for the welfare of the troops.

Pershing requested a full scale investigation of the problem and it was carried out by Raymond Fosdick of the Administrative Section of the General's staff. When the report was finally prepared, it recommended eliminating duplication of effort by competing organizations.

11

It expressed serious doubts about using private, often self-serving, organizations to provide leisure time services. The effort should be a function of the military, non-sectarian in nature, because morale is so important, said Fosdick.

Matters improved somewhat during the period between the wars. There was no shortage of baseballs and boxing gloves. What social services there were, were run most often as a sideline of the administration sections of individual military units. The relatively tiny peacetime forces were a world of their own, a world which was rapidly changing from that which Pershing knew.

Radio, for one thing, was now a major part of American life. By 1939, the young recruits were as addicted to radio as they are to television today. The twenty-one dollar a month rookie in the Philippines and Panama felt particularly cut off from home. And not being able to hear Bob Hope or Jack Benny didn't help.

This was soon to change, and the change was to fly in the face of Mr. Fosdick's cherished belief that the military should look after its own. Of course, Fosdick had never heard much radio when he wrote the report in 1919. But he had probably heard the popular phrase: "Time Marches On." Which it had.

One of the wonders on exhibit at the Golden Gate International Exposition, held on Treasure Island in San Francisco Bay in 1939, was a powerful, state-of-the-art shortwave transmitter demonstrated by the General Electric Company which had developed it. It's call letters were KGEI and its highly specialized antenna was designed to beam broadcasts either toward Asia or Latin America from a site in the appropriately named House of Magic. There was a problem, however.

Nobody had thought to put together a staff to provide programming for the station.

This rather obvious disregard for essentials was commented on by GE public relations man Buck Harris, a corporate type, who had perhaps never heard Rule Six in the *Military Handbook of Absolutely, Positively True Facts*: **"Keep your bowels open and your mouth shut."**

The state of his bowels remains forgotten, but it is known that he neglected to keep his mouth shut.

Within days he was the new manager of KGEI. His only instructions were to broadcast something, including news, to Asia four hours every day and to Latin America for three hours. Oh, yes, and build good will for GE while you're about it, said his bosses.

Although a journalist with no broadcasting background, Harris built a station which brought joy to the hearts of Americans around the Pacific rim. At last they were hearing news, music and voices from home which, in today's era of instant communications may not sound like much. Then, this was viewed as nothing short of miraculous. Many homes still had to wind up the springs in their phonographs after each three minute 78 rpm record.

The station continued to operate after the close of the Golden Gate Exposition and, as war drew closer in the Pacific, it served as an effective counterpoint to Radio Tokyo which also provided some quasi-English language programming to Asia. Ultimately, KGEI became a model for the U. S. Government's own official shortwave stations which were to come into being later as American stations targeted foreign audiences. AFRS followed a different path. It addressed only its American contemporaries wherever they might be. Still, it owes a debt to this early pioneer which opened up the airwaves for military broadcasting by opening up the minds of some of the military chiefs. Some would soon see the value of an electronic pipeline direct to the troops in the field.

Suddenly...Sunday, December 7, 1941. American battleships and aircraft were smoking hulks, fouling the blue Hawaiian skies. Within days, most of the troops in the Philippines, vastly outnumbered, and with no air cover, were being regrouped on the Bataan peninsula where General Douglas MacArthur set up his temporary headquarters for the defense of the islands.

KGEI, back in San Francisco, was now the only effective, although unofficial, link with home and loved ones.

...back after this message...

KGEI continued to provide invaluable service to the U.S. following the attack on Pearl Harbor and the battle for the Philippines. The Japanese, after the Philippine invasion, began flooding the country with counterfeit currency. At the request of General MacArthur from his Bataan headquarters, KGEI began broadcasting warnings about the funny-money and told people to refuse to accept it. The Japanese response was short and to the point. "Refuse and you'll be shot." Some did. Some were.

The American perimeter on Bataan continued to grow smaller as the Japanese, greatly outnumbering the Americans, applied ever-growing pressure. The Signal Corps people had managed to bring along a small, 1 kilowatt transmitter as they escaped Manila. This was now used to retransmit broadcasts from KGEI in San Francisco.

With the troops at MacArthur's headquarters were the Philippine President, Manuel Quezon, and his friend and fellow statesman, Carlos Romulo, who was later to become ambassador to the United Nations. They asked MacArthur if they could use the small station to send messages of hope to the Philippine population. He, of course, agreed.

The broadcasts were a disaster. So unbelieveably laced with false hope and heavy handed propaganda, they were described by one American officer as serving "no purpose except to disgust us and incite mistrust of all hopes." Romulo, perhaps with good intentions but with questionable judgment kept telling his people that the Filipino and American troops were standing firm and would soon be marching back to retake Manila as a victorious army.

Instead, the outnumbered group was forced to abandon their positions and fall back to tiny Corregidor Island in Manila Bay. With them went the transmitter which soon fell silent after an American submarine took MacArthur, his staff and the Philippine dignitaries off to safety, on orders from Washington. Soon the remaining troops were forced by lack of food, water and ammunition to surrender and they became the nucleus of the infamous Bataan Death March.

President Quezon, from exile, criticized his friend Romulo for raising false hopes and broadcasters and public relations practitioners once more had their beliefs confirmed that there is no substitute for truth, no matter how unpalatable.

It would be a nice tribute to say the Corregidor station, which MacArthur, with typical bravado, had named "The Voice of Freedom," was the first purely military station on the air. But it wasn't. While Washington dithered over who would have the responsibility for military broadcasting, strange things were happening in both Panama and Alaska.

Almost a year before, in January, 1941, and troops in both locations managed to get stations on the air, although Washington was still "studying the problem."

TOPICS FROM THE TROPICS

In peacetime, Panama was a prime assignment although at Christmas time the troops did spend some thought Dreaming of a White Isthmus. In the main, the Spanish Main that is, it was considered good duty, and important. A glance at the map makes it very clear that if the Panama Canal was taken or destroyed, ships sailing between the Pacific and Atlantic faced a considerable detour.

Troops in Panama were essentially artillery types and the major units were the 73rd and 83rd Coast Artillery Corps. Each had built additional fortified gun emplacements to protect the canal and also busied themselves setting up a number of anti-aircraft locations throughout the long, narrow country. Suddenly, they were no longer garrison soldiers. Now, many were living in small, isolated clusters of gun crews deep in the Panama jungle.

Communication between these scattered elements was essential and the military command installed a radio transmitter in the basement of the Headquarters barracks at Quarry Heights. Each unit was issued a small, portable receiver more suitable for taking on a picnic than for tactical use. The problem was, this was no picnic and the command soon learned that rather than listen to the things squawk and spit static between infrequent messages, the men turned the noisy things off.

Sergeant Wayne Woods, editor of the local Army newspaper, was told to give maximum publicity to the requirement for keeping the radios on at all times. He did, and was generally ignored.

Woods then got together with Technical Sergeant Joseph Whitehead, who ran the basement transmitter, and with Master Sergeant Paul Doster, the Public Relations NCO for the command. Their solution was to play music over the system and hope the guys would keep it turned on.

With this simple idea, military broadcasting in a form closer to what it is today, was born.

Major General Sanderford Jarman, the commanding general, approved the plan and said "go." The station called itself PCAC, for Panama Canal Artillery Command, added a couple of volunteers to the staff and went on the air. Soon the parrots in the jungle were being treated to the music of Benny Goodman and Tommy Dorsey. So were the troops who, until now, had been suffering from terminal boredom.

Newscasts consisted of someone reading stories from the local English language **Panama Star and Herald**. On Sunday, troops were treated to a sergeant giving wrap-ups of the week's news which he read directly from the pages of **Time Magazine.**

The staff felt that possibly these ingenious programs could be improved. Doster suggested that it couldn't hurt to ask for free transcriptions of the top Stateside network programs. He gave Woods the assignment of writing pleading letters to the likes of Jack Benny, Bing Crosby and Bob Hope. Benny, among others, was happy to comply and, in addition, autographed the first record. It became PCAC's first network show.

Woods apparently writes one hell of a letter. By September of 1941 NBC saluted the little station in a star-filled broadcast heard from coast to coast in the U.S. NBC also solved another of PCAC's continuing problems. One day a giant crate arrived at their doorstep. Inside, courtesy of NBC, were 2,000 pounds of transcribed NBC network programs.

Then, Pearl Harbor. The station went off the air the morning of December 7th in order not to provide a homing signal for possible Japanese bombers.

It had been fine while it lasted but it would be January 1943 before the station could return to the air. When it did, it was as a member of the yet unborn AFRS.

Now, like Sergeant Preston of the Yukon, we mush North to Alaska

It's 1941, and the military is pouring men and material into Alaska as tensions build up around the Pacific.

Among the men were included a number of pirates, who, after suffering the boredom of life in Sitka and Kodiak for a short time, decided to build pirate, unauthorized, radio stations -- homemade and strictly illegal. In Panama, the men could get away with it because the Federal Communications Commission -- the dreaded FCC -- had no jurisdiction. But Alaska, a U.S. territory, had FCC offices with officials who kept a sharp ear out for those who violate its pristine airwaves and book-length regulations.

Almost simultaneously, three stations were slipped on to the air, although "secret radio station" is something of a contradiction in terms. It is difficult to determine who was "first" and it really doesn't matter. Their efforts to fight boredom and officialdom deserve equal credit. Their efforts did much to pull the thinking of the Pentagon, as it concerned the importance of radio, into the 20th Century.

The Alaska Communications Service (ACS) handled regular military communications and was drafting all the amateur radio operators it could find. It then shipped them off to the frozen northland with little or no military training. One such, Ervin Green, enlisted in the Army and found himself in Sitka, Alaska, exactly twenty-one days later. He had worked for a radio station in Utah and ran a pirate station there at the same time, using a phono-oscillator. This was a common enough gadget in those days, being nothing more than a phonograph which legally transmitted a signal. The signal was just strong enough to reach a speaker across a normal room before its signal pooped out and disappeared forever. A clever amateur could diddle the innards a bit and make it send a signal for appreciable distances. Green brought his along.

His roommate, Jeff Boyce, was also an amateur radio operator with an FCC license and the knowledge that putting an unauthorized station on the air was considered by the FCC as ranking somewhere between high treason and, even worse, rudeness toward Government regulations.

Naturally they decided their fellow soldiers needed a radio station so they beefed up the power, added a microphone, begged, borrowed or bought records and went on the air using KRB as their call letters. Having access to the ACS, military, signal, they used the news sent from the lower 48 designed for the local newspaper and even managed to present a few locally produced radio dramas using local talent. Although the station's schedule suffered from what radio commercials euphemistically call "irregularity", it became very popular with the troops and the residents of Sitka.

Soon after it began broadcasting, boredom broke out big time in Kodiak and the finance officer, Captain W. H. Adams, hooked up a phonograph to a wired speaker system in the officer's club. Soon, everyone wanted to listen. A "radio club" was formed and two sergeants, Bill Merritt and Rule Bright, did a little moonlight requisitioning and soon had a working radio station. They called it KODK and wrote off to everyone they could think of in the States for transcriptions.

Back in Sitka, a third Alaskan station was being planned and built. Privates Charles Gilliam and Charles D. "Dowdie" Green, along with Staff Sergeant Chet Iverson, had all been on the same official Army Alaskan cruise as part of a team responsible for the communications system linking the Sitka Harbor Defense Headquarters and isolated outposts around the sound. Knowing where they were heading, and all being familiar with electronics, they managed to smuggle more than 400 pounds of electronic parts listed as personal baggage. It included a transmitter, a short-wave receiver and other essential equipment which Gilliam characterizes as "parts and junk."

At first they played music over the tactical communications system "after hours." The authorities soon brought that plan to a grinding halt. Out came the smuggled gear including a six tube wireless record player. They took turns putting records on the record changer but, because there was no microphone, they could only make voice announcements by recording them on an antique recorder using cardboard disks. These were then played over the system. They named their lilliputian contraption "GAB" -- a great name for a radio station but actually standing for "Gil and Bob."

Initially heard only within a very small radius, the pirates fiddled around with various antennas and soon come up with a heavy piece of copper cable which they attached to a large chunk of bronze plate taken off a wrecked boat. Attaching the cable to the plate at one end and the tiny transmitter to the other, they threw the plate into the salt water. That worked only too well. Their signal boomed into Fairbanks, 700 miles away. An illegal signal that powerful was bound to make pirates cautious and FCC inspectors furious. A less effective antenna was then devised which consisted mainly of a cable thrown out the window.

Other improvements followed and Gilliam even developed methods to do radio remotes, the first being the broadcast of the Easter Services from the post chapel.

At KRB, the other pirate, Ervin Green also raised power and it reached all the way to both Juneau and the quivering ears of an FCC official there. This precipitated a number of meetings between the Sitka military commander, the FCC, the Mayor and Green's officer in charge at the Alaska Communications System.

Green readily admitted he knew he was breaking the law but said it was for the common good, a theory to which the other inhabitants of Sitka readily subscribed. The FCC shrugged and there were no repercussions, particularly as Green was wise enough to take the station off the air rather than jeopardize his FCC license.

D.C. FINALLY GETS THE MESSAGE

In the spring of 1942, C. P. MacGregor, then, and for many years after, one of the most successful syndicators of radio programming, got a letter from a station he had never heard of in a place he didn't know existed. Someone in Fort Ray in Sitka, Alaska, was asking him if he could spare any old copies of his popular *Skippy Playhouse*.

He was happy to oblige but just to make sure the request was legitimate, he called the War Department and asked permission.

"Send them WHERE?" they asked.
"Your station in Sitka, Alaska."
"Never heard of it," they said.

C. P. sent them anyway. So the station got its recordings but it also got something it didn't expect -- a visit by a gaggle of Pentagon Brass who arrived shortly thereafter to find out what in hell was going on.

Back in Washington, plans had been underway for some time to provide radio to the troops and there were many -- perhaps too many -- spoons all trying to stir the pot at once. The War Department team arrived in Sitka, looked around, went up to Kodiak, looked around, talked to the troops, talked to commanders and decided that this radio was a pretty good thing after all. Maybe, they said, this is the way to go.

AFRS was about to be born. It had been one of the longest pregnancies on record. Its daddy, Tom Lewis, was with the inspection team, and he was about to go back to Washington and preside over the birth of his baby.

CHAPTER THREE

In which Colonel Tom Lewis and Colonel Frank Capra learn a great deal about bureaucracy including what Voltaire was talking about when he said, "It is dangerous to be right when the government is wrong."

Getting off the Ground...and On The Air

In spite of the recommendations of the Fosdick report, written shortly after the first World War, twenty years had now passed and little or nothing had been accomplished other than the formation of a number of committees. Fred Allen has explained it well: a committee is a group of men who, individually, can do nothing, but as a group can agree that nothing can be done.

Committees **keep** minutes. They also **lose** days, months and years.

In the case of the military trying to figure out how to improve service morale, the various committees wasted twenty years and, by 1939, local commanders still retained the primary responsibility for recreational and morale services. On most posts, if the men wanted something, they did as their compatriots in Panama and Alaska did. They just went ahead and did it, knowing full well that buried in some long-forgotten regulation was a paragraph reading, "No, NO. You absolutely **can not** do that!" They also knew that chances were good that another regulation existed that permitted it. At the very worst, a committee would be formed to study the situation.

What follows is a brief journey along the path that finally led to a workable method allowing the Military to handle such curious non-military functions as news gathering and dissemination; producing motion pictures; running radio (and eventually, television) stations. All were items that, until then, had been largely ignored in favor of passing out catcher's mitts to the troops and calling it a morale builder.

Even that simple task was not always handled well.

This is no fairy tale, but Once Upon a Time a military procurement officer was charged with buying catcher's mitts for the recreation and sports program. Being a statistical genius, although perhaps a baseball idiot, he researched how many persons were left handed. He then contracted for that same percentage of left handed catcher's mitts, not realizing that in the whole history of baseball there have only been enough left handed catchers to fill a bathtub, with room left over for Roseanne Barr Arnold.

That explained, let's toddle down the tangled path, through the thickets of bureaucracy to the creation of AFRS and its worldwide broadcasting system.

Would you believe, if told, that this man is one of the finest broadcasters ever to serve with AFRTS? Would you believe he won the prestigious Tom Lewis Award as Broadcaster of the Year? Could he really have been Commander of the Korean Network, AFKN? And is it credible that he was Deputy Commander of AFRTS Los Angeles and is now president of the Armed Forces Broadcasters Association? All true, but hard to believe.

Easy to believe: He won first prize on a Rhine River cruise for constructing and having the guts to wear the stupidest hat of anyone aboard ship.

This is an extremely out-of-uniform Colonel Pete Barrett.

15

MALICE IN WONDERLAND
or
War in the War Department

One of the most positive things to come out of the Fosdick Report following World War I was the provision to provide educational classes for the troops. As troops returned from France, the educational opportunities were continued at U. S. bases and the unanimous feeling was that this was both a genuine benefit and a morale booster.

The War Department then established a Morale Branch within the General Staff, with the mission of coordinating efforts of civilian welfare agencies and supervising internal factors affecting morale in the peacetime army. Because a stingy Congress and a civilian population anxious to forget the recent war failed to provide adequate funding, the program's effectiveness was close to nil. Consequently, local commanders continued to pass out baseballs and wonder what to do with all those left-handed catcher's mitts.

In 1939, Congress passed the Mobilization Regulation authorizing an increase in the size of the forces because of the worsening situation in Europe and Asia. The regulation defined morale largely in terms of physical welfare, leaves, food, discipline and recreation. Within a mere ten months, a Morale Division was established, this time under the jurisdiction of the Adjutant General's office. Its sections included the Army Motion Picture Service, Recreation and Welfare, War Department Exhibits, Decorations and, finally, Morale Publicity.

"Morale Publicity," in spite of its unfortunate name which calls to mind newspaper releases beginning "Private Luther Snively, chief latrine orderly at Fort Benning, said today his morale was good..." did contain the germ of the possibility that it could provide troops with information.

The regulation also required the Secretary of War to appoint a committee of civilian and military officials who were experts in welfare and community service activities. Although this was an Army regulation, other services didn't want to be either caught napping or cat napping, whichever applies, so the Navy and Marines were added to the cast of characters. This changed the title to "The Joint Army and Navy Committee on Welfare and Recreation."

This was round one.

Round two started as the Secretary of War, Henry Stimson, appointed Fredrick Osborn, then chairman of the Rockefeller Foundation, as committee chairman. The job paid a dollar a year, presumably less than the Rockefellers paid. However, he was a great friend of President Franklin D. Roosevelt who was later to appoint him to the thankless job as chairman of the Selective Service Committee.

The Joint Morale Committee did precisely what committees do best. It agreed to let someone else do it. By consensus, it turned the whole job of supervising morale functions over to the United Service Organization. The USO was then a newly formed group combining many of the private agencies which had assisted during World War I. This plan had the full support of Stimson and Army Chief of Staff, General George C. Marshall.

Roosevelt soon asked for a full briefing on morale plans and Osborn and Marshall dutifully reported to the White House, plan in hand. When they got to the part about the USO, the next round began.

At the bell for round three, Paul V. McNutt, then Director of Defense, Health and Welfare Services, could feel his empire about to crumble. He broke in and said, "We've changed that. We've set up an organization in my department to do that." A pained Osborn insisted that the USO portion of the plan be retained and morale functions be placed in non-governmental hands. The argument grew heated until Roosevelt, master diplomat, stopped it by reminding McNutt that if he were to increase his staff, he would be faced with the uncomfortable task of firing them all in the future.

"I like you too much to put you in that position, Paul," Roosevelt said. "We'll let the USO do it."

Chief of Staff Marshall, future author of the Marshall Plan, future Secretary of State, future Secretary of Defense and a man of many ideas had one more during this period. He decided to see what this morale business was all about and to find out, he called a conference in Washington and sent out orders to all major Army components requiring them to send their "morale officers" to the Pentagon. Purpose: discuss troop morale problems and solutions. Result: disaster.

Reason: when the orders arrived at the various units, many commanders discovered they had no morale officer. Solution: appoint the least needed officer immediately and ship him off to Washington as the unit's Morale Officer.

Most of the newly appointed experts had no clue what this whole thing was about and this became immediately apparent to General Marshall, a man also aware of the problems of restlessness in a peacetime Army which at that moment was best equipped to fight only among itself.

He issued a memorandum in which he said it was clear that the War Department's existing machinery was not adequate to "enable the chief of the Morale Division...to know the state of morale of the Army."

The solution, of course, was simple.

Another branch was formed. Called, logically, The Morale Branch, it was placed directly under the supervision of the Chief of Staff who appointed Brigadier General James Ulio as Chief of the Branch. Part of its charter was to develop methods to know the state of the military's morale at all times. Anything less was guaranteed to lower Marshall's morale. Ulio, who also remained in his old job as assistant to the Adjutant General, broke the Branch into four sections which included Welfare and Recreation, Planning and Research, Public Relations, and Services. Public Relations included the responsibility for camp newspapers and camp radio reception. It has since been strained through endless hands, brains, committees, supervisors, planning groups, departments and branches but remains today the great-great-great granddaddy of the American Forces Information Service and, in turn, the Armed Forces Radio and Television Service.

While the Morale Branch was forming up, Osborn's committee moved into the War Department and, by coincidence, Osborn ended up at a desk next to Ulio. General Ulio suffered a serious health problem in August 1941, and he was forced to give up his Morale Branch duties. General Marshall recommended a direct

Colonel Tom Lewis, the man chosen by the War Department to take on one of the toughest assignments any broadcaster has ever been given: to hang up his civilian suit, put on a uniform and create a network of stations reaching around the world to inform and entertain eleven-million men and women under arms. And, do it quickly!
   Somehow he did just that. AFRTS, the monument he built is still alive and well after fifty-plus years. (Photo courtesy of Dorothy McAdam)

commission as a Brigadier General for dollar-a-year-man Osborn so he could take over the Branch.

Osborn was quick to point out that he was a civilian and didn't even know how to salute, which is an absolutely minimum requirement for a General. Because Osborn was both talented and a friend of the President, Marshall made the appointment anyway. Being among the earliest of the civilian appointments, this caused no end of grumbling from long-time military officers who were still clawing for rank.

The "civilian general" was faced with a partly developed staff, little doctrine to guide him, a suspicious and somewhat resentful officer corps and field commanders who really didn't know what this morale stuff was all about. It probably had something to do with baseball, they felt.

With the bombing of Pearl Harbor and America's entry into the war some three months after Osborn took over the Branch, he changed the name to "Special Services" and added recreation and welfare as well as information and education to its operation. Information and Education, called I&E for the many following years in which it existed as a division of Special Services, was originally devised to explain to the troops why they had to fight and why they were so important to the nation and the army which required someone to do the fighting.

Marshall realized that, in spite of what the magazines said about those strange weirdos out in Hollywood, they knew what the public wanted and they knew how to get messages to masses. Millions of people spent their spare moments in darkened movie palaces or in front of their Atwater Kent radios. Sitting in Washington, he made the decision to get the best and recognizing the power of the silver screen, he immediately ordered Academy Award winning director Frank Capra to join Osborn's Special Services staff.

It was an offer that Capra couldn't refuse, in view of the fact that he had just been commissioned into the Signal Corps and the invitation came from the boss of all bosses, the Chief of Staff of the Army.

He reported to Osborn in Washington as ordered.

First impressions must be lasting for Capra, never forgot that meeting, which took place in what he described as "a sleazy, paperboard temporary building." Osborn then passed him over to Colonel E. L. Munson, Jr., the head of the Information Services section. It was to this section, which was responsible for film, news, radio and even pamphlets, that Capra would be assigned. He was not impressed, although he was to go on to produce the multi-part documentary *Why We Fight* while serving there. Many film buffs still consider this one of the finest film documentaries ever produced. During the war it was required viewing for every service member and while today its message might seem a trifle heavy-handed, in those less sophisticated and more patriotic days it provided a message that stiffened the resolve of most of the troops who saw it.

After a short orientation and a meeting with the staff, Munson asked Capra if he knew a Tom Lewis. Silly question. Everybody in Hollywood knew Tom Lewis, then vice president in charge of radio production at Young and Rubicam, one of the two or three largest advertising agencies in the country and in addition, he was the husband of screen star Loretta Young.

'dlike ta have you meet Tom Lewis

Tom Lewis is the towering figure in the AFRTS story. Born in Upper New York State in 1901, he passed away in 1988. He was graduated from Union College, Schenectady, New York, and entered broadcasting with WGY, Schenectady, soon after. WGY, the earliest NBC affiliate, was his training ground and there he first honed his skills as a writer, announcer, producer. After moving to the West Coast, he married Loretta Young in 1940, they had two children. Other than being the real founder of AFRTS and its first commanding officer, his list of credits is awesome:

...at Young and Rubicam he supervised some of the nation's most popular programs.

...he was a vice president for audience research at the Gallup Organization.

...he was founding producer of *Screen Guild Theater*, profits from which went to build the Motion Picture Relief Home and Hospital.

...he co-founded *Family Theater*, "The family that prays together, stays together."

...President of Lewislor Enterprises and Lewislor Films, producers of a string of successful films and programs.

...Creator and executive producer of *The Loretta Young Show*, which figures, and which people still remember when they see someone swoop through a door.

...Named by Pope Pius XII as a Sovereign Knight of Malta, he spent much of his time in later years serving on boards and directorates of hospitals, children's homes, drug abuse centers and councils for better schools.

Munson explained that Special Services hoped to scoop up Lewis in their net and put him in charge of radio. Capra, still in semi-shock while recovering from his recent abduction, threatened to tell Lewis to stay out of uniform to avoid becoming just another "body to be kicked around by some jerk superior."

Munson, Capra's superior, jerked back and hastily explained that "propaganda" is a dirty word to the public and that the congress was mistrustful about managed news, or propaganda, being fed to a captive audience of millions of troops. However, Munson explained, the troops had a right to know why they were in uniform and he felt that his Special Services section should use every type of modern communications media to tell them why.

Capra calmed down further when Munson explained that <u>his</u> superior, Osborn, guaranteed bi-partisan support by being a lifelong Republican while West Pointer Munson felt his contribution was to make the operation successful to "mossback Army diehards."

The conversation then revolved around what the job to be offered Lewis was all about and whether he -- or anyone -- could handle the herculean task of building a world-wide radio system from scratch that would reach every service man and woman overseas. It would be larger than all the national networks combined and would supply the military listeners everything from the Kentucky Derby to comedians to girls to football games to sports scores to the latest news from home.

Munson said that Lewis was the choice of everyone he had talked to and most had said he was a genius at getting things done. He had studied Lewis' work, talked to his associates and agreed that indeed Lewis was a genius. But, he asked Capra, "Is he lucky?"

Later, Capra was to say he replied, "I don't know about Tom Lewis being lucky, but I know damn well that you, the Army, and the country will be lucky. You've got your man, fellows -- a man who asks God for help. And gets it."

Meanwhile, back at Hollywood and Vine, the object of all this attention in Washington was considering trying to get into the Navy because he had visited the USS Arizona, now sitting on its keel in Pearl Harbor, and had been deeply impressed. While continuing efforts to enlist, the Joint Army and Navy Committee requested that he begin making a plan for a global communications operation to supply information, orientation and some entertainment to troops overseas.

Fellow advertising executive Arthur Farlow, formerly with J. Walter Thompson and now a Major representing the Joint Committee, put on an ad campaign for Lewis' benefit. He sold him on the idea of resigning from Young and Rubicam, enlisting in the Army and taking on responsibility for the operation. After enlisting, he was told, the Army would commission him a Major. Lewis wasn't totally sure where a Major stood in the military pecking order but he accepted. He resigned from Y&R and began working on the master plan. The plan was necessarily rough as obviously no one, especially a not-yet Major, had any idea where the battles to come would be fought. With broad brush strokes, Lewis sketched out a plan with rough estimates of the huge amounts of equipment needed, a guesstimate of the number of receivers required and a crystal-ball prediction as to where and when all this stuff would be needed. He also included plans for 24-hour a day broadcasting, conjecturing that the war was going to be fought globally.

As for program content, Lewis determined that the military would have to create much of the programming from scratch to fit its particular needs. Information programs, he felt, could come from essentially the same sources as the general public's. So, in large part, could the entertainment segments but he already saw the need for specially created troop-oriented material.

It became obvious to him even before he received his commission that the first part of his war was going to be fought in offices. Those "mossback Army diehards" Munson told Capra about had some sort of old-fashioned ideas that manpower and equipment and money should be used to further the combat effort. This put Lewis into what he called a "backbreaking, soul-searing" competition for manpower and materials in order to meet his commitment. His superior powers of persuasion, perhaps the result of being finely honed in the advertising business, ultimately resulted in acquiring the necessary wartime priorities to build up the organization.

By May of 1942, the plans were finished and his commission confirmed. Lewis went to Washington and presented the plan to his new bosses. It was quickly accepted and AFRS was now a reality. The only problem was, it consisted mostly of a plan. Lewis knew it had to be implemented quickly if he was to staff the organization with top people from the civilian broadcast and recording industry. This meant making his headquarters in Hollywood which, then, was the navel of the entertainment world and the home base for the finest talent and technicians.

He looked around and selected offices in the Taft Building on the corner of Hollywood Boulevard and Vine Street. Frank Capra had gotten permission to use this same location. Lewis had not, faced with the snail-like movement by the people who controlled the budgets in the Congress and at the War Department. Lewis turned to Capra for advice. Having been through it, the director told him, "Just move in."

Whatthehell, he figured. The worst they can do is shoot me. In he moved. AFRS had a home and was ready for business. Soon thereafter they both moved their business to the 20th Century Fox studios on Western Avenue which gave them a little more elbow room.

*Director Frank Capra will be long remembered by movie goers for the endless stream of great movies he turned out in his career. AFRTS owes him a debt of gratitude for being at least partially responsible for the selection of Tom Lewis to head the organization. Capra was brought into the Army as a major to head the 834th Signal Photo unit to make documentaries and was instrumental in recommending Lewis to his boss, Colonel E. L. Munson. For a time, Capra and Lewis shared crowded office space in the Taft Building in Hollywood, later moving to the Fox-Twentieth Century studio on Western Avenue. Capra's landmark documentary **"Why We Fight"** is still considered to be one of the most powerful films of its type ever made. (Photo courtesy of Dorothy McAdam)*

Right, a bright and eager young Marty Halperin reports in for the glamorous duty with AFRTS at the annex on Santa Monica Boulevard in July, 1946. Below, Once inside, it wasn't all that glamorous. This is the original recording room manned by (L to R) civilian supervisor Bud Pritchard, PFC Marty Halperin and SSgt Sid Kern. No one is sure how Halperin survived this hazardous duty. In one corner of the room was a 220 volt transformer and every time it rained, the room flooded. Halperin, as low ranking man on the totem pole, had the job of wading through the water to turn it off while trying not to turn himself into a Triscuit. (Photos courtesy of Marty Halperin)

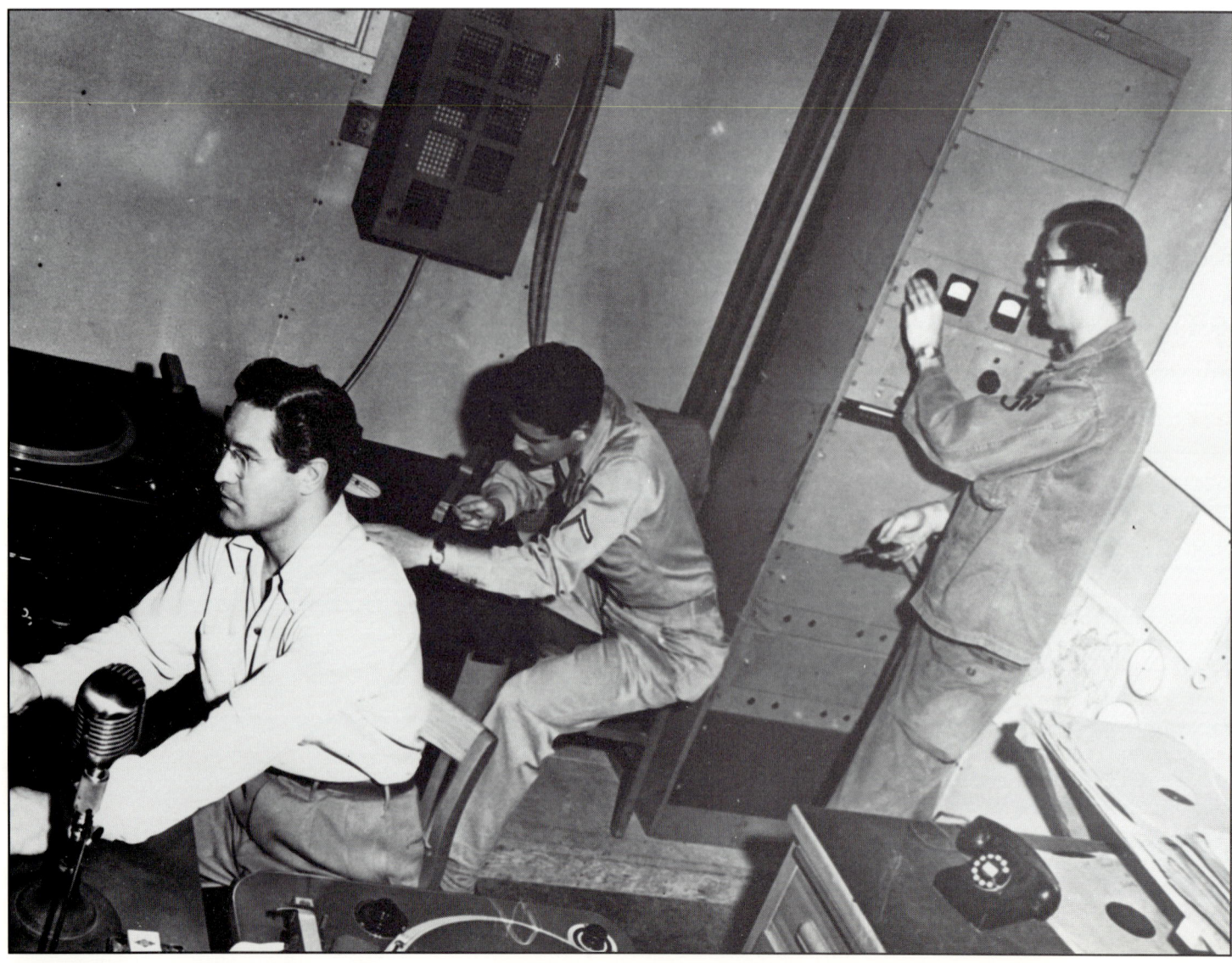

CHAPTER FOUR

In which Tom Lewis, True Boardman, and the other founders of AFRS create something out of nothing -- just the way God would have done it had God understood the situation.

"Meanwhile...back at the Flagpole"

Headquarters of AFRS in Hollywood may have lacked a bit of the physical pizzazz the public in those days mistakenly thought radio stations had and was, in truth, a semi-slum. What it had, and what radio station managers anywhere would gladly kill for, was an unlimited supply of the best broadcast talent in the country. Perhaps even better, they worked for peanuts and, better than that, there was no sales staff or profit and loss statements. Still better, there were no clients, a happy situation about which commercial broadcasters can only dream.

Lewis assembled an all-star staff of writers, producers, engineers and entertainers. Most of them were right out of the top drawer of the Hollywood broadcast industry and most of them were also grateful to continue doing the work they did best while able to go home after the work day was done. It was infinitely better than sitting in a foxhole or storming a beachhead.

Transitions were often somewhat abrupt. True Boardman, who was to become Tom Lewis' executive officer and right hand man for the remainder of the war, was sitting in a middle-eastern harem surrounded by beautiful women when they came to drag him away to AFRS. At the time he was a dialogue director on a film being shot at Universal called *The Arabian Nights* starring the sex-pot of the week, Maria Montez, and Jon Hall. He had told the director that he might not be able to finish the picture because he was supposed to report for duty. The director assured him that it had all been taken care of. Just go on about your job -- which he was doing while sitting in the harem, coaching lovelies on their lines when the military arrived and took him away.

Also coming aboard at about the same time were Charles Vanda and Jerry Lawrence from CBS and Bob Lee, a co-worker of Tom Lewis' at Young and Rubicam. Lawrence and Lee found they made a pretty good writing team and during the war turned out piles of material. After the war they continued to work together and have since given the world such plays as *Inherit the Wind*, *First Tuesday in October* and *Auntie Mame*.

As Lewis began assembling his staff, it became obvious that money and more manpower was needed although the War Department kindly met his payroll each month. Many of the departments in Washington and most of the officers with whom he dealt were something less than enthusiastic about sharing largess with a bunch of crazies sitting in Hollywood writing jokes and playing music for the war effort. Invariably a request for funds was answered with, "Don't you know there's a war on?"

One of the first to join AFRS was Al Scalpone from the New York office of Young and Rubicam. Because he had not as yet received his promised commission, Lewis sent him to Washington with orders to bring back several large sacks of money to get the new organization going. Lewis felt that because Scalpone was still a civilian, he wouldn't have to worry about offending the high brass.

Scalpone knew what buttons to push, having

Marty Halperin again, by now a corporal— presumably promoted for bravery in the face of 220 volts while standing in water. Here he proudly shows off the Presto 9N disk cutters newly installed in the recording room. The intense looking gentleman with him is Elliott Lewis, probably one of the finest radio actors who ever was. (Courtesy Marty Halperin)

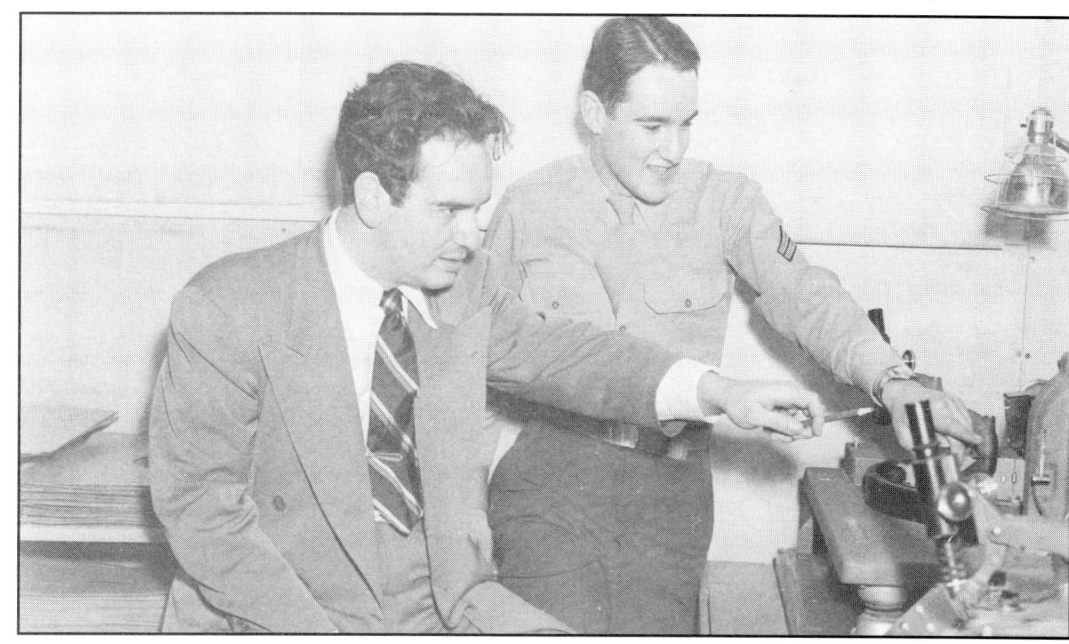

21

served as a consultant to the Joint Army and Navy Welfare Committee -- the group Osborn had headed before being anointed an instant General. Being a top-notch advertising man, he also knew how to pitch a client. He prepared a typical advertising presentation complete with charts, graphs and a nifty routine of verbal tap dancing. Being no dummy, he also ended it with a request for **twice** the money he needed.

He went through his pitch. The colonel controlling the funds said, "Don't you know there's a war going on?" No way, he said, was he about to give money for a childish thing like a radio operation.

Meaning no disrespect to the advertising profession, it must be reported that Scalpone never missed a beat and did what came naturally.

He lied.

Although he wouldn't have known Secretary of War Henry Stimson if he was trapped in a phone booth with him, he told the colonel that he was a consultant to the Secretary and if he could please be excused he would toddle down the hall and explain his problem to Mr. Stimson. The colonel thought a moment and said that perhaps he had been a trifle hasty. Would Mr. Scalpone agree to accept one-half of the money he requested? Always reasonable, Scalpone graciously accepted the amount he needed in the first place.

On his way back to Hollywood, he stopped in New York and managed to secure a number of major radio programs for AFRS use, compliments of the sponsors. Lewis was so delighted with the results of the Scalpone scam, he arranged for him to remain a civilian so he would have more flexibility in dealing with the military.

Dealing with the military remained a priority item for many years to come. As the value of AFRS became widely known through the military command structure in future years, the commands accepted it as a necessity. As it was taking on life, however, endless hours were wasted explaining to all levels of the military hierarchy why it was needed and, if it was needed, why it needed money and men. Then and now, the average viewer and listener seldom thinks about how programs get into their eyes and ears. Somehow, people often think, programs merely lurk inside the receiver and leak out when the switch is turned on.

Initially, AFRS had planned to use shortwave as its primary means of sending programs overseas. It soon became apparent that, while valuable, shortwave would have to take a secondary role. AFRS continued through the years to provide a shortwave service but in many respects it was somewhat less than a perfect solution. Propagation of shortwave signals is a chancy thing at best. Signals tend to fade or distort with the seasons, with sunspots, with time of day. The frequency response is often inadequate for music. Shortwave transmitters at this time were also in short supply and were controlled for the most part by the Office of War Information or its predecessors. This meant sharing transmitters with propaganda organizations, something Lewis and his staff were determined to avoid as much as possible.

They decided to concentrate on a new recording technology using a material then called "vinylite." In

Rosemary Clooney and Bing Crosby during an AFRTS recording session in 1944. The microphone has a CBS shield because AFRTS during these hectic days of turning out program after program was using studios any place they could get them.

Rosie's brother, Nick, would shortly be inducted into the army and be assigned to AFN in Germany where he became tremendously popular as a DJ. She later told an interviewer that at a concert she gave for the troops there, she was introduced—to thunderous applause—by the announcer saying, " Now . . . here she is . . . Nick Clooney's sister. "

1942, most recordings were done on disks using shellac as the basic ingredient. Shellac records, as anyone knows who has ever dropped a pre-World War II recording, are fragile in the extreme and trying to ship them by the thousands into warzones was obviously impractical. Besides, shellac is all imported from far away places with strange sounding names and there was very little to be had.

Prior to the start of AFRS in 1942, a Major Gordon Hittenmark, who had been a Washington radio announcer, had developed what he called a "B-Kit." "B" stood for Buddy, and the kit was designed to be issued to all units departing for overseas or isolated areas. It contained a portable long- and shortwave receiver, batteries, antennas, vacuum tubes and a hand-wound 78 and 33 1/3 rpm record player with extra needles. Also included were 48 records of popular songs and transcriptions of about 25 hours of commercial radio programs. Some kits contained a microphone and two small speakers so this bonanza of entertainment could be played to an enthusiastic audience in the mess hall or day room. And if that weren't enough to titillate any GI to write home to Mom about what a swell place the Army was, the B-Kit also contained six paperback books and seven harmonicas.

Using a grant from the Carnegie Foundation of $100,000, the Morale Branch hired a Hollywood recording specialist named Irving Fogel to help Hittenmark develop the B-Kit. It soon became obvious that shellac recordings would never make it overseas or suffer the rough use they were certain to receive. Those extra needles would probably never be necessary. Using an additional grant of $400,000 from Carnegie, Fogel developed a new type of recording. It used vinylite and packed the grooves much closer together, allowing 15 minutes of playing time to a side. It was also smaller -- 12 inches in diameter as opposed to the then normal 16 inch transcription which was too large, too heavy and too fragile.

Only a few B-Kits had been produced by May of 1942 but Lewis and his group were fascinated by the concept of recordings which could withstand hard use, were lightweight and produced excellent quality. It was decided to explore the possibility of switching to this media for the major portion of the programs to be sent to stations overseas. In the meantime, shortwave continued to broadcast the AFRS program output.

The production of B-Kits was abandoned and troops had to make do without their seven harmonicas; possibly a blessing in disguise. Lewis tapped Fogel to join AFRS and develop further his ideas for vinylite recordings for overseas use.

When Lewis had insisted earlier that AFRS be located in Hollywood to be near the center of the entertainment industry, he had no idea that the programs

Every lonely GI wants someone to dream about—and this is the woman many of them dreamed of during World War II. Things were so tough in the Pacific that some of the guys were dreaming about Tokyo Rose. That would never do, said the brass, and "GI Jill" was born. When she finished playing the latest records, reading mail from all over and whispered in the sexiest voice imaginable, "Gooooood Niiiiiight," sighs could be heard from New Guinea to Okinawa.

This is the picture that went out to the thousands who requested one. What was never mentioned was her real name—which was Martha Wilkerson—or that she was married to Mort Werner who continued to work with Pat Weaver at both NBC and McCann-Erickson Advertising. Mort was to become program Director of NBC while Pat was President of the TV Network.

would be going out to the world on small vinyl circles with a hole in the middle. Nor did he consider that most record pressing plants were on the East Coast. There were such plants in or near Hollywood but all of them were working around the clock filling orders for both private industry and the large number of government agencies which were providing recorded programs or announcements to radio stations around the country to sell War Bonds, urge people to enlist, pay their taxes or save fat.

Sending the AFRS master recordings back East for processing was obviously out of the question, what with wartime transportation difficulties and the necessity to get the material out to the field in the shortest possible time. Lewis and Fogel decided to work with private industry and were able to obtain facilities by helping secure building permits, equipment, manpower and scarce building material to allow existing pressing plants to expand enough to meet AFRS' needs.

Those needs were considerable. As the war expanded and flared around the globe, AFRS stations proliferated and the requirements for programs grew daily. Within a year, AFRS was shipping each station 106 different shows totaling 42-hours of broadcasting each week. About half were programs received from the major networks and the remainder were produced by AFRS program staff members.

In the very early days, AFRS had no studios or recording facilities of its own. Instead, it would rent facilities from CBS, Mutual or NBC. The programs would either be recorded at the source or sent over specially balanced telephone lines to recording studios which would prepare the master disk for pressing and shipping.

Commercial programs presented a particular problem in that the commercials had to be removed prior to being sent overseas. In those early pre-tape days, this was something of an art. Two identical 16" disk recordings would be prepared. Then they would be so cued up that the commercial references were skipped. Recording number one would play up until the commercial. Then an engineer, who had cued up disk two after the commercial, would switch from one to the other. All this was being recorded on to a third, sub-master disk. Then the engineer would quickly find the next edit on disk one and cue that up. Sounds simple enough, except that no mistakes were allowed. If he goofed on the very last edit, it was necessary to go back to the beginning and start the whole tedious process over again.

Many of the standard techniques used in the recording process and the editing procedures were developed by the AFRS technical staff. At that time, big time radio meant CBS and NBC which had two networks, the Red and the Blue. Neither allowed recordings, except for sound effects, on their premises. All programs were done live and those heard across the country were done twice, once for the East and once for the West to allow for the time difference.

AFRS developments in the handling of recordings proved that audiences really didn't care if an entertainment program was a live broadcast or not. All they asked was that the quality be good and the material entertaining. When Bing Crosby's contract with NBC was up for renewal in 1945, the Blue Network had just been sold and it didn't care whether programs were recorded or not, so anxious was it for business. Crosby held out renewing with NBC and demanded the right to record his program in advance. NBC refused, Crosby moved to the Blue, now renamed ABC, and the dam was broken.

Now that the engineers had perfected cutting material out of transcriptions, it remained for program people to figure out a way to fill the holes up again so the programs would run the proper length. Lewis enlisted the aid of two of the most talented radio broadcasters around. One was a namesake but no relation: Elliot Lewis, actor and producer, who, along with his wife Kathy, were then, and remained, two of the most acclaimed performers in radio for many years. As his assistant, Dresser Dahlstead joined AFRS. Dahlstead, then a top announcer on many network programs and later a producer with the Ralph Edwards organization, joined Lewis in what was called the Domestic Rebroadcast Subsection, a forgettable name if ever there was one.

Its job was to fill holes left by editing engineers.

They built a backlog of musical material, using whenever possible the orchestra usually heard on the program. For example, they had a library of Phil Harris music for use on the Jack Benny program. Then they would get the program's announcer to record a special introduction to the music and insert it into the program.

Openings and closings of programs also presented problems because they almost invariably contained the name of the commercial sponsor. Lewis and Dahlstead, over time, were able to have most of the major program casts and orchestras record special openings and closings for use on the AFRS version. It was even necessary at times to rename the program. *Lux Radio Theater*, for example, lost its suds and became *Hollywood Radio Theater*.

Also prominent in the program section were three of Hollywood's leading talent agents, all now in uniform. They were Lester Linsk, George Rosenberg and Barron Polin. Their job was to obtain the talent for the AFRS produced programs. It was almost too easy. The cream of the Hollywood talent crop made themselves available for broadcasts to the troops with almost embarrassing regularity. It got to the point where it was necessary to limit the appearances of some of Hollywood's top names in order that the troops didn't get tired of hearing them.

Sylvester "Pat" Weaver, who later became President of NBC Television, who invented the "special," developed the *Today* and the *Tonight* Shows and is considered one of the geniuses of the business, arrived at AFRS in 1944. He said many years later that "this was the zenith of American radio. AFRS programs may well stand as the highest expression of American radio broadcasting."

✣ ✣ ✣ ✣ ✣

AFTER THIS BRIEF PAUSE TO TURN THE PAGE, WE'LL TAKE A LOOK AT WHAT THESE PROGRAMS WERE. THERE HAS NEVER BEEN ANYTHING LIKE THEM SINCE.

✣ ✣ ✣ ✣ ✣

"G. I. Journal" was still another of the popular weekly programs featuring Hollywood and Broadway personalities. Here, at rehearsal, are three of the best: (L to R) Mel Blanc, the man of a thousand voices, the incomparable Lucille Ball and Alan Ladd. (Photo courtesy

Flanked by two members of the AFRTS production staff, comedian Jack Benny, his foil and singer on his own program, Dennis Day and movie star Rita Hayworth laugh it up in the dressing room prior to recording a "Command Performance" program. Dennis at this time had already joined the Navy, but continued to perform whenever he could get leave. (Photo courtesy AFRTS)

25

Left, In the true spirit of honest journalism, and because of the firm belief that truth shall prevail, we report that we don't have the faintest idea whom it is that is carrying on a conversation with Bob Hope. All that can be honestly reported is that he certainly is a spiffy dresser, that his lapels could be used for flying jibs during the America's Cup and his necktie is handpainted. Below, The audience—exclusively military—lines up outside the CBS studios at Sunset and Gower waiting for the doors to open for "Command Performance," which, thanks to the unlimited supply of the world's most popular talent might well have been the world's most popular radio program. (Photos courtesy of AFRTS)

CHAPTER FIVE

*In which we learn that war is hell when the **Brown Derby** is your mess hall and sitting around throwing jokes at Bob Hope can be dangerous if he decides to fire back.*

Ready...Aim...Talk

AFRS produced hundreds of titles and thousands of programs during the War years but the jewel in the crown, and unarguably the program most remembered and most requested at the time was *Command Performance*. Oddly, it had its start elsewhere, although it was soon to become the showpiece of the fledgling network.

The Bureau of Public Relations of the War Department had a Radio Division, the mission of which was somewhat unclear. One day they received a request from the office of Lieutenant General Lesley J. McNair, Chief of the Army Ground Forces, passing on a letter from troops in Iceland asking for sports broadcasts. Although this did not seem to be part of their job, the Radio Division agreed to take on the task.

Edward Kirby, a civilian broadcaster from Tennessee, was then on assignment to the group and was handed the job of developing "radio to troops." By late spring of 1941, he had put together a small team of production people and was able to get sports programs on shortwave to the troops. Kirby's staff included Robert C. Coleson and Glenn Wheaton. Wheaton, a former producer for the popular *Quiz Kids* program, and Coleson had both volunteered as dollar-a-year-men.

Soon requests for musical programming started to come in and the staff began to develop ideas. Early plans included getting popular orchestra leaders and their singers to record voice tracks introducing their own songs requested by overseas troops. Wheaton wrote a number of scripts but the project never got off the ground or on the air.

December 7, 1941 interfered, as Japanese aircraft bombed Pearl Harbor. Now the U. S. was at war and Wheaton expanded his concept of a recorded musical show. He knew that, because of the war, he would now be able to get all the talent he wanted. His new idea, first broached at a Radio Division staff meeting, was to have the troops "command" their favorite performers to appear. This, he felt, would make them feel they had a direct connection with the program -- and with home. The title *Command Performance* fell into place naturally and Wheaton was given the assignment to form a production staff and get the as yet unborn child toddling.

Wheaton, as producer, and recently assigned Lieutenant Rankin Roberts who was charged with getting shortwave broadcast coverage for the program, went to New York to arrange for talent and try to persuade Vic Knight, then the producer of the *Fred Allen Show*, to join the team as director. They did...and he did.

Knight, who was a perfectionist, tended to be somewhat dictatorial, rather than directorial, when we was in a control room. Many feel he was the inspiration for the Clark Gable character in the book and film *The Hucksters*. He accepted the job at the usual fee and, because the talent on the program was not paid, became the only person associated with the show who got a raise. Starting at a dollar a year, the Army showed its appreciation of his talents later by drafting him as a private at $21 a month -- a raise of 2,520 percent.

Wheaton assumed overall direction of the writers and, with Knight, began lining up talent, more writers, studio space, equipment and air time. All of it was obtained, and at no cost to the government. Within three months, the first program was ready to go and it set the standard for the programs to follow for the war years and several years thereafter.

Troops on March 8, 1942 were treated to program number one. It was transmitted over 11 shortwave transmitters and featured Eddie Cantor, Dinah Shore, Danny Kaye, Merle Oberon, Bert Gordon (the "Mad Russian"), Bea Wain, the Cookie Fairchild Orchestra and, as a treat for the guys in Iceland, a replay of the Joe Lewis-Buddy Baer heavyweight championship fight was included. Harry Von Zell was the announcer.

The first six programs were produced in New York but it soon became apparent that it would have to move to Hollywood if the major talent pool in the country was to be drained. Besides, Los Angeles was home to both Wheaton and Knight. The move was made and Robert Coleson borrowed office space from CBS on Sunset Boulevard and arranged to use their largest studio when required. Talent was obtained through the newly formed Hollywood Victory Committee and its radio director, George Rosenberg. He was soon to put on a uniform, be assigned to AFRS and continue doing the same job.

Although not originally planned as a regular weekly show, it soon became obvious that with all the talent clamoring to do their part for the war effort, it would turn out that way. And it did. The first show from Hollywood, in mid-April, featured Gene Tierney as mistress-of-ceremonies, Betty Hutton, Gary Cooper, the

Andrews Sisters, Ray Noble and his Orchestra, Edgar Bergen and Charlie McCarthy, Ginny Simms and Bob Burns. Future shows featured the same kind of dynamite casts and, luckily for the War budget, they donated their high-priced talent.

Shortly after the move to Hollywood, Kirby replaced Vic Knight as director with Maury Holland due to what he called Knight's "increasing sense of his own importance." The incident which provoked the change came after Knight demanded the appearance of a singer, even though she already had a commitment to appear at a bond rally in Boston. He flatly turned down a suggested replacement, a young lady named Judy Garland. Later, still under the control of the Radio Division, top J. Walter Thompson producer Cal Kuhl became the third director.

By now, AFRS was becoming a reality and, as is the nature of a Public Relations organization, the War Department Public Relations people were beginning to use the program for their own purposes rather than that for which it was intended. One program was done in Cleveland at the National Association of Broadcasters Convention. Another was done at the National Theater in Washington for an audience of top government and military officials. Appearing were the Kay Kyser Orchestra, Bing Crosby, Jimmy Cagney, Hedy Lamarr, Abbott and Costello, Dinah Shore, Paul Douglas, Ginny Simms and others. It ran so long it was possible to edit it into three programs.

The final program produced by the Radio Division was a special Christmas show which broke tradition by being broadcast on domestic commercial networks, on the BBC and on shortwave around the world. It featured Bob Hope as MC, The Andrews Sisters, Red Skelton, Spike Jones and his Orchestra, Bing Crosby, Edgar Bergen and Charlie McCarthy, Jack Benny, Fred Allen and the entire 20th Century Fox orchestra. Nice cast. It also featured Dinah Shore who was to become the champ with more than 30 appearances on the program over the years. This program was the only *Command Performance* ever broadcast in the U. S. When it aired, it was under the control of AFRS, which took command of *Command Performance* on December 15, 1942.

The Army Takes Charge

Although only about six months old, AFRS was now in the radio production business big time. Stars were knocking at their door begging to appear on the air. The draft was scooping up writers, directors, technicians and AFRS was becoming adept at capturing them before they were assigned to a motor pool in Texas.

Working out of cramped quarters, working for military pay and turning out more programs than the major networks, this group was uniquely capable of getting the job done. It was also totally incapable of convincing the more hidebound elements of the military that they were anything but a bunch of escaped lunatics. Perhaps in some cases they were, but they were inspired lunatics. Much of that inspiration went into getting the job done in spite of the obstacles inherent in a military situation, such as rank. Rank was generally ignored. Many of the men had been long-time friends of Colonel Tom Lewis and had difficulty calling him anything but Tom. Privates directed programs and gave orders to officers such as Major Meredith Willson, later to write *Music Man* but then leader of the AFRS orchestra.

Military inspections often led to strange results

"Mail Call" ranked right alongside "Command Performance" in popularity with the troops. A comedy show featuring top names, it answered requests from the troops overseas. The program is seen here recording inside the CBS studios in Hollywood. In a few weeks it will be on disks and be on its way overseas.

including the inspection team walking away shaking its collective head at this peculiar organization. During one such inspection, a general officer asked a private, who was sitting staring into space, what his job was.

"I'm a writer, sir," said the private.

"Nonsense. I've been watching you for five minutes and you haven't written a word."

Attempts were made to put a little military fiber into the backbone of the AFRS team, but it wasn't always successful.

Sherwood Schwartz recalls a number of incidents. As a top writer for Bob Hope, he was a highly respected and highly paid member of the broadcast community one day and a private in the army the next. After completing basic training, his outfit was slated to sail to the Aleutians the next day when he was miraculously reassigned to AFRS back in his old home town. He spent the remainder of the war doing more good with his typewriter pounding out *Command Performance* and *Mail Call* scripts than he possibly could have done with a rifle on a remote island in th Bering Straits.

Happy Wartime Memories

Schwartz remembers the day someone somewhere decided to shape up the outfit. Frank Capra's film unit had a rough, tough adjutant, a regular army captain named Virginal Petito. Obviously anyone named Virginal HAD to be tough. Capra, being a kind man and a friend of Colonel Tom Lewis, kindly lent him to AFRS.

Petito's first order was that men would report for roll call and inspection each morning at zero six hundred. This meant that they were at their desks, typewriters or microphones at about 6:15 each morning. As Schwartz explains, "No one on earth can be funny at 6:15 a.m." The men then got in the habit of strolling off to breakfast for a couple of hours before pitching into the day's work.

Petito, not to be outdone, issued the order that anyone found away from his work place after morning inspection would be punished. Sure! said the group, and they strolled off for breakfast the next morning. When they returned, they found they had been put on grass-pulling duty.

Austin Peterson, one of Tom Lewis' earliest recruits and now a ranking AFRS officer pulled in to see his talented treasures barbering the bermuda. He asked the men what was going on and one explained that they had been naughty and were being punished. Calling in Petito, he countermanded the order and explained AFRS logic.

"Look," he said, "we've got shows to get out. It's okay to make guys who can write jokes pull grass for punishment as long as you can replace them with grass pullers who can be punished by writing jokes."

Among the more popular programs, according to the mail received at AFRS headquarters, were *Mail Call* which was a joke-filled extravaganza featuring top names, and AFRS's answer to Tokyo Rose, *AEF Jukebox*, which was hosted by a sultry-voiced charmer who used the name **GI Jill.** In actuality, she was Martha Wilkerson, a popular radio actress. It would have crushed her thousands of fans sitting on bleak Pacific islands to know that she was to marry another of Lewis' merry men, Mort Werner. Werner worked with "Pat" Weaver at AFRS and their relationship continued after the war. Weaver became President of NBC Television and Werner became Television Program Manager.

Schwartz wrote a number of *Mail Call* scripts and his recollections provide insight into the type of problems the Brass Button Broadcasters faced.

He wrote a script featuring Bob Hope and Clark Gable, then the he-ist of he men. The comedy sketch called for them to set each other up with dates from their respective little black books. It climaxed with a double date, Gable having fixed up Hope with Dame May Whittey, then 80-years old, and Hope having gotten Gable a date with Margaret O'Brien, 8. The problem was that he-man Gable was terrified of microphones. He had to be talked into appearing but finally consented, only to come face to face with a microphone and start to shake. His script rattled so loudly they had to stop the recording.

Hope took him aside and tried to convince him that the audience, and the troops who would hear the show, loved him. It took some doing on Hope's part but he got him back on stage and held both Gable's hand and his script throughout the program. It turned out to be a very funny show. Except to Gable.

More touching was the appearance of Jerome Kern on a *Mail Call* salute to prominent composers. Also appearing were Jimmy McHugh, Johnny Mercer and Sammy Kahn. At first Kern begged off doing the program, saying that the young people in the audience wouldn't know who he was. This from the man who wrote more than a thousand songs the audience had grown up with.

Actor Robert Young rehearsing for a "Command Performance" performance. At that time he was a romantic lead. Later as Kindly Dr. Welby on television, you can bet the farm that he would have never let himself get photographed with a ... sssshhhh ... cigarette in his hand. (Photo courtesy of AFRTS

Schwartz went to see him personally and pointed out that it would be impossible to do a program on America's greatest composers without Kern. Kern, a truly shy man, was terrified at the thought of saying anything. Schwartz assured him that he would write only a couple of lines which Kern could read to introduce himself playing one of his songs. Kern finally agreed, only to change his mind at the last minute. He said he was so afraid no one would know him that he had developed a large fever blister.

After much negotiation, Kern finally appeared, fever blister and all. The show was written so that he appeared last and by the time for his entrance he was a nervous wreck. Finally he made his appearance in front of a predominantly soldier-sailor audience and they went crazy with applause as they all came to their feet in a standing ovation.

The man who was afraid no one would remember him was unable to say even "thank you." He just walked over to the piano and played his "Smoke Gets in Your Eyes" with tears streaming down his cheeks.

Meanwhile *Command Performance* continued and became what old time radio buffs, and "Pat" Weaver, consider the greatest program of all time. It was on this show that the fake but funny "feud" between Frank Sinatra and Bing Crosby was dreamed up. It was a gimmick they were to use for years on their own programs.

After fifty years, those who have heard the "Wedding of Dick Tracy" episode still remember it as the absolute ultimate in radio comedy. It could hardly miss with a cast like this:

Dick Tracy	Bing Crosby
Flat Top	Bob Hope
Shakey	Frank Sinatra
Tess Truehart	Dinah Shore
Vitamin Flintheart	Frank Morgan
Snowflake	Judy Garland
The Mole	Jimmy Durante
Summer Sisters	The Andrews Sisters
Police Chief	Jerry Colonna

The program started at a point Chester Gould, the creator of Dick Tracy, never reached -- at the wedding of Tess and Dick. It took two hours for the event to occur, it being interrupted by bank robberies, murders, mayhem, a kidnapping, a holdup in which Tracy shoots 13 innocent bystanders, as well as some major interruptions.

Although these shows certainly deserve the fond recollections most people have of them, AFRS also turned out a full schedule of musical, religious, sports, news, classical and every other kind of program imaginable for the use of the stations overseas who were insatiable in their demands for more and more entertainment.

Shortwave broadcasts continued from both the East and West Coasts with news, sports and entertainment. Vinylite disks rolled out in a steady stream to the

Backstage at CBS just before going on for another AFRTS show (L to R) Patty Andrews, Jimmy Durante, Maxene . . . or is it Laverne . . . Andrews, Gary Moore and Laverne . . . or is it Maxene . . . Andrews. The Andrews Sisters, Ladies and GENTLEMEN—Patty, Laverne and Maxene. (Photo courtesy of AFRTS)

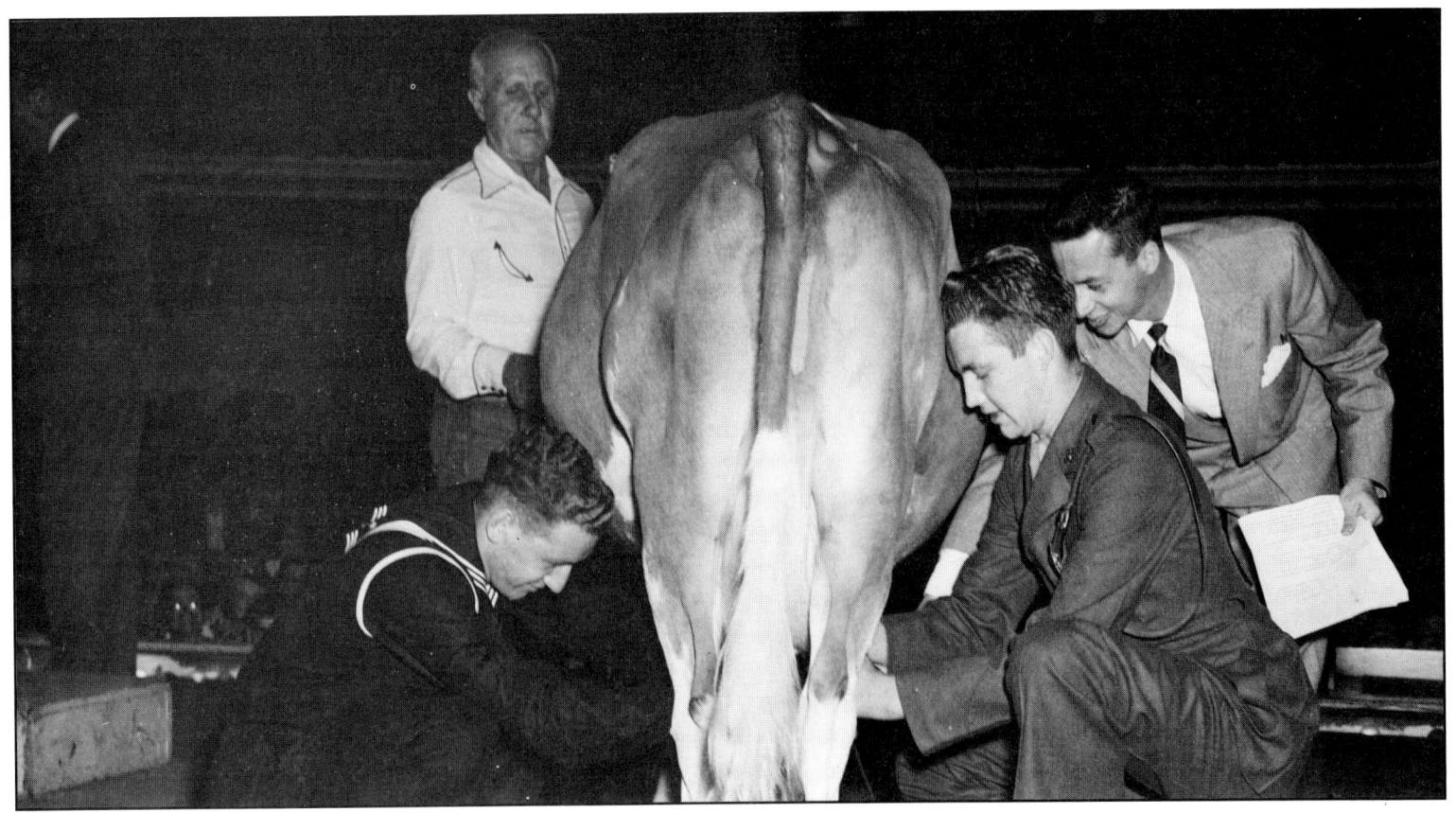

For this "Command Performance," a homesick farm boy far from home wrote in that he was nostalgic to hear the sound of fresh milk going into the milk pail. Writer Vance Colvig, far right, was able to pull it off.

He also managed to make it fun for the military audience by turning it into an udderly ridiculous contest between the marine and the sailor, shown here trying to drain Old Betsy's transmission and get the most fresh milk in his own bucket.

The audience went wild as the contest ended in a tie, but the cow from then on tended to get nervous everytime she heard Irving Berlin's "The Yank's are Coming." (Photo courtesy of Vance Colvig)

stations overseas and by 1945, at a special ceremony, Colonel Tom Lewis was presented with the one-millionth record.

Still, it's *Command Performance* that people remember. One who remembers is Vance Colvig who was a writer on the program toward the later part of its run. Bob Lehman was the director, Major Tom Smith who was to civilianize and stay with AFRS for years was also a writer as was Bill Norman. This is the way Colvig remembers the show:

"We opened the program with the Star Spangled Banner but on shows where there were foreign dignitaries, we also played the anthem of that country. One day we had an ambassador from a South American country whose anthem had a sort of Latin beat. Danny Kaye was guest host that day and he did a sort of dance to it. The State Department wasn't amused and demanded an apology.

One GI wrote in and asked Ann Miller to tap dance in Army shoes. She did, too. We built a hollow wooden box for her to dance on and the sound was deafening.

Kay Kyser had a different kind of shoe problem. His didn't fit. His price for appearing was a pair of navy low quarter shoes. We had to send a courier down to San Pedro Navy Base to get a pair and he happily led the band in his new shoes. As far as I know, he was the only performer to ever get paid, even with a pair of shoes.

The biggest laugh I recall was someone wrote in and said he missed fresh milk. I got a cow wrangler to wrangle in a cow and we had a milking contest between a sailor, a soldier and a marine. The sound of milk hitting the pails was wonderful. It ended up a tie although they each tried to outdo the udder.

Very often a person who had promised to appear would get a paying job they couldn't turn down. That was always fun. We would merely go to the bar at the Brown Derby on Vine Street and grab any actor who happened to be sitting there. I can recall walking out with both Ken and Wendell Niles, with Marvin Miller, June Foray, Howard Duff and Jack Kruschen. And we found Willard Scott there quite often.

The question of whether all these fun and games were a serious contribution to the war effort was answered by Tom Lewis in a speech he made before the Advertising Club of Hollywood in 1945 commemorating the 25th Anniversary of Radio. He read a letter he had received earlier from a chief steward aboard a merchant vessel off the Normandy beaches. The man wrote:

It's not a pretty sight to see American ships blown up by a floating mine or a direct hit from one of the shore guns. And believe me, this was really a test of nerves. Can you imagine how we felt when we came into the mess room for a cup of coffee and to sit down for half an hour. The radio was on and there was a rebroadcast from the American Forces Network of **Command Performance**. *It seemed good to hear the other crew members laugh at the jokes of the radio stars and hum the same tunes that the vocalist was singing. For a moment the grim business of carrying on the war seemed so remote. I imagined that I was at home safely listening to my bedroom radio. Listening to their radio here in the ship's mess seemed to give a lift to all the crew -- and when we were called on deck to abandon ship -- the radio was still playing.*

And it <u>still</u> is.

STATION BREAK NUMBER ONE

We pause now for Station Edification

As all the world knows, radio and television announcers, technicians and anchors are super-human. They are cool under pressure. Their minds continue to function flawlessly while those about them are losing theirs. In addition, their highly trained tongues continue to perform grammatically, dramatically, intelligently and constantly while worlds crash about them.

Right?

Wrong. Their brain is just like ours -- it starts working perfectly when they first get up in the morning, but often marginally stops functioning when they get to the office and into the studio. Some on-air personalities are more adept at hiding their natural tendencies to say and do absurd things but it would be a safe bet that there isn't anyone in broadcasting who doesn't admit to occasionally committing Public Stupidity.

AFRTS personnel are no exception to the rule. In spite of the fact the military strives for perfection, it is an ideal but unobtainable goal. Military people who remain out of trouble often do so by following the excellent maxim: *A Closed Mouth Gathers No Feet*. Unfortunately closed mouths are not in demand in broadcasting as military broadcast supervisors are often reminded. Consequently, Brass Button Broadcasters often find themselves with several feet of foot protruding from their mouths.

Why? Who knows? Former AFRTS staff member and now the brilliant columnist for the New York **Times**, William Safire, puts it sagaciously:

Is sloppiness in speech caused by ignorance or apathy? *I don't know and I don't care.*

AFRTS can take a prideful bow for being among the first to break radio's color barrier. They did it with a show called **Jubilee***, which, for the first time in radio, featured black performers. It seems ridiculous now, but it ranked with Jackie Robinson joining the New York Yankees when the show first began. Here at a rehearsal for the groundbreaking program are (L to R) Bandleader Stan Kenton, the Glorious Lena Horne and Eddie "Rochester" Anderson (Photo courtesy of AFRTS)*

☆

One of the very earliest AFRS-produced programs was called **The AFRS Ranch House Party.** It, like most programs in those early 1940ish days featured live orchestras, live talent and, supposedly, even a live announcer. On this particular day, the program had gone perfectly. The orchestra didn't miss a beat. The vocalists never forgot a lyric. The program timed out perfectly. Everybody was mentally congratulating themselves on a splendid job as the announcer stepped up to the microphone to close out the program and proclaimed, **"This is the Armed Faces Rodeo Service."**

☆

Sometimes it is difficult for the management to know whether a semi-crazed or simply bored announcer makes a mistake on purpose, or whether it is merely a case of forgetting to put the old brain in gear. No one will ever know what caused an AFN announcer to introduce the National Anthem with a solemn, **"Ladies and Gentlemen, The Star Spangled Banana."**

☆

Some inappropriate remarks can only be put down to cases of simple stupidity, whether permanent or temporary. It really doesn't matter; once said, there is no retrieving them. One such came during the height of the Watergate hearings and the depths of President Nixon's popularity. Sergeant Barry Bennett, outwardly normal and otherwise talented, suddenly lost control of his judgement during a major program on AFN one Sunday afternoon and ad libbed, **"President Nixon saw the movie 'Deep Throat' but hasn't been able to get it down Pat."** Instantly, every supervisor at home listening headed for the studio to have the honor of (choose one) killing him or relieving him on the spot. So quickly did they leave that no one noticed he followed his irresponsible remark with a recording of the then current hit, "Let's Come Together."

☆

Still at AFN, it should be noted that one announcer verbally discovered a medical miracle when, in the middle of a newscast, he described a patient recently admitted to the local hospital as suffering from **eternal bleeding.**

☆

Don Busser, who preceded Adrian Cronauer (the man on whom the Robin Williams role in "Good Morning Vietnam" was based) as the morning DJ in Viet Nam, had his tongue slip one morning following a night of wild revelry in the fleshpots of Saigon. As he did the morning newscast, he misread the phrase "automatic weapons" and announced that the Army was testing **"atomic weapons in Vietnam."** This might not sound like a horrible error, but to those in Vietnam it was every nightmare come true, although it is NOT true, as Busser later said, that the commanding general jumped off a bridge and pulled the river up over his head.

☆

Maintenance engineer Mike Cleary at Lajes in the Azores wasn't an announcer but his mouth did cause the station some embarrassment. Mike was a native Britisher who had gone to the States and joined the U. S. Air Force in order to expedite American citizenship. One day the British royal yacht "Britannia" pulled into the harbor and the station was honored to be able to give a tour to the British Ambassador to Portugal, Sir Charles Sterling, who was accompanying the royal party. Station Manager John Bradley introduced the fellow Brits and the suave diplomat asked Cleary, "And how is it you find yourself in the American Air Force?" Cleary showed why he was in the Air Force rather than the diplomatic corps. He replied, "My sister married an American and that gave me a chance to get out of the bloody country." Later the local General told Bradley that it might be a good idea to see that Cleary was otherwise occupied during any future VIP tours.

☆

The standard station break at the affiliate in Kaiserslautern, Germany, is "Serving you West of the Rhine." During the annual Christmas party, the duty announcer broke away to make his announcement which came out: **Serving you the rest of the wine."**

☆

President Nixon came in for it again, although not so vulgarly, when an AFRTS announcer in Europe said he had just gotten word that Nixon had safely arrived in Paris and had been greeted with a 21-gun salute. "Unfortunately," he ad libbed, "they missed."

☆

One evening in 1950, Max Lash was the announcer on duty at FEN Tokyo. Through the glass window across from him was newscaster Phil Lenhart waiting to do the 6 p.m. evening newscast. Lash read his final announcement and introduced the news. The first story concerned the arrival in Tokyo of the British Chancellor of the Exchequer, Sir Stafford Cripps. He was to be the guest of General of the Army Douglas MacArthur during his visit.

As Lenhart was reading his story, the door behind Lash opened and, reflected in the window separating the studios, he could see MacArthur who was accompanied by a man in civilian clothes. Lenhart glanced up from his copy and paled.

"Ladies and gentlemen," he said, "General Douglas MacArthur has just walked into our Tokyo newsroom with uh...Sir Stifford Crapps, the British Chancre of the Exchequer. Uh...I mean Sir Clifford Scraps...uh..." At this point he remembered he had a microphone switch and he turned it off while he took a deep breath and, finally, recovered and got it right.

Sir Stifford...ah, STAFFORD...to his credit, took it with good humor and laughed heartily. And Lenhart, proving the thesis that *everyone* in broadcasting drops his or her brain and tongue into neutral once in awhile, went on to spend many years with ABC in New York.

☆

...We return now to our regularly scheduled program.

Right, this view of the radio tower at the Navy-run AFRTS station in Keflavik shows graphically why the country is called Iceland. Below, inside the Kevlavik television studio, weatherman ensign Ernie Martz also shows why. Note the temperature—22 degrees. (Photos courtesy of AFRTS)

CHAPTER SIX

In which we learn that, if a Little Knowledge is a Dangerous Thing, both the Army and The Navy are in danger because they have little knowledge about what the other is doing in North Africa.

Getting Tunes to the Dunes

While he and his staff were getting the programs underway in Hollywood, Lewis was careening back and forth between both coasts like a berserk yo-yo. Not only was he trying to find more money, more people and more respect for what he was trying to do, he was trying to wrest control of military radio away from the entrenched organizations which already had a piece of the action. His prime target was the Radio Section of the Bureau of Public Relations which, in mid to late 1942 was not only producing *Command Performance* but was beginning to build an broadcast empire of its own.

The USO had a number of carrier current stations in Alaska and they were delighted to turn them over to AFRS. The Bureau of Public Relations, and other government departments producing radio programs, continued to balk. In June, President Roosevelt signed an executive order creating a new agency, the Office of War Information, and appointed highly regarded newsman Elmer Davis to run it.

Davis showed absolutely no inclination to give up any control of radio. The problem was that OWI was designed to propagandize America's role in the war effort. Its broadcasts, of necessity, were filled with slanted and somewhat less than subtle propaganda which, of course, was what it was supposed to be doing. AFRS was trying to bring a touch of home to men serving around the world and that, to Lewis and his staff, meant avoiding blatant propaganda at all costs and to not only entertain but present the news quickly, honestly and without bias. Consequently, Lewis found himself in frequent "extensive and heated negotiations" involving Dr. Milton Eisenhower, General Osborn and members of Congress.

At this early stage, AFRS had no powerful shortwave transmitters on which to put their programs. These were controlled by OWI. Negotiations finally ended when Davis ordered OWI to cease producing broadcasts for the troops and to share time on its shortwave stations with AFRS. Lewis was happy with this temporary solution, but not so happy that he didn't take a few precautions. While AFRS was broadcasting from the OWI studios, armed guards were placed at the door and no OWI personnel were allowed in. Lewis explained in a speech years later that he merely wanted to insure that the GI overseas was hearing the same news as his folks at home; not news filtered through the sometimes biased brains of OWI writers.

The Bureau of Public Relations also came in for its share of arm twisting by Lewis and General Osborn and, as explained, caved in and turned over control of troop broadcasts to AFRS in December 1942.

Lewis realized that shortwave was only a partial solution to meeting troop needs. Initially he and his staff felt that it would be possible to use, either full or part time, existing stations in countries to which U. S. troops were being sent. He and the OWI opened negotiations with a number of foreign governments and individual radio stations and succeeded in getting air-time in such far flung places as Australia, Hawaii, the middle East, South Africa, India and China. As an integral part of the agreements, the foreign broadcasters had to agree to protect the rights of the performing artists, musicians, writers and composers who were donating their services to the U. S. government.

Although the plan was undeniably a good quick-fix, it had a number of limitations. One was obtaining shared time at hours the troops could or would listen. The more serious limitation was there were few American troops who were patient enough to sit through an unknown number of hours of Farsi, Hindustani or Chinese while waiting for **GI Jill** to come on and play a Kay Kyser record. And because most foreign stations didn't follow the American tradition of starting programs on the hour or half hour, it was impossible to tell the troops when they could expect to hear anything they could comprehend.

Lewis and staff now realized that, although originally conceived as primarily a program source, AFRS was going to have to get technical about things and start developing its own stations. Even though the War Department didn't assign this mission to AFRS until early 1943, Lewis sent Lieutenant Martin Work on a cross country trip at the end of 1942 and Work bought twelve 1,000-watt transmitters, eighteen 250-watt transmitters and a batch of associated equipment. Development was also started on producing portable stations which could easily follow the troops. It wasn't long before AFRS had one which would fit into five suitcases, one of which was the music library. They only put out 50-watts of power but this was exactly what was needed for coverage of limited areas.

In the confusion of war, however, it sometimes takes time for the word about who is doing what to whom to get around. Not every group involved with broadcasting understood that AFRS was now the sole supplier of troop broadcasts. If they did understand, they perhaps ignored it. It was particularly confusing at this period because the U. S. was about to make its first landing on foreign soil. Masses of men on masses of ships were heading for the massive messes on the beaches of North Africa. It was to turn out to be a mighty melee among the various broadcasters. Three individual groups of them were ready to set up shop on shore, and none was aware of the others. It was also the first chance AFRS had to prove that it was ready to do its job near where the shooting was going on. It almost got shot down.

Enter André Baruch

The phrase "radio announcer" as used today has an entirely different meaning from the phrase **RADIO ANNOUNCER** as used during the 1930-45 era. Today's often scruffy radio announcer bears approximately the same relationship to a **RADIO ANNOUNCER** as a three-chord electric guitar player in a high school rock band does to the concert master of the Chicago Symphony.

André Baruch was a **RADIO ANNOUNCER**. His voice was among the most recognizable in the country. He was an integral part of countless coast-to-coast broadcasts and his early credits include *The American Album of Familiar Music, The Shadow, Your Hit Parade* (for 25-years,) *Second Husband, Myrt and Marge* and the *Kate Smith Show*. He had learned his trade in the school of live radio during the glitter days when announcers were stars, by gawd, and acted like it. At six o'clock across the country, **RADIO ANNOUNCERS** put on tuxedos, if only to read station break announcements.

On the morning of November 8, 1942, Baruch was not wearing a tuxedo. He was wearing an army lieutenant's suit and standing on the gun deck of the U.S.S. Texas, which in turn was standing off the North African coast, preparing to support the first major amphibious invasion by American troops in the war. He was a part of a group of broadcasters known as the "First Broadcast Station Operating Detachment." It consisted of 11 officers and 19 enlisted men and had been formed by the Adjutant General's office at Camp Pickett, Virginia, in October and attached to the 3rd Infantry Division, a part of the Western Task Force which now found itself ready to invade a part of North Africa. Its assigned task was to broadcast instructions in French to the civilian population ashore to help avert civilian casualties when the assault started.

Baruch's orders were to begin the broadcast over the 5000-watt transmitter aboard the *Texas* at 5:00 a.m.

The time came and he opened his microphone switch.

"*Bon jour, Mesdames et Messieurs, ici André Baruch,*" he said, which was a far as he got. Coastal batteries ashore, with absolutely no concern that Baruch was a **RADIO ANNOUNCER**, opened fire and knocked down the transmitter antenna aboard ship.

Realizing that what he was seeing was a part of history, he and his group quickly set up their recording machine, a monstrous lathe which cut spiral grooves into a glass-based disk faced with acetate. Again he began:

"Good morning. It is the morning of November 8, 1942, an historic time..." The giant 16-inch batteries aboard the *Texas* fired a salvo, the recoil sent the recorder into temporary oblivion and effectively ended Baruch's career as a combat commentator.

All members of the First Broadcast Station Operating Detachment finally joined forces in Casablanca on the 14th with orders to begin operating existing stations in Rabat and Casablanca. They discovered that the French had already taken over the Rabat station and that there was no station in Casablanca. For weeks they pondered and wandered. One day Baruch's executive officer, Lieutenant Houston Brown, suggested they start their own station. All agreed it would be possible providing the local commanding General gave his permission. His name was George S. Patton, and Baruch was not particularly anxious to meet him.

After some prodding, Baruch went without an appointment to the General's headquarters and asked to see the General. Patton, well known for his fabulous imitation of George C. Scott, greeted him with a polite and cheery, "What in fuck do you want?"

Baruch explained and Patton, perhaps remembering the successful experimental radio station AFRS had built for him while both the broadcasters and Patton's troops were training in the California desert at Camp Young, signed a memo giving Baruch permission to start a radio station.

At first they borrowed equipment from the French who gave them a 250 watt shortwave transmitter. This the Americans rebuilt to broadcast on standard AM frequencies. When they signed on the air, the station was sparsely equipped to say the least. It boasted one turntable, one amplifier, a borrowed microphone and 17 10-inch records. By December 15 it was on the air using a long wire as an antenna and calling itself "The Army Broadcasting Service -- the Voice of an American Soldier and Sailor."

Two weeks later, Patton issued orders to give everything back to the French and vacate the premises. This meant resorting to what was then called "moonlight requisitioning" although some called it scrounging and the scroungees called it stealing. Chief scrounger was Lieutenant j.g. Ralph Carson. He provided the unit with food, radio scripts and at this point, a 300 watt transmitter and enough equipment to make a radio station. By January 15 they were back on the air, reading gossip items from magazines, reporting what news they could get their hands on and playing their 17 records in monotonous rotation. The record library continued to grow as sailors cruising off shore became so tired of the same music, they would track down the station when they came ashore and donate their own record collection.

One day Baruch heard about an officer who was supposed to have a fabulous collection of recordings. Hoping to appeal to his sense of patriotism and duty to his fellow servicemen, Baruch tracked him down in the officer's billets.

He entered to find the man dreamily listening to a Bea Wain record. The walls were plastered with pictures of Miss Wain. The record collection consisted entirely of Bea Wain recordings.

When Baruch mentioned that he understood the man had some records, the man began to sing the praises of Bea Wain. Baruch then interrupted to explain that he was from the radio station. The man jumped to his feet.

"You want to borrow my Bea Wain records! Forget it, buddy. Get your ass out of here!" He acted as indignant as if he had been asked to donate his youngest child to the local scrap drive.

Baruch backed out without mentioning that he, too, was a Bea Wain fan. Nor did he mention that he was married to her.

And he happily remained that way until his untimely death in 1991.

To fill the air time without endless repetitions of recordings, the station put on live dramas and skits. Most of them were humorous, or attempted to be, and one day a Colonel suggested Baruch and gang "stop fooling around" and do something serious. When questioned, he suggested a program about the infantry, a subject of dubious interest to a man hiding behind a sand dune being shot at by German troops. Still, a suggestion from a Colonel merits consideration, especially if you are a Lieutenant.

Fortuitously, both Humphrey Bogart and Frederic March were in Casablanca on a USO tour and, being friends of Baruch, agreed to do a show if he could come up with a script. With the staff, he produced one called "The Infantry -- Queen of Battles." During the broadcast, they used live ammunition for sound effects and as background music, a recording of Stravinsky's *Firebird Suite* which fit the mood perfectly.

Later the Colonel visited the station and praised the staff for a wonderful job. He admired the actors and asked who they were. He particularly admired the music and asked who did it.

"Stravinsky, sir," answered Baruch.

"What's his rank?" asked the Colonel.

Somehow Baruch avoided telling him that Stravinsky was then a fire fighter in Moscow.

At this time neither AFRS nor the Special Services Division had a clue that a radio station even existed in North Africa. Enter, then, Special Services to fill a void already filled.

At the request of the General in charge of Special Services in the States, Major Charles Vanda and an enlisted technician procured a 250-watt and a 1000-watt transmitter and took off for North Africa. The men arrived in good shape but the transmitters as well as turntables, recorders, a record library and other necessities had disappeared into a martial black hole. At about the same time Vanda discovered his equipment was absent, he discovered Baruch's group was present. He considered them a "freelance operation" and continued to try to find his missing station.

Suddenly Baruch's Broadcast Station Operating Detachments got orders transferring them to the Psychological Warfare Branch. Not wanting to wait until Vanda found his equipment, General Arthur Wilson, commander of Special Services, arranged to have four officers and 13 men from Baruch's outfit remain and continue operating the station. Four days later, orders were received placing all radio stations providing troop entertainment under Special Services Division.

Vanda finally found his stations and an agreement was reached to provide Baruch with the 1000-watt transmitter and all the other equipment except the small 250-watter, which was sent to General Mark Clark's Fifth Army, then headquartered in Ouijda, Morocco.

It is now March and word has reached Los Angeles and AFRS about what is happening in North Africa. Lieutenant Martin Work from Armed Forces Radio Service is dispatched to the dunes to investigate. The first thing he did was manage to get all of the Baruch group released from the Psychological Warfare unit and returned to Special Services. He then arranged for another 1000-watt transmitter to be sent from the U. S. to Algeria for the use of the armies now pouring East to meet the British troops pouring West with Germany's General Rommel in the middle.

Dwight Eisenhower, then a three-star General, but soon to become Supreme Allied Commander, ordered that in the future he wanted radio stations installed in his commands whenever and wherever practical. Baruch was named Chief of the American Expeditionary Stations in the North African Theater. Troops invaded Sicily and fighting ended on August 7, 1943. By August 13, a station was on the air. By January, there were a total of eight stations operating in the Mediterranean stretching from Casablanca to Naples. And they were under the wing of AFRS, receiving regular weekly packages of programs from home and news by shortwave for retransmission without censorship. They were also supplied with the latest musical hits including a good supply of Bea Wain records.

Baruch cherished a letter he received from a hospitalized soldier. It read: *Last week we had a tent full of boys who were treated during the Sicilian Campaign. When we turned to your swing program, smiles of joy and complete forgetfulness of their pain came over the faces of the boys. Words can't express the happiness that your programs have given to the American soldiers. So thanks for giving us a few hours of happiness each day. Good luck!*

This is John Hayes who pretty much controlled radio in the European theater during the war, rising from Captain to Colonel. Hayes selected the original staff and generally charted his own course as commander of AFN. After the war he returned to the States to head up WTOP in Washington and was later appointed Ambassador to Switzerland. (Photo courtesy of AFRTS)

CHAPTER SEVEN

*In which we learn that Samuel Johnson erred when he said **When a man is tired of London, he is tired of life.** Our guys were merely tired of having bombs drop in on them while they were in London.*

The European Vocation

AFRS had backed into its first major overseas operation in North Africa. Now it was high time to get into the big time and begin to build stations in the United Kingdom and get ready for the big push across the English Channel which everyone knew was to come. When it was to come remained, along with the recipe for the atom bomb, the best kept secret of the war.

Convoys which managed to evade the German submarine wolfpacks were delivering untold tons of material and fighting troops by the tens of thousands to the U.K. The common saying among the Brits was that the only things keeping the bloody island from sinking were the tethered barrage balloons. Americans, living and training in the cold, wet British climate or flying day after day through miserable weather, were in many cases sorry the balloons were there.

If ever a group of people needed a voice from home, this was it. Tom Lewis knew from his trip to Alaska in June 1942 how important local radio stations could be to men and women isolated from home and most creature comforts.

Of course there was the BBC, but to most of the young American troops it might as well be broadcasting from Mars, so foreign to U. S. tastes were its methods and programming. "Boring" was a word often heard. It was used to describe the heavy diet of classical music, two-day cricket matches, twelve part adaptations of obscure Thomas Hardy novels, news stories about Yellow Toed Sapsucking Titwillows being spotted by bird watchers in the Midlands and other broadcasts designed to tweak the interest of the British psyche. To the newly arrived olive drab guests, it was -- well -- boring.

It is true that after the Dunkirk debacle in 1940, the BBC had initiated "The Forces Program," filled with military news. A common complaint by the Yanks was that it never mentioned them. Nor were they interested. They wanted the kind of rapid-fire, frenetic-paced entertainment they had grown up with. Besides, they had difficulty understanding the language which Trevanian, in his book *Shibumi*, described as BBC's own peculiar pronunciation to which is added a clipped, half-strangled sound which the world audience has long taken to be the effect of an uncomfortable suppository. True or not, to the average American, it was a pain in the fanny.

Murray Brophy, of the War Department's office of Coordinator of Information, had accompanied Lewis to Alaska and recognized the need for American radio. He and Lewis reached an informal understanding and Brophy departed for Britain to negotiate as a representative of COI for permission and frequencies. The plan was that COI would provide the low-powered transmitters, temporary personnel and connecting land lines. AFRS would then take over the operation as soon as it had the capability to function properly overseas.

Technically, the plan was simple. Politically, it was a can of squirming worms. Stripped to essentials, the BBC controlled the British airwaves, had always controlled the British airwaves and, by Jove!, always would control the British airwaves.

General George Marshall had been instrumental in creating AFRS in the first place and Brophy met with him in England. Marshall quickly gave his support. Following the meeting, and a request that the Office of War Information prepare a study, Brophy sent Brewster Morgan, the OWI's Chief of Broadcasting, to England to prepare and deliver the report to General Eisenhower. With his approval in November 1942, permission was granted to begin on an experimental basis with twelve transmitters.

The outcome of the OWI negotiations with the BBC was that the little network would be considered an additional BBC service. Once approved, the BBC would supply technical facilities such as land lines and studios as well as two frequencies. On the other hand, AFRTS would have to abide by the same program restrictions as the BBC, including restrictions on questionable lyrics and writings and compliance with British copyright regulations.

Eventually, the BBC and the British counterpart of the Federal Communications Commission -- the Wireless Telegraphy Board -- produced a waiver of the BBC's monopoly and operating agreements were signed between the OWI, the U. S. Army and the BBC. The OWI ordered 28 50-watt transmitters and, as AFRS assumed control, it ordered an additional 25 through the Signal Corps.

The OWI turned to the Army Signal Corps to install the transmitters. The Signal Corps had earlier sent newly commissioned Captain Loyd Sigmon, (one L of a guy), to England along with 16 radio engineers, although he had no orders as to what to do after arrival. Sigmon, who later became one of America's most innovative broadcasters at Gene Autry's KMPC in Los Angeles, is the man who developed the "Sig Alert" which tells Los Angeles drivers where to go to get themselves caught in a traffic jam. After arriving in England, his group split up and headed in different directions doing other work.

Eventually the OWI was able to run them all down and assign them the job of putting the transmitters on the air. Sigmon was told the first four had to be operational by July 4, 1943. The BBC, although willing to cooperate, was not willing to wholeheartedly enter into the spirit of the thing and refused to let AFRS put a transmitter in London. They were okay out in the boondocks but not in the center of the British universe. Oxford was as near as they could come. For the most part, they were installed at air bases and large training camps. To reach the Americans in London, the Sigmon group, including Bill Pickering, installed a wired sound system similar to Musak which would feed loudspeakers in American quarters and office buildings throughout the city.

It is one thing to have approved paperwork, frequencies, transmitters and landlines. What was now missing from the equation was programming and a staff to get it on the air.

Morgan drew up plans for such a group. It called for OWI personnel to be in charge but with Army personnel providing the programming, both locally produced and from AFRS in Los Angeles. A board of directors was selected consisting primarily of the Chiefs of Special Services, Public Relations, the Signal Corps, the Press Officer and Engineers. Morgan selected Lieutenant Colonel Charles Gurney to be the Radio Officer for the European Theater of Operations and, as such, Gurney became the first chief of AFN, the American Forces Network. Captain John Hayes was appointed his assistant and by May 1943 they had assembled a staff of four officers and 13 enlisted men. This being long before the days of computers, finding this staff meant spending endless eye-straining hours manually going through stacks of personnel files, hoping to run across people with a former radio background.

Their mandate from General Eisenhower's headquarters was to "supply American military personnel in Europe with a radio programming service for the information, education and entertainment of such personnel." In addition, the service was to "be as much a duplication of American broadcasting at home as it was possible to achieve overseas."

This was a fine idea, everyone agreed, but the realities were that AFN would be functioning in a war zone. Hayes spent considerable time discussing with the OWI, the BBC, the Canadian Broadcasting Corporation and the various U. S. services the problems of sounding like "American broadcasting at home" when faced with the reality of censorship. The final result was a consensus with Eisenhower's staff that the only censorship would be in the area of military security. Otherwise, it was directed that the network would exercise self-discipline in so far as taste, tact and regard for the sensibilities of the Allies were concerned. They were to practice the normal restraints which the American system of free broadcasting customarily adopts for itself. No arguments there. In fact, AFN was still quoting Eisenhower's dictum to irate generals a half-century later.

July 4, 1943, was selected as the target date to sign the new network on the air. The new staff busied itself preparing plans for programming which was to come from its own productions, from AFRS in Los Angeles, from the BBC and a few programs done by local stations in the U. S. A news staff was formed and arrangements completed to receive the major wire services.

Meanwhile, Pickering, Sigmon and group were installing and hooking together the first transmitters and the OWI, still nominally in charge of the operation, began hiring engineers to prepare the London studios from which the programming would originate.

A young engineer named John "Jack" Boor was in London, trying to get back to the States. In 1939 he had heard Britain's Lord Beaverbrook explain that Britain stood on the brink of disaster and could use the world's help. He wrote Beaverbrook, then running the London *Times*, and obviously impressed him. After being screened by British Intelligence, Boor was asked to come along although he had no inkling of what he was to be doing. Next, in violation of the Neutrality Act of 1939, he found himself in the British Royal Air Force. When America entered the war, the British felt the Americans who had entered British service would be happier in their own and released them. Boor failed the U. S. physical and found himself an unemployed civilian in the middle of a war. Hearing about the OWI from a friend, he applied for a technician's job and, because of an extensive radio background, was snapped up. He remained a civilian throughout the war and stayed with AFN until 1946, leaving as Chief Engineer.

First order of business for Boor and friends was to get the studios ready. The BBC arranged for AFN to take over small studios at 11 Carlos Place, just off Grosvenor Square. It was from here also that Ed Morrow did many of his famed "This...is London" broadcasts. The facilities consisted of a small studio, a control room, two turntables and a rack of distribution amplifiers feeding the outgoing lines to the various London public address systems and the on-air transmitters in the hinterlands.

ON THE AIR...AT LAST

Stars and Stripes, the Army newspaper, announced in a front page story that the network was to sign on the following day. And they were right. It did. General Eisenhower threw the switch. The National Anthem played and enlisted announcer Syl Binkin introduced the first program. Programming the first night included *The Bing Crosby Music Hall, Edgar Bergen and Charlie McCarthy* and *The Dinah Shore Show*.

At first the network could only sustain a limited amount of programming. Sign on was a 5:45 each afternoon and sign off came at ll:00 p.m. The staff was so small initially it called on *Stars and Stripes* staffers to read the 7:00 p.m. Sports and the 10:00 p.m. News. As new people joined the staff, these hours increased. Within several months they were able to provide luncheon music for their listeners. From noon until two each day, AFN broadcast what passed for a tricky title in those days. It was called *Concert for Chowhounds*. An even trickier title was that given its first remote broadcast which originated at a Belfast enlisted men's club and featured volunteer singers, musicians and comedians. The snappy title was: *Uncle Sam's Boys Entertain.*

Although some of the shows may have bombed, it was the real thing that bothered the staff at ll Carlos Place. The British, with typical stiff upper lips, referred to the frequent buzz bombs sailing in from the Continent as "a spot of bother." More than once, bombs exploding nearby caused the tone arm to jump off the record being played on the air. Standing orders were not to explain to the audience what had happened, just in case German Intelligence was listening and would thus be able to pinpoint the accuracy of the bombs.

After a few months, the staff had grown to respectable proportions. Lieutenant Colonel Gurney had departed and his former assistant, John Hayes, was now commander of AFN and was destined to lead it through the early growing pains and the most traumatic experiences it would ever encounter. Captain Bob Light, soon to be Major, had come on board from Tom Lewis's office in Los Angeles and Jack Boor was appointed Chief Engineer, the only civilian on the management team.

Winter came to Great Britain and it was the worst in memory. The famous London fog was so thick that the beam of a 3-cell GI flashlight wouldn't reach to the ground. Boor recalls getting caught in such a fog one night while walking with a friend. They stumbled along and luckily found an open door. Actually they were VERY lucky. It turned out to be the door of the White Horse Distillery. Still another time they became so totally lost they asked a passerby to please direct them to 11 Carlos Place. The gentleman took off at a brisk pace and they were forced to ask him to slow down. "Sorry," he said, "I'm blind and I do this all the time." He took them directly to the door.

Although D-Day, the invasion of the Continent, was months away, allied preparations were frantic. Training intensified, material piled up in depots, bombing raids increased. AFN also began both training and preparing mobile units which would be assigned to each of the U. S. Armies once they reached the European mainland.

During the miserable winter, the staff had just about had it with the buzz bombs and it was decided to move from central London which was the prime target of the bombs. The BBC offered them space in its 80 Portland Place facilities. This being slightly outside the bullseye, AFN accepted and Boor and his staff began preparing the new home.

At first things progressed quickly and the British technicians assigned to help were completely cooperative. Then it was noted they were slowing down. Next, a delegation showed up to explain that it was a treasured British custom to stop twice daily for tea. There being no places nearby where tea was available, and because coffee breaks were unknown to the Yanks, Boor told them no. Three days later they were back and even more insistent. He then hired a tea lady who appeared daily, dispensed her tea and biscuits, and went on about her business. There was never another word of complaint from the British. The Americans, however, raised all kinds of hell when "their" tea didn't arrive on time.

Work was completed in time to meet their June 1 deadline which, although they didn't know it at the time, had been imposed to prepare them to begin broadcasting to the continent following D-Day, June 6.

Among the new members of the programming staff which joined AFN during this period was Johnny Vrotsos, who called himself Johnny V and who remained with the network for the next twenty years, becoming perhaps the best known voice in Europe. Keith Jaimison, who had been a top network announcer and the official announcer for the President, dropped his duffle bag in the billets one day and he remained a staff member for a few years. Sergeant Broderick Crawford, already a well known actor, was a staff announcer. Another sergeant to arrive was Vic Knight who had been the original producer of *Command Performance.*

Stars and talent of all kinds used AFN as a gathering place while in England entertaining troops. Marlene Dietrich agreed to participate in a drama program being produced at AFN one day and Sergeant Broderick Crawford was dispatched to pick her up in Leicester Square. They somehow missed connections and each spent several hours waiting for the other in separate pubs. When they finally returned, they were holding each other up and the recording session was delayed until their headaches were better.

General Jimmy Doolittle dropped by and told about bombing Tokyo with his B-25s flying off the deck of carriers. Dinah Shore was a frequent visitor as were both Bing Crosby and Frank Sinatra. Crosby sat at the piano one afternoon and wrote a song that began, "Sinatra...Sinatra...Clear from Rangoon to Sumatra." Unhappily for the world, the remainder of the lyrics are lost forever.

Other visitors included Kay Kyser, Spike Jones, Judy Garland and Fred Astaire. For a time, the station break signature was "AFN" tapped out in morse code. What the audience didn't know was it was tapped out by Fred Astaire with his tap shoes, a board floor and a low microphone.

Tommy Dorsey arrived with his entire orchestra to entertain the troops. Along with Glenn Miller, it was perhaps the most popular orchestra of its day. AFN arranged to have him record a special program and borrowed the biggest and best studio at the BBC Broadcast House for the event.

It was a perfect example of British-American cooperation. Hands across the sea and all that, don't ja know.

Dorsey was world famous for his skill with the trombone and his absolute insistence on getting his own way. He decided that the broadcast would require 22 microphones. The BBC chief engineer understood the effect Dorsey was trying to achieve musically and suggested 18 microphones would do the job perfectly.

Dorsey threw a small fit, ranted, raved, shook fists and demanded 22.

In the spirit of inter-allied cooperation, the British engineer caved in and placed 22 microphones on the bandstand. Afterward Dorsey said it sounded exactly the way he wanted, thanks to the 22 microphones.

"Absolutely, Mr. Dorsey," said the engineer. Then came a long British pause followed by, "Of course, only 18 were turned on."

Hi Ho, Hi Ho, It's Off To War We Go

Although much of the programming was coming from the AFRS headquarters in Los Angeles, John Hayes was determined that AFN should retain as much identity as possible. He had the network admit that it was affiliated with AFRS only at the beginning and end of each broadcast day. He also removed the AFRS identification from the end of each of their programs. Furthermore, although the OWI was involved in the operation and furnishing much of the staff, Hayes refused to broadcast any of their material in order that AFN could maintain its credibility as a voice of home -- not a voice of Government. It was the start of AFN's lengthy attempt to retain as much autonomy as possible within the framework of AFRS. It was also the start of a generation of AFRS executives making one word out of "that damned AFN." So ingrained did this mind-set become at AFN, and so determined was the staff to maintain its own identity, that it was not until the formation of the Army Broadcast Service in 1980 that the attitude began to appreciably change.

That's the future. The present is the short period immediately prior to D-Day, June 6, 1944. John Hayes is told that his precious and very independent radio network is going to have to join forces with the BBC and the Canadian Broadcasting Corporation when they move on to the Continent with the troops.

Damn! Is that any way to run a war?

SOUND EFFECT: OMINOUS ORGAN CHORD.

ANNOUNCER: Wellllll! Will the Brave Americans have to go to war with a bunch of foreigners? Will AFN have to start answering its mail from headquarters? Will cricket matches replace baseball on the radio? These and other unasked questions will be answered in a future episode. But first...this word from our sponsor -- World War II.

Will the Real Corporal Stuart Please Step Forward!

Pictured here are, right, M. K. Kan, born Kan Man Loh, and left, Corporal (?) Stuart—or, perhaps, Stewart. The picture was taken in 1944 or 1945 near Kunming, China, home of the Flying Tigers and the terminal for those Gooney Bird flyers dragging war materials to China over the Hump.

Kan was adopted at the age of three months by an American missionary, Cornelia Morgan, whose grandfather was a senator from Alabama. For years, the two of them toured China until the war broke out and Kan found himself in Kunming. Here he met his first real American buddy, a disk jockey with the Kunming AFRS station. His name was Corporal Stuart (Stewart?) and the two became great friends. When Stuart returned to the States in 1945, Kan was determined to follow.

Unfortunately, his address book was stolen and Stuart's address was gone forever. In 1948, Kan came to the U.S., joined the Air Force and became a citizen, living in California. Since that time he has spent much of his spare time trying to locate his first American friend.

United Press International and other news media have carried his story but to date no one has come forth with information about the missing AFRS DJ. If anyone recognizes him from the picture or served with him in China and knows anything about him and will contact the author through the publisher, the two can be reunited. Kan you help? (Photo courtesy M. K. Kan)

CHAPTER EIGHT

In which we learn how AFRS once broadcast to the Japanese and speculate that they must have enjoyed it. Otherwise, why would they now be making so many radios for the Americans?

Terrific in the Pacific With Jingles in the Jungles

While John Hayes and his merry men were getting ready to move to the Continent with the armies massing in England, Colonel Tom Lewis and his group back in Hollywood faced a Pacific predicament. Hayes, with the backing of SHAFE, Eisenhower's headquarters, could utilize the Signal Corps to obtain equipment, requisition person-power or supplies from various sources and count on cooperation from field commanders.

In the Pacific Theater, the situation was less clear cut. Troops were island-hopping up the island chains toward their eventual goal of the Japanese home islands. Stations would, because of the huge distances, be thousands of miles apart and totally dependent on their own resources. Much of the potential audience was Navy and could only hear AFRS when in port or when sailing near an island with a station unless powerful transmitters were obtained. Linking the far flung stations into a network was clearly impossible in the early days of the Pacific war while stations were being set up, broadcasting, torn down and moved almost weekly as the war moved ever closer to Japan.

Further complicating an already complicated logistical problem of distance was the fact that the War Department had carved up the Pacific map into two areas. The Southwest portion was under the command of the Army and General Douglas MacArthur. The Central Pacific was essentially Navy country under the command of Admiral Chester Nimitz, although once territory was captured, Army troops were generally stationed on it along with sailors and marines.

Each area, for a number of reasons, was handled somewhat differently. Lewis and his group realized that finding trained personnel in the middle of the world's largest ocean would be difficult. Even trained dolphins aren't that smart. AFRS assigned a group of officers and enlisted men with a background in radio to prepare for overseas duty running radio stations. With logic that is slightly cloudy to this day, the men were sent to the desert at Camp Young near Indio, California to train. This area is frequently used to simulate the depths of the Sahara in motion pictures and occasionally wild camels are still seen. After completing their desert training, the men were sent to the steaming jungles of the Southwest Pacific or to Hawaii to set up an additional training facility.

Groups operating in Navy country were referred to officially as the Pacific Ocean Network, or PON. Once in place and operating, they referred to themselves as The Mosquito Network. Those farther south in MacArthur's territory were spread among the rainy, mosquito infested, slimy jungle islands and called themselves The Jungle Network. These were fine names, if somewhat inaccurate, in that they were in no sense networks at all. Instead, they were struggling little entities, often understaffed, always overworked and more often than not living in deplorable conditions.

The staffs of the Jungle and Mosquito networks were never up to authorized strength although AFRS did everything possible to supply qualified people. Much of the work, of necessity, had to be done by volunteers and there was never a shortage of those. It was much more fun than sitting around waiting for the next invasion and slapping mosquitoes for recreation. The higher headquarters in Hawaii had a personnel office at Pearl Harbor and tried to keep everyone supplied with people. In spite of their best efforts, qualified staffers were usually in short supply.

All of the stations had similar operational problems once they hit the beach. For the most part they were operating on tiny dots of land which had been formerly owned by people with a highly developed native culture but to whom such taken-for-granted amenities as electricity were unknown. This meant the use of portable generators in order to whip out the watts. It also meant sharing the generators, always in short supply, with other users. It became a fact of life that as the sun went down, so did the coverage area as more and more lights came on and voltages dropped. This also caused turntables to slow down, making Dinah Shore sound like a basso profundo in the throes of a drunken fit.

Even stations with their very own generator found that it never put out enough power to supply both the station and their other needs. Everyone soon learned exactly how many lights could be turned on each night before Dinah started going into her fit.

The newspapers back home would have the public believe the enemy was Japan. Not to the AFRS staff. It was Dear Old Mother Nature.

Among the little tricks DOMN loved to play was to infest the stations with "juice ants," a generic term for the hordes of insects living in the constantly damp and rotting vegetation. After a billion generations of eating abundant rotten vegetation, they were delighted to change their diet and eat electrical insulation. This caused many short circuits of long duration.

Fungus was another serious problem. It not only grew in shoes, toes and crotches, it grew in equipment such as microphones. The ultimate in radio listening pleasure is **not** listening to a man scratching his fungus-filled crotch while being cheerful through a fungus-filled microphone. The station manager at Noumea, New Caledonia, reported back to AFRS that it was necessary to twice-daily blow out the fungus from the microphones with a bellows. He made no mention of how he solved the other fungus problems.

Fungus, along with frequent periods of 99.999 percent humidity, limited the useful life of radio receivers to about four months before their innards simply rotted away. Radios were already in short supply and were most often placed in the mess tent or other common-areas so they could be enjoyed by the greatest number of troops. It would be much later in the war, but the military eventually developed a radio receiver sprayed with plastic that was virtually waterproof.

The station at Noumea became the headquarters of the Mosquito Network although it was far removed from most of its command. It got a station, commanded by Captain Spencer Allen, on Guadalcanal three months after it started broadcasting in New Caledonia. Within a year, Spencer was promoted to Major and became Chief of the Armed Forces Radio Service in SOPACBACOM, an acronymn for something no one now knows or cares. What is known is that he now commanded stations in garden spots recently denuded by Naval bombardment, Air Corps bombardment and both infantry and marines who weren't always too tidy as they fought their way off the beaches. His command included names familiar to every household in America then but which have been mostly forgotten by anyone not now claiming Social Security. They include New Caledonia, Guadalcanal, Munda, Bougainville, Espiritu Santo, Nandi and Tutuila.

In one letter back to AFRS, a year after opening the first station, he reported that generally the initial problems had been solved and most stations were no longer working out of leaky tents and scratching their crotches. He also reported what every program director and station manager since the first day of broadcasting anywhere has noted. He wrote:

> Where once the listener was happy if we played nothing but Harry James recordings all day, he has now become critical. We're in for abuse if we clip the last two minutes of the NBC symphony to join the short-wave news; or if "Your Radio Theater" (Lux, to you) is canceled; or if we play "Rum and Coca Cola" too many times. In short, the longer we're in operation, the more demanding the GIs become in their listening tastes.

Programming consisted of local disk jockeys playing "Rum and Coca Cola" too many times, the package of Stateside programs supplied by AFRS, as well as the programs produced especially for overseas outlets by Armed Forces Radio. Some of the Mosquito Network stations were able to produce other local programs including church services and local boxing matches. Nor did they forget their mission to provide information. Although commercials had been removed selling soap, cigarettes and laxatives, there were other products and ideas that needed selling.

One, reminding troops in this malaria infested area to use mosquito repellent was a copy of a typical perfume commercial. A sexy voice each night read: "Are you repellent? No? Then use Toujours Gai. It keeps the mosquitos away. Remember to rub it on your delicate skin each evening as the sun goes down. Thank YOU."

The station on Guadalcanal used subtle sell to remind the troops to take their atabrine each evening to prevent malaria even if it did turn the skin a nauseous yellow. Daily at 5:30 the station presented the Atabrine Cocktail Hour. Each evening the program originated from a different glamour spot on Guadalcanal and troops would tune in to listen to music from "The Fungus Festooned Fern Room" or the "Starlight Roof high atop the Hotel DeGink in lovely downtown Guadalcanal." Just the mention of the program title several times throughout the program was enough to make the men in the chow lines remember to take their medicine — which was always waiting for them at the end of the line.

Because a number of Americans were stationed in New Zealand, permission was obtained from the New Zealand Government to open a station near Auckland. They supplied technical assistance and the U. S. furnished four broadcasters. It was the southernmost outlet and remained on the air for almost a year. When it closed late in 1944, a columnist for the Auckland Star wrote: "You've taught us how to smile, a lesson we so badly needed. The old town won't seem the same when the American station is off the air, and many a radio will want for use when you are gone. Mine especially."

Returning to the Jungle...Network, that is

Concurrent with the growth of the Mosquito Network, was that of the Jungle Network. It's area of operation stretched across the central Pacific from Hawaii to the Philippines. Under the control of the Navy, which in those days was both very busy and seldom ashore, AFRS stations faced a certain lack of priority in the minds of the commanders. AFRS was rarin' to go but getting the necessary cooperation from local commanders, busy with other things, was as difficult as trying to shave with a marshmallow. And as slow.

Colonel Lewis sent his deputy, True Boardman, to the Pacific in an effort to convince the various island commanders, and the top brass, of the necessity for providing a radio service. His journey may well be the first example of shuttle diplomacy.

He, too, island hopped. Jumping from one location to another, he visited places he felt would benefit the most from an American radio station. Many of them were under the joint control of both the Army and the Navy although the high headquarters was Navy, located back at Pearl Harbor, Hawaii. His first task was to convince both the Army and Navy subordinate commanders of the need. He outlined the benefits they, as commanders, could expect in the way of improved morale of their troops and the advantages of having a quick and direct pipeline to disseminate local information and policy. After selling the commanders of one location, he would move on to the next. It was a slow and laborious process. When it was completed, he then had to return to headquarters of the Navy in Pearl Harbor and obtain a joint working agreement that allowed AFRS to begin establishing stations.

He started his shuttle in May 1943. By February 1944, the first of the Jungle Network stations signed on the air from Port Moresby, New Guinea under the command of U. S. Army Captain Edgar Tidwell and Australian Captain Robin Wood along with a staff of six men. This was a joint American-Australian operation because Port Moresby was in the half of the island administered by Australia. The next station to open was in former Dutch territory, at Nadzab, and because the Dutch had other things to occupy them — such as the German army — this was a purely U.S. venture. Next to open was Finchhaafen which Tidwell chose as the flagship station of the group.

A new commander in the person of Ted Sherdeman, a highly experienced and highly regarded radio executive, was appointed. He arrived at the time General MacArthur and the Navy were busily invading island after island on the way north to the Philippines and, eventually, the Japanese home islands. Under his direction, the stations would island-hop right along with the invading forces.

There was no chrome-plated furniture or snazzy blond receptionists as these stations. The station at Finchhaafen was at the top of a muddy hill overlooking the harbor. It occupied a 20 by 45 foot prefabricated hut which could only be reached by four-wheel drive vehicles when it rained, which was all the time. Surrounding it were tents providing less than luxurious accommodations for the staff, both officer and enlisted. The precious generator rated its own tent. Because reaching the location was so difficult, many of the locally produced shows were done from the bottom of the hill in a studio in the Engineers' recreation hall.

This was one of the more elaborate stations, although the staffs of the various outlets showed supreme ingenuity. Many of the studios were in 18 foot squad tents. By putting the center tent pole through the center of a supply-drop parachute, and spreading it out from the peak, a double ceiling provided both insulation from the sun and outside sounds. The bright red or yellow silk ceiling undulating in the breeze gave something of the impression of being in a Turkish harem. The effect was somewhat lessened knowing the nearest person of the opposite gender capable of making a harem a harem was usually six islands and 3,000 miles away. Nor did the olive drab GI blankets hanging from the walls to improve the acoustics add to the ambiance.

One station was on the side of a muddy hill and in constant danger of changing location drastically during the frequent heavy downpours. Another was located strategically between a fuel depot and an ammunition dump. Although not the intention in selecting this location, it had the effect of keeping the staff alert.

As in many wars, the war in the Pacific consisted of lengthy periods of sheer boredom between invasions, followed by relatively short periods of terror and tension. These men had little to occupy themselves with other than a few tattered paperback books and radio. There were no towns. There were no clubs. Food was generally terrible and monotonous. Movies were shown outdoors, rain or shine, and were often years old and in terrible condition. Mail from home was a chancy thing and it was not uncommon to have mommy send a picture of the baby's six-month birthday party which arrived before the recipient knew he was a daddy. These guys were not happy campers. They just wanted to get this thing over with and get out of the slime and muck and go home.

Sherdeman knew this when he wrote to Tom Lewis back in Hollywood about programming from AFRS:

Don't, to the slightest degree, wave the flag. It's sure death. I'd suggest you have whatever script is prepared carefully gone over by somebody who's been over here — I mean somebody who's been stuck in New Guinea for awhile. The brief inspection trips can't reveal the truths of what men feel in this country. Only those who've lived in it and gone through the loving and hating it can tell whether what is written is honest or not. If it isn't honest; if it doesn't ring true to them — they'll jeer it down.

To make sure the stations under his supervision followed this philosophy, he made sure the stations were operated by the enlisted men. It was a rule followed by AFRS stations everywhere in years to come with certain exceptions such as chaplains, base commanders or general officers. Officers often managed the stations but the enlisted personnel ran them. The rule — frequently neglected by some over-enthusiastic officers — was "guide but don't lead."

In 1944 the U. S. military took off in earnest and names no one had ever heard of before were suddenly in the headlines. As the Japanese fell back, places like New Britain, New Ireland, Kavieng, the Admiralty Islands, Rabaul, Tarawa, Makin, Kwajalein, Eniwetok, Saipan, Iwo Jima, the Marianas, Peleliu and Ulithi were being discussed in knowing tones across America's dinner tables. AFRS opened a station on each of them.

Because the military strategy was to by-pass many other locations harboring enemy forces, AFRS was called upon to provide a service which violated Lewis' firmly held conviction to make the organization exclusively a voice for the American serviceman and woman. On the other hand, AFRS is in the military too and sometimes it's necessary to break the rules in order to get the job done. So, for a short period, some of the stations such as Peleliu began doing broadcasts in Japanese addressed to those holdouts on by-passed islands. Many of them, cut off from sources of accurate news, refused to believe the war was over for them. AFRS would play some Japanese music to get their attention and Japanese speaking Americans would tell them the truth about what was happening to their forces.

In what might be the most unusual radio promotion scheme ever used by AFRS, aircraft flew over and dropped leaflets on Japanese positions announcing the time and frequency of the next Japanese language broadcast. Although it was impossible to measure the success of these broadcasts, captured documents found after the war revealed that they were extremely successful in convincing Japanese commanders that future efforts were futile. It's also possible that the Japanese, their ears accustomed to a five tone musical scale, surrendered rather than listen to any more American jazz.

Still another deviation from policy was the inclusion of coded messages to "coast watchers." These were most often British, Australian or Dutch nationals who had been living on islands taken by the Japanese earlier. Many of these brave souls took to the jungle and, with small two-way radios, reported Japanese activities going on around them. AFRS broadcasts sometimes included coded information requesting specific information or setting times for their next transmission. Few survived their years in the jungle; if discovered they could expect to quickly become casualties.

Austin Peterson, chief of programming at AFRS, was sent to the Pacific by Tom Lewis to determine whether the AFRS programs were hitting their target. Peterson determined they were and found that equipment rather than programming was the most serious problem facing AFRS outlets. Even more so was the shortage of radios for the troops. Early directives of the War Department had prevented troops from bringing their own radios with them. And now, with the U.S. getting ready for their final push toward Japan, radios were in short supply. Peterson found that the Saipan station had distributed 381 of them to hospitals and other listening points. When he got to Manila and told one GI that a station was about to go on the air, the reply was, "What the hell good is a station if there are no radios?" He agreed the man had a good point.

He also found that there was a critical shortage of equipment at many of the stations — just as there was a critical shortage of most things both overseas and domestically. Many of the stations were becoming adept at getting equipment by "finding" it. Providing it wasn't welded down, it was often possible to find something before it was lost and put it to good use. On Saipan, the station found a shortwave receiver in the OWI office and on Guam the station found a receiver before the base information office knew it had lost it.

As he had promised when he left Corregidor in Manila harbor in early 1942, General MacArthur returned to the Philippines as the island of Leyte fell in December 1944. A station quickly went on the air followed by one on Samar, the next island north. By February, Manila had fallen and AFRS was broadcasting from there. So quickly did it go on the air that perhaps the staff failed to notice that the building in which it was installed only had three walls, the result of an artillery hit earlier. Soon realizing this was not an ideal situation for a radio studio, the station moved into the Sugar Building and, for the first time in the history of the Jungle Network one of its stations was in what could be considered adequate quarters. Not only were there hard walls, floors and ceilings, there were toilets and light switches.

Close to the site of the famous photograph of the Marines raising the flag after the capture of the Iwo Jima, AFRTS picked out the site for the radio station there. This place doesn't look like much now, but just wait until the station is built.

This became the headquarters of all the remaining Pacific stations and, now that the war had moved this far north, it was decided that they were now out of the Jungle. The Jungle Network, by mutual agreement of Hollywood and Manila, took on a new name.

The stations were now a part of The Far Eastern Network. Japanese troops who were still holding out throughout many of the Philippine islands and who heard this new voice had no way of knowing the network's next headquarters would be Tokyo and call itself more simply The Far East Network.

They should have known. AFRS Manila and Mindinao also directed a number of programs to them, as they had done from the more southern islands. To help, the military, which had trouble finding receivers for American troops, managed to find a number of them to drop by parachute to the hold-out Japanese sitting on isolated islands. This may be the last recorded instance of the U.S. sending radios to the Japanese.

STILL MORE ORIENTATION ABOUT ORIENT STATIONS

The war in the China-Burma-India theater — the CBI — never got the publicity that the fighting in Europe and the Pacific got. Some say that's because it was so nasty, no correspondents wanted to go there. Much of it was fought in slimy jungles filled with leeches and other creepy-crawlies. All of it was fought in rain and mud with short supplies and back breaking labor. Early on, General Stilwell had been pushed out of the jungles by the Japanese troops. Literally. Thousands walked or crawled back to what passed for civilization in Burma and India. Mostly today what is remembered is the airlift over "The Hump" to supply the Chinese with essential war materials using rickety C-46s and C-47s. People remember the Ledo Road, scraped through hundreds of miles of jungle and mountains from Ledo in Northern India, across Northern Burma to Myitkyina where supplies could be moved forward to Kunming in China.

Rear echelon headquarters for the CBI theater were in New Delhi, India, when the first station went on the air. Named VU2ZY, it boasted 50 meager watts of power and a giant staff of four. The three program people were Captain Lee Black, Chuck Whittier and Bill Stulla. Eskil Holt was the engineer. Black later became manager of KIMA in Yakima, Washington. Bill Stulla returned to Los Angeles where he was one of the first to explore Children's programming on television. He got his feet wet at the local NBC station, KNBH, with *Bill Stulla's Parlor Party* and then became "Engineer Bill" hosting a kid's show in Los Angeles for many years.

Eventually the "network" in the CBI, which was never an actual network in that the stations all operated independently, served such unlikely locations as Karachi, Agra, Calcutta, Lalmahirhat, Bangalore, Gaya, Tezgaon, Jorhat, Misamari, Kandy, Ledo, Chabua, Shingbwiyang, Myitkyina and Bhamo. Captain Black was the theater radio officer and it's doubtful he ever learned to spell all the places under his supervision. Announcers trying to

And here it is! It doesn't look like much now, either. If the Academy of Television Arts and Sciences ever gives an award for a bare bones operation, this is a sure winner. (Actually, soon after this early photo was taken, AFRTS managed to put more equipment than that lonely microphone in place. (Photos courtesy AFRTS)

say, "This is AFRS Lalmahirhat" didn't have an easy time of it either.

Other staff members added as the stations increased included Les Damon who, before the war, had been doing The Thin Man Series. Also on staff was a perpetually smiling Bert Parks who roamed the theater doing interviews which contained considerably more substance than those he later did with potential Miss Americas. On the other hand, his interview subjects in the CBI were usually covered with mud and leeches and never once inspired him to sing, "Here He Comes, Flying Tiger Pilot Stanislaus Wojiwinski..." His writer and partner was Major Finis Farr. Parks went on to do *Break the Bank, Stop the Music, Double or Nothing, the Miss America Contest* and appeared on Broadway in *Music Man*. Farr continued writing and has written well received books on Frank Lloyd Wright, Jack Johnson and Margaret Mitchell. So good was the material the team turned out, the NBC Blue Network made a regular Sunday series of it called *Yanks in the Orient*.

Staff Sergeant Gene Kelly (no, not THAT Gene Kelly), was chief-everything at the Bhamo station. Prior to entering the service, he had been a minor league ball player and, afterwards, became the voice of the Philadelphia Phillies. He broadcast a number of live baseball games in the CBI between such teams as the Cobras and the Mudcats but surely must hold the record for doing the most unusual baseball remote — and possibly the dumbest — of all time. He wanted to broadcast the game, but he also wanted to play. He did both. With an extra long microphone cable, he played first base and described the action at the same time. Made for an unforgettable broadcast.

Private Bob Greene of the New Delhi station, is remembered still by those who heard him for his subtle but stinging ad libs. A former CBS announcer, he reported on a parade in Tokyo led by Emperor Hirohito. "He sat astride his mount like it was a part of him," Greene told the audience. "And in my opinion, it was."

By August of 1945, the CBI stations stretched from India to China and everybody wanted to go home, including the station staffs. As rumors of the war ending

The station in Myitkyina, Burma called itself "Half Way House" because it was exactly half-way around the world from Hollywood headquarters. It was on the Irawaddy River, on the other side of which is China. It was built under emergency conditions to solve a major morale problem following the battle of Myitkyina in which Allied Troops suffered fifty percent casualties.

Among the listeners in the area were the famed "Merrill's Marauders," the "Mars Task Force," and the "Flying Tigers." Generals "Vinegar Joe" Stillwell and Dan I. Sultan were there along with Flying Tiger Chief, Colonel Claire Chennault. Allied forces were also listeners including the First Chinese Army under General Chiang Kai Shek and British, Australian and Gurkha troops under General Lord Montbatten.

Part of the staff seen here are (L to R) Technician Bud Sawyer, Station Manager Charles Purnell, Engineer Max Fink and Commanding Officer Al Davis. The dog is named "Spam." He did the opening bark every morning for "Dog House News." (Photo courtesy AFRTS)

increased, the radio people manned their short wave receivers for the latest word. The Bangalore station was first on the air with the official word. In fact, they were the first station in India to broadcast it, after Sam Boyd had monitored shortwave for 96 straight hours.

They were not, however, the station which is best remembered by those who were there on that great day. That honor belongs to Chabua and station "Victor George." Colonel John Virden was there and remembers it well. He heard the whole thing from Ledo, which he describes as "a stinking town." It was the night of August 10, 1945 and, because Ledo averages 300 inches of rain a year, it was pouring buckets. His group was gathered around a radio in a leaky tent drinking rain water and a hideous concoction fermented from green bamboo which the Indian manufacturers labeled "Old Fighter Brand." It packed a giant wallop, if only you could get it to stay down.

Reconstructing what went on that night at the radio station, the staff apparently just opened the microphone and let the good times roll. They also had a fine supply of "Old Fighter Brand." Gurgles and slurred words began coming over the air. Then the announcers, who had for much too long been playing music requested by the audience, decided THEIR time had come. The three most requested songs for a year had been Blue Grass Roy singing *Strawberry Roan*, Roy Acuff's *Great Speckled Bird* and a lagubrious ballad named *There's A Star Spangled Banner Waving Somewhere*.

"We're all going home soon," one announcer said. Glug...Glug...Glug. "And what am I going to tell my grandchildren I did in the war? I'm going to have to tell them I worked at a stinking jungle radio station and my audience's repertoire of requests consisted of *There's A Star Spangled Banner Waving Somewhere*." (He had a lot of trouble with repertoire but got it right after three tries.) "It's about some poor goon who wants to go to war but can't because he's got a twisted leg. I wish to God I had a twisted leg."

There was a sudden crash as he smashed the record on the microphone. "That Star Spangled Banner that waved somewhere ain't gonna wave here no more," he said, followed by a series of glugs. By now he was completely in the bag, but continued manfully. "While I'm at it, I'm gonna pull every damn feather out of the damn tail of that damn Speckled Bird," he said. CRASH! Another disk gone. More glugs. Then he killed the Strawberry Roan.

By now, the entire staff had gotten into both the Old Fighter Brand and the record library. The sounds of records they didn't like were heard for the last time as they were smashed to the accompaniment of "take that's" and maniacal laughter. Soon most of the records, and all of the staff, were smashed and the only sound heard as dawn approached was the gentle snoring of the young fighters sleeping off the Old Fighter.

The Army felt obligated to order an investigation but some officer who had heard, and enjoyed, this unique radio program, assigned an investigator who already had orders in his hand to go home. The investigation died before it began.

❖ ❖ ❖ ❖ ❖

With the war over, stations opened throughout Japan and Okinawa while stations began closing elsewhere in the CBI and the Pacific. Troops now were more interested in how fast they could get home and AFRS faced a gigantic information job preparing them for the transition from military to civilian life. AFRS prepared a series of programs explaining the process of how men would be returned to the States for discharge and, while today that subject might sound dull, it was the most exciting thing the exhausted troops had ever heard.

Back in Hollywood, many of the writers, producers and technicians were trying to get out of khaki and back into grey flannel as quickly as possible. Tom Lewis left in October and Martin Work assumed the job of trying to provide programming with a staff which was shrinking rapidly. Even many of the entertainers who had happily provided their talents during the war were losing interest in contributing time and effort now the war was over.

Luckily for everyone, many of the staff also decided to put on their grey flannels but remain with AFRS. Marty Halperin was one such. The guru of the recording section, he went home in Khaki one day and reported back to work the next day as a civilian. He was to remain a keystone of the huge recording operation for ten years.

Another was Jack Brown. At that time he was a navy enlisted man with a PT boat squadron on Samar. As the war ended, he heard the AFRS station was in need of personnel. Hunting down the station, he was hired on the spot. From the junior man blessed with the worst shifts and the dirtiest jobs, he quickly rose to program director and then station manager. Failing to get a civilian position there when it came time to return to the States, he remained in the Navy waiting for orders which, he was told, would be to either China or Alaska. When they arrived, he found himself assigned to McCadden Place in Hollywood, AFRS Headquarters.

It was one of the best things that ever happened to AFRS. Starting as a Navy news broadcaster on AFRS shortwave, he became progressively a civilian and the chief of broadcasting guiding the entire worldwide program operation for many years until he retired in the 1980s.

SOUND EFFECT: Loud, reverberating Gong

ANNOUNCER: We'll soon return to the mysterious Orient for the exciting continuation of our story. But first, through the magic of the printed word, we return to the foggy British Isles for the conclusion of the exciting battle for independence by AFN. Join us now as we travel to England...1944.

The first in a long series of ever improving mobile units make their way north as the Army advances through World War II Italy. At war's end, these became the nucleus of the Blue Danube Network in Austria and later versions would bring radio—and later television—to the troops wherever the fortunes of war took them. (U.S. Army Photo)

CHAPTER NINE

In which we learn that nothing much has changed in the 1,900 years since Petronius *said, "We trained hard...but every time we were beginning to form up into teams, we would be reorganized. I was to learn later in life that we tend to meet any new situation by reorganizing...and a wonderful method it can be for creating the illusion of progress while producing inefficiency and demoralization."*

"The Camp Followers"

When last we left England, the young but growing American Forces Network was facing up to its first major crisis. Commander of AFN John Hayes was absolutely insistent that his little network would maintain objectivity in its presentation of the news and other material. He was determined that it would avoid at all costs any material prepared by the Psychological Branches of the military or by the Office of War Information. Much of this material was prepared with the highest of national objectives in mind but, in large part, was so heavy-handed and outrageously partisan that even the most unsophisticated listener instantly recognized it as containing large gobs of barnyard effluvium. He considered it basic that AFN should confine itself solely to programming for American military and civilian-attached personnel and should ignore the fact, other than breaching security, that both allies and the enemy could hear the American stations.

In a statement made later, Hayes said, "American personnel in the European Theater should never be subjected to any broadcast heard over the facilities of AFN which might be construed to be propaganda or commentary or news analysis or hard news which would be prepared by other than personnel of the American Forces Network. Any other action would have led the AFN audience to the conclusion that the American Forces Network was being used as a propaganda arm of the Government."

Generally, succeeding commanders over the years, both at AFN and at other AFRTS stations around the world, have been able to keep this article of faith alive and well. Soon, however, Hayes and his staff were going to be called to the battlements to fight to maintain the wall of credibility they were working so hard to strengthen.

After having been on the air eleven months since the sign-on of the network on July 4, 1943, AFN was visited by Colonel Tom Lewis who arrived in the United Kingdom to supervise preparations for American broadcasts to the troops following D-Day, the invasion of Europe by the Allies. Few knew the date but if they did they kept their mouths tightly closed rather than face a court martial or a firing squad. Still, it was obvious from the preparations going on all around them that the date could not be far off.

Lewis arrived in May 1944 and found that Hayes had replaced Charles Gurney as commander and was in the process of moving the AFN headquarters from Carlos Place to 80 Portland Place, a move designed to get the studios outside the bulls eye of the buzz bombs still plaguing the city of London. He was met at plane side by General Osborne at Prestwick, Scotland. Osborne immediately stressed the necessity of secrecy concerning the dates of D-Day. Lewis' reply was something less than serious, in that he didn't have the vaguest notion of what that date was. He soon became serious when Osborn told him about a two-star general who, in his cups, talked too much and too loudly and suddenly found himself a Lieutenant Colonel under house arrest.

The explosions caused by the buzz bombs were insignificant compared to the explosions Lewis found going on at Supreme Allied Headquarters in London. They were caused by the controversy raging between British and American planners on the method to be used to furnish radio broadcasts to the troops after they hit the beaches of Normandy and began slogging toward the German heartland.

Lewis was in complete agreement with Hayes that each of the allies should provide its own troops with a radio service tailored to their particular needs and desires. The British were in agreement as well. It did not take Lewis long to find out that his orders from "a civilian general," General Osborn, didn't carry much weight with the established military bureaucracy. Almost immediately Lewis, "a reputed communications expert," and an instant Colonel found himself bucking heads with General Ray Barker, General Eisenhower's communications officer.

THEME; ESTABLISH AND FADE TO ANNOUNCER;
We'll be back in a moment to find out whose head was hardest after these brief messages:

While the high-level fighting was going on, broadcasting continued and some of it was slightly on the bizarre side. In spite of AFN's justifiable pride in avoiding, sometimes with great difficulty, the inexplicable demands of higher commands, there were times when they simply HAD to go along with them.

One example, recalled by John Hayes, was an occasional request from the highest of high commands to play some totally ridiculous song at a specific time of day. Down would come the order: "Play **Lily of Laguna** at 2:43:25 p.m." And play it they would.

Once, he recalls, London was ordered to play **Sur le Pont d'Avignon** fourteen times in a single day. It was obvious to all that someone was telling someone else something that someone somewhere needed to hear. Best guess was that someone was telling the French Underground something important and that AFN was not just supplying music for a disco party in Avignon.

❖

Right in the middle of the blitz in 1944, AFN took a poll. Nothing strange about that. While broadcasters in some distant future describe the events of Armageddon, others will be busily taking a poll to determine which station is getting the highest rating.

This poll determined what the most popular songs of the previous year were according to AFN listeners in England. If you remember all of them, you'll find the number of your nearest Social Security office in the front of your phone book.

1. Long Ago and Far Away
2. I'll Be Seeing You
3. I Love You
4. I'll Get By
5 Amor
6. I'll Walk Alone
7. It Had to be You
8. San Fernando Valley
9. Besame Mucho
10. Trolley Song

We return now to our story...

Stripped to its essentials, General Barker's plan was to combine the separate British, American and Canadian broadcast operations into a single Allied Expeditionary Forces Program which would serve all the troops of these nationalities and services. This was in direct opposition of everything Lewis and Hayes had been fighting for. According to Lewis in a statement made after the war, "General Barker believed he had created something new in a field so new to him, and he held to it despite all opposition." Lewis was frustrated in that Barker had the ear of Eisenhower but at the same time was making decisions in spite of the fact he didn't have a clue as to what AFRS was all about -- that it was designed to keep **American** soldiers and sailors in touch with the sounds and news of home, by means of free, undoctored public communications. Lewis said that Barker negated everything AFRS was all about by changing the whole concept into a "joint British-American mish-mash."

He made a major effort to change Barker's decision, something not always easy to do to General Officers who sometimes shoulder God-like attitudes along with their stars.

Lewis prepared a plan of action for AFRS to be followed after D-Day. Along with Arthur Page, a civilian consultant to Secretary of War Stimson, he took it to the general. Dated May 20, 1944, several weeks before D-Day, it called for AFN to use one of the high-powered BBC transmitters on the English Channel coast to broadcast programs to the American Troops. The BBC would use a second transmitter for its General Forces Programme. (Purists please note that Americans broadcast **programs.** British broadcast **programmes.** The British often find programs incomprehensible. Americans tend to find most programmes dull.)

AFN also planned to attach mobile broadcast vans to each of the invading American armies and set up permanent stations in newly liberated major areas.

Barker read the plan. Lewis explained it required the General's endorsement and Eisenhower's signature. Barker read it again and said, "This, I take it, is a complete reversal of the plan I am in favor of." Lewis responded that it was a restatement of a plan already outlined to Barker and one upon which AFRS was founded. He said he could not endorse a plan which, in his professional judgment, was a mistake.

General Barker tried to suppress his anger, and said, "I do not approve, Colonel Lewis. Career army men like myself sometimes wonder why it was necessary to give such rank to civilian specialists like yourself when men older than you have worked their entire lives for such rank." Lewis tried to explain further but Barker broke in: "Only one thing wins wars, Colonel Lewis: Leadership! Not radio leadership, not Madison Avenue leadership, not Hollywood leadership. Military leadership! General Pershing did not need radio speakers in the first World War. Remember that!"

Lewis noted that radio had not existed during World War I and that the issues of that war had been comparatively simple. He explained that he and Frank Capra's film unit were trying to tell the United States Armed Forces why they were fighting, who was leading them, the nature of their enemies and the nature of their allies. He said their campaign was intended to "bring General Eisenhower close to the many thousands of men of different nationalities that he now commands." He said that, partially because of radio, the men knew him, liked him, called him "Ike" among themselves and knew that he was with them.

It was a losing battle. General Barker was not impressed although he prepared a letter of commendation for General Osborn for Lewis to hand-carry back. When Lewis asked what it would take to make Barker change his mind, Barker refused to answer.

As Lewis boarded his plane to return to the U. S., Barker called him at Prestwick, perhaps to make sure he was leaving and getting out of his hair. Besides, Barker knew D-Day was only days away. Lewis didn't. Nor did Lewis know that while his opposite numbers at the BBC were on his side, Eisenhower had placed high priority on integrating all aspects of British, Canadian and American forces under his command.

Lewis' battle had been lost before it began.

In a final skirmish, Barker met with the British Minister of Information, Brenden Bracken, the director-general of the BBC, and others on May 19. The purpose was to obtain permission to use the BBC transmitter at Start Point on the Channel Coast for the combined broadcast service. Bracken wasted no time in informing the group that the BBC Board of Governors were unanimous in their opinion that the Barker plan was impractical. Barker reacted in the same way he had with Lewis and stated that Eisenhower and his staff had already approved the plan and that the purpose of the meeting was to arrange terms of a lease for the transmitter, not to debate the plan.

Bracken continued that it was impossible to meet the interests and needs of the different armies with one service. "Let's be practical," he said. "You have that fellow Bob Hope. Very funny to your people. Not to ours. We have (British Comic) Tommy Trindler. I doubt if your people can understand what he is saying, much less laugh at him." He ended by stating that the BBC board was absolutely unanimous in its opposition.

Barker would not accept this but told the group that he would inform General Eisenhower of the opposition and ask for a decision at the very highest level. He got it.

Eisenhower wrote to Winston Churchill asking for a formal acceptance of the combined broadcast operation. Churchill wrote to Bracken giving him two choices: either the BBC take over the entire combined operation or give Eisenhower the transmitters so he could undertake it through his headquarters. Churchill was too much of a politician to ask the parliament to change the broadcast laws during wartime and change the BBC mandate. The BBC, knowing the kind of leverage Churchill had, quickly agreed to reconvene, rethink, reconsider and retreat. In short order, the BBC caved in to political pressure and, with a few last mumblings that the plan would never work, accepted the inevitable. Effective with the still unannounced date of D-Day, the combined Troop Broadcasting Service of the Supreme Headquarters, Allied Expeditionary Forces would become a reality.

AFN continued to broadcast to the hinterlands over low powered transmitters located at various air bases and training areas throughout the United Kingdom. The BBC's chokehold on broadcasting within the British Isles continued to prevent over-the-air transmissions in London and other high density population areas. London was naturally the center of activity and all major U. S. commands were located there as were thousands of support troops. AFN supplied service to these people by means of a wired system which fed loud speakers in all the major clubs, offices, housing areas and other gathering places.

Such a system provided the opportunity for a

The contents in this unknown Private's mess kit doesn't look very appetizing, because it isn't. It does, however, provide food for thought. Radio receivers were in such short supply during World War II that troops built their own using any materials at hand. This man built his in his mess kit and strapped a battery around his waist. What some people won't go through to hear Kay Kyser records! (U.S. Army Photo)

number of non-governmental broadcasts directed toward specific individuals. One such was to a young officer who would disappear frequently with a young British female employee for lengthy "lunches" at his apartment which, because he was an AFN officer, was wired into the system. The engineering staff, with impeccable timing, waited until the crucial moment and plugged a microphone into the amplifier feeding his apartment. In the throes of passion, the two lunch partners heard a sepulchral voice intone: "This is God. Aren't you ashamed of yourselves?"

On May 24, just weeks before D-Day, William Haley, Director General of the BBC, informed General Barker that he had appointed Maurice Gorham, a long-time BBC executive, as director of the new Forces Broadcasting Service. American Colonel Edward Kirby, newly arrived from Washington, was to be head of the SHAEF Broadcasting Services which was an excellent match as both men had worked together previously. Tom Lewis turned over his files to Gorham who discovered that his operation was to commence on D-Day. This presented him with something of a dilemma because no one knew when that was to be. At the same time, the BBC turned over its large transmitter on the Channel coast to Gorham who later discovered that the coverage area precisely matched the landing zones on the French Coast. He was promised a staff to be partially supplied by the Americans and the first to come into place was John Hayes, until now commander of AFN but now assigned as second in command to Gorham. Also arriving was Bob Light of AFN who became AFN commander in place of Hayes. Showpiece of the combined staff was Lieutenant Colonel David Niven who had left Hollywood lotus land to return to do his bit for his native country. Writer Dorothy Parker's husband, Captain Alan Campbell, also joined the staff.

The first official meeting of the combined staff, which by now included Gerry Wilmot of the Canadian Broadcasting Corporation, quickly degenerated into a donnybrook. Gorham announced that the BBC had decided the official name of the combined operation was "The Allied Expeditionary Forces Programme of the BBC." The Americans and the Canadians didn't object to the AEFP call letters but "of the BBC" was more than they could take. The whole negotiations started to unravel and it took General Barker to step in and solve it.

He did so by ruling in favor of the BBC and "of the BBC" continued to be a source of irritation and confusion to staffers and listeners alike throughout the life of the AEFP because AFN, the Americans, supplied 50 percent of the programming and the Canadians another 5 percent.

ACROSS THE CHANNEL

The morning of D-Day, June 6, 1944, allied troops established a tenuous beachhead on the coast of Normandy. They were a little too busy to listen to the radio that day but on the morning of June 7, Margaret Hubble, a popular BBC personality, signed on the AFEP transmitter with the following announcement:

WE ARE INITIATING TODAY A RADIO BROADCASTING SERVICE FOR THE MEMBERS OF THE ALLIED EXPEDITIONARY FORCES. WE SHALL CALL THIS SERVICE THE AEF PROGRAM. IT IS TO BE A SERVICE EXPECIALLY PREPARED FOR YOU AND WE SHALL TRY TO MAKE IT OF A CHARACTER SUITED TO YOUR NEEDS. ITS PURPOSE IS THREEFOLD: TO LINK YOU WITH YOUR HOMES BY MEANS OF NEWS BROADCASTS FROM THE UNITED KINGDOM, CANADA AND THE UNITED STATES; TO GIVE YOU THE LATEST NEWS OF OUR OWN AND OTHER WAR FRONTS, AND THE WORLD EVENTS; FINALLY, TO OFFER YOU DIVERSION AND RELAXATION DURING THOSE PRECIOUS FEW MOMENTS OF LEISURE FROM THE MAIN JOB AT HAND. FOR THIS LATTER PURPOSE, WE SHALL BRING YOU THE BEST ENTERTAINMENT THAT CAN BE SUMMONED FROM OUR ALLIED NATIONS. THE BBC HAS GIVEN GENEROUSLY OF ITS RESOURCES AND SKILLED PERSONNEL. THE AMERICAN FORCES NETWORK AND THE CANADIAN BROADCASTING CORPORATION ARE WORKING CLOSELY WITH THE BBC IN THIS PROJECT, MAKING IT A TRULY INTER-ALLIED EFFORT. WE SHALL GO FORWARD TOGETHER TO VICTORY AND ON THE FORWARD ROAD. THE AEF PROGRAM WILL BE CONSTANTLY WITHIN YOUR REACH, SERVING YOU, WE HOPE, IN A MANNER WORTHY OF YOUR DEEDS.

Never one to give up easily, AFN continued to broadcast on its own transmitters a full schedule to the troops remaining in the United Kingdom. In addition, it still had the mobile broadcast vans which Tom Lewis and AFRS had envisioned earlier as supplying the American programs to the troops once they started moving across the continent. These were assigned to the various American armies on the move and they provided the entire entertainment package supplied by AFRS in Los Angeles to listeners within their range. Although the AEF program, or to be precise, programme, continued throughout the war, AFN was there as well. Should the American troops get bored with listening to day-long cricket matches, which they invariably did, chances were they could find AFN playing the latest hits from home.

In all honesty, the AEFP had a few of these as well. Glenn Miller, leader of the most popular orchestra in America, had joined the Army with many from his orchestra whom he augmented with top musicians from other bands. Sent to England, he became the American Band of the Supreme Allied Command with the primary duty of providing musical programs for the AEFP. The 40-piece group supplied weekly programs and also broke up into smaller aggregations to broadcast specialized programming such as "Strings with Wings" featuring the string section and a jazz program headed by famed pianist Mel Powell, formerly with the Benny Goodman band. Johnny Desmond was the vocalist who came to the group from the Gene Krupa band. This was Miller's first use of a string section and it was an innovation he hoped to incorporate into the postwar Miller organization.

It wasn't to be. After the liberation of Paris, Miller decided to move the group there for the broadcasts in order to be nearer the troops. The band arrived as scheduled. Miller, flying in a single-engined aircraft with a Colonel who had offered him a ride, disappeared into the mists over the English Channel and was never found. The band continued to broadcast, however, under the baton of Sergeant Ray McKinley who had led the dance orchestra from behind his drums.

As the allied armies continued to roll westward, Tom Lewis dispatched his chief trouble shooter, True Boardman, to visit the fronts and find out how the combined operation was functioning and what the reaction of the American audience was.

In a lengthy report, Boardman wrote that the acceptance was shaky at best by the Americans. They frequently referred to the AEFP as "that BBC thing" and never did quite understand what those British comics were talking about. They expressed interest in the Miller music shows but were somewhat less enthusiastic about the British band, George Melachrino. Boardman's report praised the British professionalism and expressed understanding for the need to initially show a sense of solidarity between allies. But now that the war was entering into what seemed to be its final phases, he ended his report by saying, *"The joint operation now in effect provides a program service less than satisfactory to most American listeners and one which works against, rather than for, friendly relations with our British allies."*

He found that the most common complaint among the American listeners was a definite tilt toward British victories on the news, to the detriment of American accomplishments. The U. S. was not getting due credit, soldiers felt, and public information officers tended to agree with them. They pointed out that although the program was supposed to be 50 percent American, this was unfair in that the American troops far outnumbered those from Britain.

MEANWHILE, AT AFN

With mobile units accompanying each army across the Continent, and still broadcasting to troop areas in the U.K., AFN continued to chart its own course in spite of the hated combined operation. It was this spirit of total independence that was to persist for many years to come and which would cause more than one AFRS commander back in the States to go home and snarl at his wife while he kicked the dog -- or vice versa.

Building a program schedule on a solid base of U.S. programming supplied by AFRS, AFN built a news staff headed by G.K. Hodenfield, late of *Stars and Stripes.* Other staffers now included Ford Kennedy, Marty Smith, Johnny Vrotsos (who got wild applause from the German members of the audience who still remembered him when he was brought back to be master of ceremonies at the AFN 40th Anniversary party,) and former presidential announcer Keith Jamieson. Russ Jones also joined the group. John Hayes was temporarily attached to the combined AFEP group and Bob Light, now a major, was commander. John "Jack" Boor, the wandering civilian, was chief engineer. Using primitive disk or wire recording equipment, the network produced hundreds of documentaries including actual descriptions from bombing raids. One dramatic broadcast was done by Captain Jack London and Major Bob Light who climbed up on the roof and recorded the sounds of a buzz bomb approaching, the silence as the motor shut off and, seconds later, the explosion as it ripped apart buildings a short distance away.

By the time D-Day arrived, AFN had a network of 60 stations throughout the British Isles, including six in Northern Ireland fed by cable under the Irish Sea. Although the BBC still would not permit a transmitter

Everyone has to start someplace. This someplace is Samar, Philippines, and the redoubtable Jack Brown, doing a remote. (A "remote" is a broadcasting term used to describe a broadcast from someplace like Samar which is about as remote as it gets.)

Those with the fortitude to keep reading will find Brown doing his thing from much nicer places at a later date. (Photo courtesy of Jack Brown)

anywhere near London, it was estimated that some 5-million Britons listened to AFN regularly.

AFN never assigned any appreciable staff to the AEFP and tensions remained strained between the groups. Although AFN had agreed not to establish any additional stations in Britain after the beginning of the AEFP, it did begin a shortwave service to the China-Burma-India theater of operations. In October following D-Day, Eisenhower's goal of a single unified service was again ignored as AFN began to establish permanent stations in France under AFN control. Bob Light was given the assignment to set up the first station -- in Paris.

Light dispatched Chief Engineer Jack Boor and a crew of eleven men to Paris with the equipment to begin setting up a station in the Shell Building on the *Rue de Berri*. Boor got the truck and the equipment but could only find two men so they set off with the vague orders: "Find a ship or something at Southampton and go." They found a ship or something but sat in the Bay of Biscay for three days. The LST, loaded with troops, went up the Seine to Rouen where they proceeded to back the truck off into axle-deep mud. Getting pulled out, they discovered that all road signs had been removed to confuse the Germans. Because none of them spoke French, they picked up two French-Canadian hitchhikers. The French-Canadians spoke French but the French couldn't understand a word of their accent.

Finally arriving on the outskirts of Paris, they got on the road circling the city and proceeded to do just that for a number of hours. When they stumbled across the Shell Building, the building janitor kindly helped them unload and the two enlisted men proceeded to string a long-wire antenna between two chimneys. It turned out to work just fine -- if you were receiving by sky wave in London. Unfortunately, it could only be heard for a six block radius within Paris.

Bob Light soon arrived and managed to obtain a 15 kilowatt transmitter from the French and to find a better location for the station. From then on, AFN Paris was the center of operations on the Continent although network headquarters and the news operation remained in London.

Interestingly, the French loved the station. Jumping ahead, when the station closed in June 1946 following the departure of most American troops, the French citizens petitioned the American ambassador to reopen it. For diplomatic reasons, he did so and the station was fed from the new AFN headquarters in Germany until late 1947.

The AFN mobile units continued to move along with the rapidly advancing armies although it must be admitted that the troops they were addressing themselves to were sometimes too busy to listen. When they did, they heard some amazing things such as the day General Courtney Hodges, commander of the First Army, stopped by the station and politely asked if he could someday "speak over this." The awed staff assured him he certainly could. He showed up that night and announced that his army had just captured Cologne, the first indication the world had of this victory.

A small portion of the AFN Munich transmitter liberated following World War II by Bob Light, Walt Cleary and others. The tubes stand more than six feet high and generate so much heat they must be water cooled. Output power was 100,000 watts. This transmitter fed the tallest tower in Europe—850 feet tall. Because the tower base insulator was broken, Cleary ordered two from the famed Rosenthal China factory. One was damaged during installation. With one left, the Army engineers attached three cranes to the tower, gently lifted the monster off the mounting and placed the last remaining giant porcelain insulator in place before lowering the tower to it. (U.S. Army Photo)

At least two AFN staff members were killed during this period. Sergeant Jim McNally became the first casualty when he was killed during a strafing attack on the Seventh Army's station. Shortly afterward, Sergeant Pete Parris, an AFN newsman, was killed while accompanying an airborne attack in France.

While the AFN mobile units were heading East toward the Rhine, the Fifth Army station which Andre Baruch had begun in North Africa and which he had shepherded through Sicily and ashore in Italy was now headed north toward the Alps. These broadcasters never worked out of a building until they reached Rome, thousands of miles from their start. The experience they developed in mobile broadcasting was invaluable to Light and his group.

To take care of the rear encampments, stations were set up in headquarters areas at Cannes, Nice, Marseilles, Dijon, LeHavre, Reims and Biarritz. None of them were exactly hardship posts and the staffs were quite content, especially the one at Reims who discovered their building was headquarters of France's most prestigious champagne bottler and that there were several million bottles of the stuff neatly hidden away in their basement.

By April 1945 it was obvious that the war was winding down. German armies were on the run, crushed by the Allies moving in from the West and the Russians from the East. John Hayes contacted Commander Harry Butcher, Eisenhower's aide and himself a broadcaster, and asked for a recorded message from Eisenhower to be played thanking his armies on the day of the Armistice. What arrived was a recording of Eisenhower thanking the brave workers of Britain for standing by their machines during the war. Hayes called Butcher and explained that there would be a riot among the troops if this were played and they weren't given credit for their sacrifices. Butcher explained that he was sorry but this was what Eisenhower wanted. Hayes told him he would not under any circumstances broadcast it and thus AFN, Eisenhower's own network, failed to provide a victory message from the Supreme Commander. Hayes even went so far as to call Kay Sommerville, Eisenhower's driver and extremely close friend, to see if she could whisper in his ear but even that didn't work.

In the last stages of the war, Light, Boor and Ben Hoberman took off in a jeep to make their way to Munich which was about to be captured by the Americans. The trip was sheer hell. Boor, a civilian, kept getting arrested because he didn't have a uniform. The solution was simple. He put one on, although without insignia. They kept getting lost. Once they got to Nancy, France, they were recalled to Paris. It seems they were in John Hayes' jeep and he wanted it back. They got another and wended their way through France and Germany by way of Pfungstadt, Frankfurt, Augsburg and, finally, Munich which had just been captured by General Patch's Seventh Army.

Light arranged to take over two 100,000 watt transmitters from Patch's Psychological Warfare Section. One was in Stuttgart and one in Munich. They found a perfect studio location in Munich. It had been the home of the Munich *gauleiter*, or district leader, during the war and previous to that the lavish mansion of famed German artist Kaulbach. Although luxurious, one of its main attractions was that it had transmission lines already installed leading to the transmitter. Chief Engineer Boor went to work installing the necessary equipment and the station was soon ready for sign-on.

On sign-on morning Bob Light proudly opened the microphone for the first time and announced, "Good Morning. This is AFN Munich, the voice of the Seventh Army."

Wrong!

They had been so busy they didn't know that Patch's Seventh Army had moved on to other pastures and terrible tempered General George Patton and his Third Army had moved in. Patton, who had never been a friend of radio, AFN, or much of anything else, was shaving with a straight razor when he heard the announcement. His hand jerked, blood flowed from a nasty neck nick and he screamed for an aide to find that stupid son of a bitch and court martial him. Soon finding himself surrounded by MPs, Light managed to convince a sympathetic cop that he was an innocent man. But, in view of what he knew about Patton, he took the easiest course and by evening was on his way back to Paris.

With war's end, the AEFP went out of business and Margaret Hubble, the woman who had signed it on 417 days earlier, signed it off. This time Eisenhower had prepared a thank you to the troops. Hubble's, and the AEFP's last words were: *This is Margaret Hubble saying goodnight to you for the last time in the AEF programme, and I can think of no better way of saying goodbye to you than by repeating the words of General Dwight D. Eisenhower, which were...."goodbye -- and wherever your new activities may take you -- good luck and God Speed."*

The war was over. Johnny was about to go marching home to an uncertain postwar future. As for the future of military broadcasting, it, too, was very much in doubt. Stay tuned for the next exciting episode and find out what happens next.

*Jack Brown, who earlier had been an unsung staff member on the Philippine island of Samar, and later ended up as an enlisted sailor at AFRS in Hollywood, talks to Donald O'Connor.
In the picture above, taken in 1951, Brown interviews O'Connor at a movie premiere in Hollywood. Things were to go well for both of them.*

Picture right, some twenty-five years later, Brown has moved through the AFRTS Hollywood system and for many years been Chief of Programming for the worldwide network. He also has been elected President of the Pacific Pioneers Broadcasters, whose membership embraces every name remembered fondly by radio and television buffs. Seen here following a PPB tribute to O'Connor, the two seem amused at the program featuring the pair during much earlier days. (Photos courtesy Jack Brown)

CHAPTER TEN

There is nothing permanent except change.
-- Heraclitus (540-457?) B.C.

When Johnny Came Marching Home

Now that the war in Europe was over, troops were being warehoused in encampments across the Continent waiting transportation home. AFRS was perhaps more important than ever as it played a major role in trying to prevent boredom among the hundreds of thousands of troops impatiently waiting to board ships to return them to their former lives. Because the war in the Pacific was still being fought by fanatical Japanese troops determined to prevent foreign troops from invading the home islands, rumors were flying through the European camps that their inhabitants would soon be taking a tour of the Orient rather than going home. Morale among the troops was at an all-time low.

Ben Hoberman, who had returned to AFN Paris after helping Bob Light and Jack Boor install the station in Munich, was officer of the day on August 12, 1945. The rumor that Japan was about to surrender had been floating around for days and he decided to find out if it were true. Not wanting to be accused of communicating with the enemy, he picked up a phone and called the Japanese military attache in Bern, Switzerland. Using somewhat cloudy logic, he felt he could get away with that because Switzerland was a neutral country.

Hoberman recalls that "the timing was marvelous." The attache, Hoberman says, "spoke marvelous English. He was crying on the telephone in response to my question of whether it was true the Japanese had surrendered. He said he had just gotten off the phone after speaking with the Japanese Minister of War, who had confirmed that it was true." Hoberman immediately relayed his story to headquarters, which was still in London, and AFN broke the story to the world.

The war, and the primary reason for AFRS's existence, was over. Tom Lewis and his staff in Hollywood realized that although the mission had changed, the need for military broadcasting remained primary in keeping the troops fully informed of what was happening, what their future was and what they could expect. He and the staff knew they could do a great deal to provide information and morale-building entertainment while at the same time helping smooth the transition back to civilian life.

AFRS continued to send out a full schedule of entertainment programs from the commercial networks while changing the emphasis on its own in-house produced material from entertainment to information.

Lewis knew that the organization would suffer major changes as the hostilities ended. Immediately after the Japanese surrender, AFRS prepared a "Forward Plan" which set forth the premise that, since the number of troops being served would undoubtedly be smaller, AFRS would be needed in a proportionately reduced volume. The plan called for a gradual reduction in program services to begin on July 1, 1946.

The "Forward Plan" concluded with a masterpiece of bureaucratic gobbledegook. After July 1, it said, a new plan would be drawn up "based upon the crystallization of the present fluid National policy in regard to the activities of the War and Navy Departments." Although few people are qualified to translate that, it most likely means that "we don't know what in hell is going to happen and neither does the government. When they decide what to do, we'll write another plan."

The plan was very specific about one thing. It pointed out that with massive demobilization of military personnel, the AFRS staff was rapidly bugging out from its wartime home in the tacky buildings supplied by the military and heading for the chrome-plated offices of NBC, CBS, ABC and the various commercial production companies. It was suggested strongly that replacement personnel be selected from among those who still had a reasonable amount of time left to serve and that as many civilians as possible be hired.

For a short period, the question as to whether AFRS would remain in operation at all remained unanswered. Some felt it should be mothballed until the next war came along. Others, perhaps stunned by the sheer verbosity of the three volume "Forward Plan," prevailed and the Army Information and Education Division instructed Tom Lewis to implement it. This was as close as AFRS ever came to expiring. Although it reverted to a somewhat smaller operation than during its wartime heyday, it continued its information, education and entertainment activities including the shortwave function.

AFRS had taken over many shortwave operations

from the Office of War Information in 1943 and continued to send out news, sports and special events using OWI transmitters from studios in New York, Los Angeles and San Francisco. Some entertainment programs were also included, such as *G. I. Jive, Melody Roundup* and *Command Performance.*

It was an article of faith that sports should be the backbone of the shortwave service and it came as something of a shock when the Army News Service advised in 1945 that "interest in sports is variable and spotty...the longer men have been away from home, the less interest they seem to have in stateside sports..." Later, as the old timers overseas rotated home and were replaced with fresh troops, interest in sports again peaked.

Implementing Lewis' "Forward Plan" turned out to be impossible to do in any sort of orderly manner. He and his staff were charged with continuing to broadcast three hours of programming a day over four transmitters to Europe, the Caribbean and the Middle East from the New York studios. San Francisco was closed, leaving Los Angeles to transmit eleven hours daily of AFRS material to the Pacific area.

Also continued -- and enlarged -- was the little known **Bedside Network.** During the war, AFRS had supplied military hospitals throughout both the U. S. and overseas with the basic music library plus special information programs, totaling about 17 hours a week. The hospitals would play this material over public address systems. Now the War Department directed that AFRS also provide technical equipment to improve the quality of the hospital systems as well as train the staffs to operate them.

This all was happening while the highly qualified staff at AFRS, which Lewis had developed so laboriously during the war, was heading out the door and returning to jobs paying slightly more than the 50 bucks a month a private received.

Lewis himself understood their desire to walk up to Sunset Boulevard and catch a bus home for the last time. In an internal newsletter in September 1945, he wrote:

> *Your job in the war just passed has been one of the greatest. You have suffered hardships, undergone many privations, yet you have done this willingly because you know what Armed Forces Radio Programs and services have meant to our troops. Before we all return to that life we have dreamed about for so long, let's be sure that one big challenge that we have prepared AFRS to meet is still before it -- the challenge of assuring the continuation of the AFRS as long as there is a need for it.*

Lewis left in October 1945 and returned to Young and Rubicam as Vice President in charge of radio. Major Martin Work, who had served as Lewis' executive officer, assumed command of the organization pending assignment of a regular army officer. AFRS within months was beginning to resemble a civilian organization as new civilian personnel were hired or former military staff members hung their uniforms in the closet and civilianized. Command of AFRS remained a joint Army-Navy function under the control of the Troop Information Branch of the Department of the Army.

During the war years, Lewis had successfully fought to keep control of AFRS, including policy making, in Hollywood. With his departure, Washington quickly assumed overall control and policy and command functions moved there. The Los Angeles headquarters now had only the responsibility for executing basic policy relayed to it from the Pentagon, while still remaining the production and distribution center for programming and for technical advice and supply. Although the policy-making centers would change names and functions a number of times in coming years, this is essentially the same command and control system which is in effect today.

With the cut-back in personnel, some of the popular AFRS programs were impossible to keep in production or were canceled due to their exclusively military nature, no longer appropriate in a world at peace. One program which continued was *Command Performance*, still the showpiece of the AFRS schedule. It continued its wartime format of answering letters from troops asking for particular performers.

Larry Gelbart, whose name now leads the list of credits on innumerable programs including M-A-S-H, joined the writing staff as a private in 1946. He recalls that for a year he wrote for *Command Performance* exclusively and says that there were no problems getting performers although the program became more and more like commercial radio due to the lack of wartime tensions. He says the writing staff directed its efforts "more to the boys away from home rather than to the boys in danger or who wouldn't come back." He also recalls that at AFRS, the military veneer was extremely thin. Not caring much for the issue enlisted man's uniform, he had his own tailored more along the lines of those worn by officers. No one seemed to object. True, there were occasional nights when he had to pull guard duty although he couldn't figure out what he was needed for except "to prevent people from stealing jokes."

At the end of the war, AFRS had been turning out about 20 hours a week of original programming. By early 1947, this had dropped to 14 hours. The remaining 41 hours a week sent out in the weekly transcription package came from the commercial networks. Much of the AFRS programming changed emphasis and devoted itself to emphasizing pride in service, American heritage and historical aspects of the military. One such example was a series called *Medal of Honor*, each of which dramatized the story of a Medal of Honor winner.

As more and more civilians joined the staff, payroll became a consideration as no longer were inexpensive employees in khaki supplied by Uncle Sam. Costs of production consequently rose, as did budgets. Concurrently, as the urgency to contribute to the war effort waned, so did the willingness of the entertainment community to contribute unlimited time and talent as their contribution to the AFRS mission. Bob Hope, who had been a keystone of so many AFRS programs through the war years and who had contributed so unselfishly of his gigantic talent, announced in 1948: "I will do one show per year for the army -- that is all. I feel obligated to do other benefits -- just as important to my country, like cancer, Red Cross and so forth." No one could blame him and the military continued through the years to call on him and he continued to respond.

Throughout the war years, it had not only been the performers who had contributed their talents. The various crafts, guilds and unions, without whose skilled personnel there can be no programs, had also provided their expertise. Among these were the American Federation of Musicians (AFM), AFTRA, the American Federation of Television and Radio Artists and ASCAP, the American Society of Composers, Authors and Publishers which controls through its membership the majority of the rights to copyrighted music.

The relationship between AFRS and these groups, which later included music-rights group BMI, has been outstanding throughout the entire course of AFRS' existence. During the war, members of these groups had received only a token payment under agreements worked out earlier. These arrangements continued until 1947 when the AFM and AFTRA asked for a review of the earlier concessions. Both asked for a slight increase in the absurdly small fee then currently being paid in view of the rapidly rising living costs and lack of military emergency.

Negotiations were carried out in a spirit of mutual understanding. A compromise established a number of concessions and provided for slight, stepped increases in compensation over a five year period, at which time the members would receive a fee pegged at one-third of the prevailing rate. AFRS agreed to two major conditions. It would base its rate on a minimum number of musicians and actors expected to be hired during one year's production operation and unemployed actors and musicians would have preferential hiring rights.

Although these additional charges caused some budget difficulties, the guilds and unions continued to be firm in their appreciation of, and dedication to, military radio. Many members of the entertainment organizations noted that while manufacturers of planes, tanks and guns enjoyed profits from their government contracts, AFRS did not and was entitled to every possible consideration in an effort to save tax dollars. This warm and understanding relationship is perhaps unique in the entertainment industry and has continued to the present day. Without it, there could be no AFRTS as it now exists.

Most of the wartime Johnnies had now marched home and the audience overseas was in garrison and in a training status. With the change in requirements for specialized programs and the sharply rising costs of production, AFRS phased out much of its in-house production and by 1950 the majority of the programming was decommercialized network material and disk jockey programs. This marked the end of AFRS Hollywood as a center of creative programming and from now on it would operate essentially as a procurement, duplication and distribution center for programs obtained from outside sources or recorded by contract employees using AFRS facilities. Radio in general was going through a period of growing panic as television began to dominate the U. S. broadcast scene. Radio entertainers moved from the microphone to the sound stages and radio programmers searched for new formats with which to hold their audiences. For the most part this consisted of disk jockey programs which were cheap to produce. AFRS was carried on the tide and its weekly bundle of programming

In 1946, the war was over but Command Performance continued for the benefit of the occupation troops remaining overseas. Seen here, the post-war production staff. Left to right, writer Vance Colvig, writer (Major) Tom Smith, director Bob Lehman and writer Bill Norman.
Colvig became a popular television personality and continued to write for numerous comedy shows. Smith remained with AFRS as a civilian and served in Hollywood and at stations in Europe. (Photo courtesy of Vance Colvig)

With the war over and budgets being slashed, and with Hollywood stars still patriotic but not anxious to donate valuable talents, AFRTS cleverly got around the problem by covering the many events such as movie premieres at which stars gathered. Many of these assignments went to Jack Brown.

Here, nattily attired in his foreign correspondent's trench coat, he introduces Humphrey Bogart and Lauren Bacall to the AFRS audience. (Photo courtesy of Jack Brown)

for the overseas stations soon consisted primarily of 25-, 30- and 55-minute record shows featuring all types of music presented by the best DJs available. Like their wartime predecessors, they worked for minimum pay.

While the excitement and creativity which marked the wartime operation was gone, the programming output was exactly what the new audience needed. Rather than receiving large doses of sometimes ponderous, flag-waving inspirational "message programs," overseas listeners now heard the same type of shows as their contemporaries back home. It was more directly in tune with the new audience which was generally younger, and less experienced in both military and overseas life, than the wartime group it had replaced.

They wanted, among other things, their favorite kind of radio shows and mama's home cookin'. AFRS was able to supply one but not the other. AFRS Hollywood furnished many of the shows. The overseas affiliates aired them, along with other material needed by the troops, and by so doing became that vital "voice of home."

AFRS had its own set of problems during the transition period following World War II. The overseas outlets had a different kind. Here is what happened at some of the overseas stations and networks when peace broke out.

ANNOUNCER: NOW, THROUGH THE MAGIC OF THE PRINTED WORD, WE TAKE YOU TO THE DEPTHS OF THE PACIFIC. ACTUALLY, NOT THE <u>DEPTHS</u>. JUST TO THE PACIFIC.

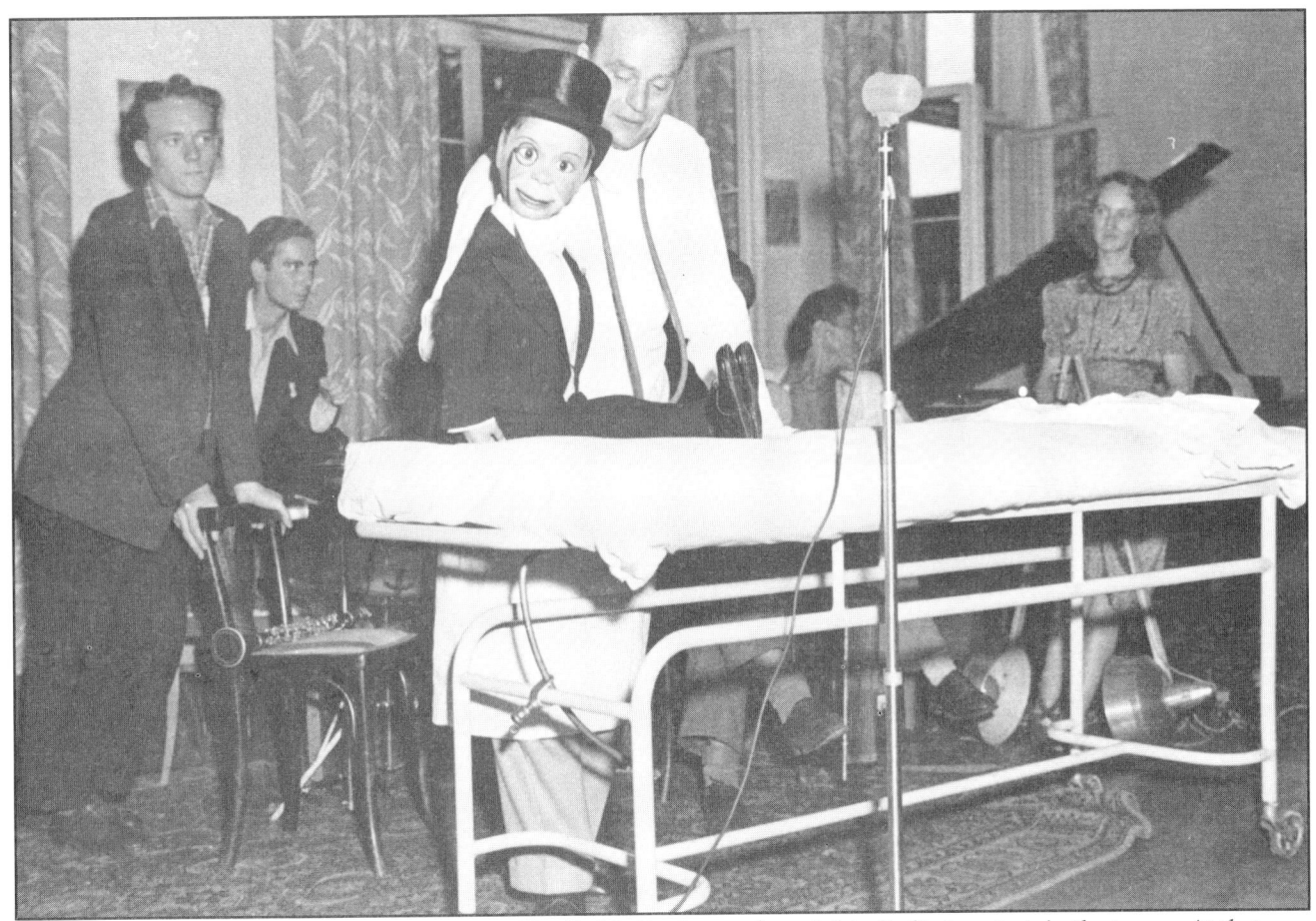

In 1948 Edgar Bergen and his sidekick, dummy Charlie McCarthy, made a visit to Berlin to entertain the troops. At that time he was one of the country's most popular personalities although today he is perhaps remembered more for being "Murphy Brown's" father.

During his Berlin visit, Bergen is interviewed by AFN Berlin's TSgt Ray Cava. Notice the heavy, old-fashioned microphone. It took two hands to hold it steady. Bergen, on the other hand, is having no trouble holding his drink steady with one. (Photos Oscar Skocik)

Three views of the Von Bruening Castle, the oldest part of which dates back to 1357, at which time the builders made no provisions to make it a radio headquarters. AFN moved in when Eisenhower moved his headquarters to Frankfurt, Germany, and AFN, wanting to be slightly away from the flagpole, took over this castle in the suburban village of Hoechst. Above, the oldest part of the castle towered over the city and served to house the unmarried staff members. Surrounded by a moat, it failed to keep out an occasional young lady of the town intent on intensifying German-American relations and international goodwill. The moat also acted as a magnet for staff members, awash in good German beer, who from time to time convinced themselves they could jump across. Right top, The courtyard and gardens perched on a cliff overlooking Main River. This portion of the castle complex held the offices and studios of the network as well as the club and music library. After twenty-one years AFN moved to a new facility in Frankfurt and the castle was renovated by the I.G. Farben chemical combine. This portion became a posh guesthouse for VIP visitors, the tower portion became a museum. Right bottom, The main entrance leading into a courtyard. This is the "new" part of the castle, having been built in the 1600s. The authentic renaissance-style structure, as can be easily seen, has been greatly enhanced by the tasteful signs designed by the Army Corps of Exterior Decorators. Just to the right through the gate is the entrance to the AFN club which contained such military necessities as a bar and slot machines. (AFN Photos)

CHAPTER ELEVEN

AFRS has survived the war. Now the question is: can it survive the peace? "War is only a cowardly escape from the problems of peace."
-- Thomas Mann

While Johnny <u>Waited</u> To Come Marching Home

The beginning of the AFRS operation in the Pacific was in the Mosquito Network and the Jungle Network, both of which had leapfrogged from island to island as American and allied troops had secured them on their move toward the Japanese home islands. The Jungle Network had moved its headquarters from Hollandia, New Guinea, to Manila in February 1945. Manila at that time was, quite literally, a wreck. Heavy fighting had destroyed much of the city. Unrecovered bodies lay festering under tons of rubble in the old city center. A few discontented artillerymen had made a political statement by shooting large holes in the upper floors of the Manila Hotel after hearing that General MacArthur planned to use it as his home. The infamous Santo Tomas concentration camp still held a few gaunt, skeletal Americans who had managed to survive the war and who were now enjoying all the food they could eat. The newly arrived troops were delighted to find the San Miguel Brewery had survived intact and was quite willing to part with as much of its product as a person wanted, unlike the Army which issued only a niggardly two cans a week to the thirsty troops.

The station temporarily set up in a bombed out building which, because it had only three walls, proved unsatisfactory. It quickly moved to other quarters. It also changed its name. Major Graf Boepple, the officer in charge, explained in a memo to MacArthur's headquarters that "Jungle Network" was a misnomer now that it had moved to the comparative civilization of the Philippines. Henceforth, he explained, his group would be known as The Far Eastern Network.

At that time, Boepple's group was scattered from Milne Bay, New Guinea, to Manila. Although not physically a network, his Manila headquarters provided a number of transcribed programs to the affiliates as well as other assistance. Procurement was also begun on six to eight mobile radio stations to be used during the forthcoming invasion of Japan. These were 400-watt, self-contained soundproof studios mounted on the frames of the army's ubiquitous deuce-and-a-half trucks.

They weren't needed. Something called an "atomic bomb" was dropped on someplace called Hiroshima, according to the AFRS newscasts on August 6 and on Nagasaki on August 8. It looked like the war was finally coming to an end. And so it did, a week later.

When General MacArthur and his staff landed at Atsugi Airport in Japan on August 30 to complete arrangements for the surrender, AFRS was right behind him. The surrender on the deck of the Battleship *Missouri* on September 2 was broadcast to the world and, within days, semi-permanent stations were placed in operation.

To this day, no one really knows which station went on the air first. In the chaotic days following the surrender, troops fanned out throughout Japan and AFRS did its best to go where they did. A portable transmitter operated by the 2nd Marine Division in Northern Kyushu went on the air the day before the surrender, on September 1. During the next two weeks stations in Kure and Osaka were operating.

Boepple, meanwhile, met with the U. S. Signal Corps and representatives of the Japanese government. He explained that he wanted to use one of the two Japanese government stations in Tokyo for 16 and one-half hours a day along with the second Japanese network which included outlets in Hiroshima, Nagoya, Yamaoto, Sopporo, Osaka, Sendai and Tokyo. In a face-saving gesture, the Japanese asked that the request be put in writing but it was a foregone conclusion that it would be granted. Boepple didn't even bother to wait for formal approval. He and his group marched into Radio Tokyo and began broadcasting days before approval was received on September 23.

After several years of operating out of tents, shacks, quonset huts and bombed out buildings, this was heaven. Instead of cello-tex soundproofing so common in American stations, the walls of the main studio in Tokyo were upholstered with embroidered silk. The equipment looked familiar as well. The control boards, amplifiers and even the microphones had been copied from RCA specifications right down to the red dot trademark -- which contained a Japanese character instead of the familiar "RCA."

Both of the networks -- one now being used by the Americans, the other by the Japanese -- originated programs from the same studios. It became commonplace to see an American jazz group waiting outside a studio while a Japanese samisen player finished his program inside.

Hy Averback was one member of the Tokyo staff as was actor Hans Conried. Averback had been announcer on the Bob Hope show in civilian life. Conried, determined at all costs to retain some sort of individual identity in spite of being in uniform, sported red socks to the constant dismay of the command group. Perhaps as punishment, he was assigned to do a program at ll p.m. each night consisting of his Barrymore-like voice reading poetry over a background of solemn organ music. It was designed to help people fall asleep and all agreed it did a magnificent job of doing just that.

Although hardly anyone noticed, the network again changed its name and *The Far Eastern Network* became *The Far East Network* the day it went on the air from Tokyo. Although now located outside the bustling center of Tokyo, at Yokota, it is a name it carries proudly to this day.

Troop consolidation began almost immediately in occupied Japan and AFRS stations, as always, followed suit. From a peak of 39 FEN stations in 1946, the number dropped to 16 at the beginning of 1947 as bases closed and troops went home. By 1949 there were 11 located in Japan, Korea, Okinawa, the Philippines and the Marianna Islands.

Although the Japanese are traditionally extremely nationalistic and proud of their unique culture, AFRS and the American style of music and of broadcasting made quick inroads among the Japanese listeners. It wasn't too long before one manufacturing company produced a radio which could only pick up the Far East Network. Baseball, first broadcast over AFRS shortwave and on FEN, enjoys tremendous popularity today although it was almost unknown in Japan previously. American music, seldom heard in Japan prior to the arrival of FEN, is now heard everywhere and American performers, known there through their recordings, draw overflow audiences when they play live concerts.

❖ ❖ ❖ ❖ ❖

ANNOUNCER: NOW OUR STORY TAKES US FROM THE INSCRUTABLE EAST TO THE SLIGHTLY MORE SCRUTABLE WEST. SO LETS BE SCRUTING ALONG, BACK IN TIME AND PLACE, TO EUROPE IN 1945 WHERE THE WAR HAS JUST ENDED AND THE WAITING TO RETURN HOME HAS JUST BEGUN.

❖ ❖ ❖ ❖ ❖

With the end of the war, the AFN Headquarters was still in London although most of the programs were being fed out of the Paris studios. The Southern operation which had begun with the invasion of North Africa and which had followed the troops through Sicily and north through Italy was still operating out of vans. Some of them had marched as far north as Austria and the Danube where they were getting ready to hunker down and start a network of their own.

At this point, the man who was to become one of the key players in the entire military broadcasting system was a rag-tail army lieutenant named Robert Cranston who was sitting with his First Army Unit near the Elbe River in Germany. With little going on, he wangled a few days off and headed for Paris. Having been a listener to AFN's First Army mobile station, and having been an assistant program director at WBAP in Fort Worth, which was managed by his father, he was interested in seeing the station and meeting John Hayes who was a friend of his father. He was given a tour, treated very nicely and sent on his way back to his unit. There he was told that he was going to be given command of a company and sent to Japan.

Shortly thereafter his slightly upset battalion commander called him in and handed him a set of orders. It was signed "By Order of General Eisenhower" and assigned him to AFN Paris. It was the start of a long military career devoted to military broadcasting and

The war just over, Lieutenant Bob Cranston and a couple of wartime buddies do a bit of hammed-up celebrating. Cranston is the center slice of ham. A couple of weeks later, he is to receive orders to report to AFN and does—still wearing his grungy combat uniform and getting picked up by the MPs in Paris. Hard to believe, but this disreputable-looking character went on to be the hub around which the entire AFRTS system revolved. (Photo courtesy of Robert Cranston)

public affairs during which Bob Cranston became perhaps the best known practitioner of these arcane arts.

That all happened later. At this time he was a scruffy line officer who had just arrived in Paris wearing a dirty field uniform after a miserable flight in a C-47 filled with displaced persons and Jerry cans of gasoline. After being picked up by the MPs who explained that in Paris an officer was expected to look neat and tidy, he found a hotel and cleaned up. He also met by accident the Executive Officer of the Office of Information and Education, Lieutenant Colonel Hy Miller, in the hotel bar. This office supervised AFN and Miller arranged to take him to his new assignment the next morning. There he met the I&E boss, Colonel Paul Thompson, later to become a General and still later Editor of *Readers Digest*. The AFN Headquarters turned out to be in the former home of the now unemployed German Ambassador to France, Otto Abend, at 19 avenue d'Iena. The station was about a mile away.

Others on the staff included Major Bob Light, who had been the AFN commander while John Hayes worked with the combined AEFP broadcast group. Ben Hoberman was there and was later to become a vice president of ABC Radio. Newsman Baxter Ward, later known to Los Angelenos as a long-time anchor man who ended up on the Board of Supervisors, was also a member of the staff, as was actor Broderick Crawford.

Hayes assigned Cranston as Executive Officer and his first job was looking for replacements for the people who were leaving. Working with the personnel offices, he found Lou Edelman who was to remain with AFN for many years and whose first assignment was to the station in Biarritz. Cranston also found Lieutenant James R. Lewis whom he assigned to Frankfurt and who turned out to be responsible for one of the most unusual and legendary locations ever used by an AFRS network.

During the hectic and constant shifts in personnel in these chaotic post-war days, it was decided to close down the United Kingdom operation and move headquarters to the Continent. Cranston had now become an expert at closing, opening and moving stations and it fell to him to move to London temporarily to get that job done.

Returning to Paris, and now a Captain, he continued closing stations and handling all the jobs that naturally fall to an executive officer. The job called for a major and one soon appeared. Cranston was reassigned as the administrative officer and put in charge of program distribution. Most of the programs still came from AFRS in Los Angeles although it was difficult to tell sometimes. Hayes still insisted that AFN was an entity unto itself and ordered that the AFRS announcement at the end of each program be covered with an AFN identification. Tom Lewis and his successors back in Hollywood were never happy about this but they had their own set of problems to worry about and the independent attitude of AFN persisted for years to come.

Offered the opportunity given to men who had been overseas for lengthy periods, Cranston took a 35-day leave in the States. It proved the widely known fact that if you value your job, take short vacations. When he returned he learned that because of a rumor that persons taking this leave would not be coming back, his job had been given to someone else. He was then assigned to the Information and Education section of AFN. Worse, while he was gone, the entire operation had moved to Frankfurt, Germany. The French operation was being gradually closed down so it was off to Germany for Cranston.

What he found was that one of the men he had gotten assigned, Lieutenant James Lewis, had more than paid his way. When Lewis had been sent to Frankfurt, he discovered that the studios were in a house which had been commandeered earlier. This was fine except it was in the shadow of the flagpole and the flagpole cast a mighty big shadow -- it belonging to the Supreme Allied Commander, General Dwight D. Eisenhower. Eisenhower had taken over the I. G. Farben headquarters building which was at that time the largest office building in Europe. Springing up around it were all the accoutrements of American civilization in Europe -- post exchanges, billets, snack bars and offices of all types.

Being nobody's fool, Lewis realized that the nearer the flagpole, the easier the interference from high ranking military types. He and the rest of the staff knew that *everyone* considered themselves to be experts on radio programming. The best thing to do was distance themselves as much as possible from the daily harassment of listening to self-styled experts expound on their theories of how to improve the radio programs. Lewis got in his jeep and began scouring the nearby countryside for a home. He found it in the small city of Hoechst, about ten miles outside of Frankfurt off the road to Wiesbaden on the Main River.

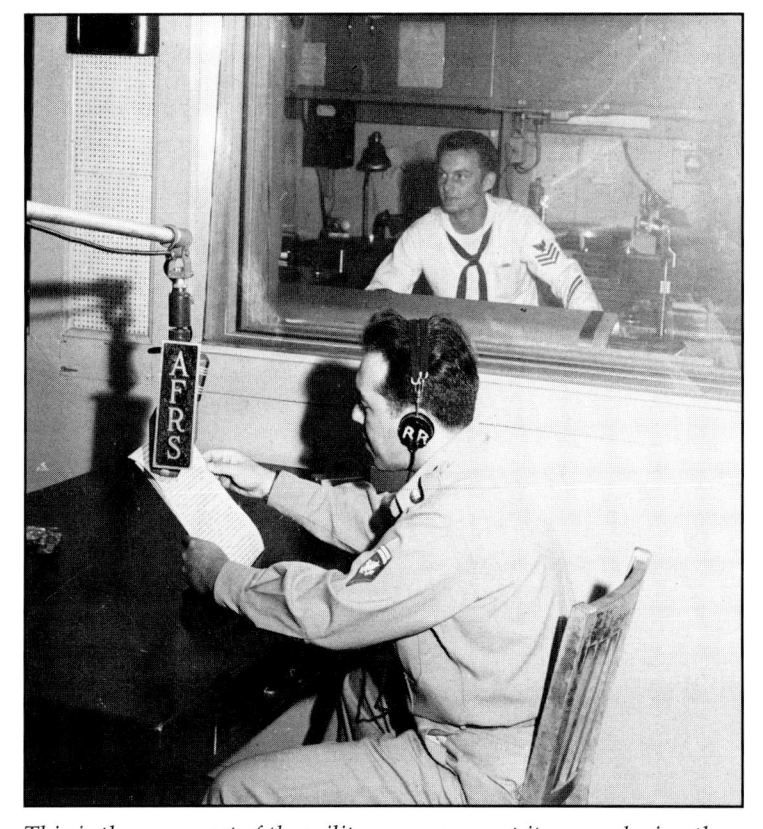

This is the way most of the military overseas got its news during the 40s and 50s—by shortwave from the McCadden Place studios of AFRS in Hollywood. SP1 Martin Wax is reading the news; RM1 William Eggers is the engineer. (Photo courtesy of AFRTS)

What he found was a castle, parts of which dated back to the 12th Century. It was occupied by the Von Bruening family who were asked to find other quarters immediately as the victorious U. S. Army which had won the war with words, public service announcements, interviews and big band recordings required it. What they took over was a very large 16th Century building complete with ball room, chapel, and all the other necessities of castle living. In addition, they got a moat and a 13th century building with a tower several hundred feet high. Both buildings were surrounded by lovely gardens complete with fountains and statuary, all perched on a bluff overlooking the river. The tower building was converted into billets for the troops and the elegant addition across the moat became studios, record library, offices and a club complete with bar, food service, slot machines and friendly frauleins. In short, just a normal radio station.

The engineering staff, under the direction of the new Chief Engineer Walt Cleary, installed the equipment and AFN began broadcasting from the castle in late 1945. Other AFN stations in Germany at this time included Munich, Bremen, Berlin, Bayreuth and Kassel. It took a lot of doing in war-ravaged Germany but the engineering staff managed over time to interconnect the stations and even install a clock system which worked from a master clock at headquarters so it could switch originating points on the split second. Cleary also found a strange piece of equipment in a warehouse. A German engineer explained it was a "Magnetophone" and it recorded and played back sound on a roll of what looked like paper. That sounded pretty stupid but Cleary managed to find people who had worked with the things, got some of what the Germans called "tonband" and what was later to be called tape by the Americans and -- whaddayaknow? -- it worked. AFN started using it and thus became the first American broadcaster in history use audio tape.

Because there were still large numbers of troops left in France, stations were maintained in Rheims, Marseilles, Cannes, Nice and Biarritz with headquarters in Paris.

On New Year's day, 1946, the London operation closed its doors for good and Frankfurt became the network headquarters. Paris remained a major station and supervised the French group. The 100,00 watt Munich station was run by Lieutenant Bruce Wendell who stayed with the network and with AFRS Los Angeles until his retirement many years later. Lieutenant Ben Hoberman was Chief of Operations in France.

Bob Cranston, was still a Captain and because his job as Executive Officer had been taken over by higher ranking Major Oscar Stegall, found himself Commander of Troops, Supply and Transportation. Earlier in the year. John Hayes had returned home to reume his distinguished career, which included being named Ambassador to Switzerland, and Lieutenant Colonel Ernest Sanders was now Commander. A former regional director of the National Association of Broadcasters, Sanders was a natural selection for the position. After about a year, Cranston was notified that he had been overseas too long and it was time for him to come home. While he was packing to return, he learned that the job of Commander of the Blue Danube network was open and it was his if he could get the War Department to extend his European tour of duty. He did, and was soon Vienna bound -- which is where we will next catch up to him as he continues his climb to the top of the military broadcast heap.

There may have been more unusual broadcast studios and network headquarters, but it is difficult to figure out where they might be. The Von Bruening Schloss (Castle) was commandeered by AFN in 1945 and remained the network headquarters for twenty-one years. It came complete with such non-broadcast accoutrements as a chapel, a moat, a portcullis, a dungeon and a watch tower. It even had a Watergate 500 years before Nixon had one.

Drafty and poorly sound-proofed, crowded and chilly in winter, the staff loved it, except for the newest member. He or she always got the room at the top of the tower in order to build character and leg muscles.

One of the former guests was Napoleon Bonaparte, who spent time here after being unceremoniously tossed out of Russia. (Photo courtesy Trent Christman)

Back in Frankfurt, Roy Neal was station manager. Neal is familiar to U. S. television audiences today as a long time former NBC newsman based in Los Angeles and the anchor for NBC's coverage of the early space shots. During this early period, there was no assurance that military broadcasting would continue into the post-war period. Much depended on the way in which it satisfied the needs of the new audience. That audience consisted of troops still waiting rotation back to the States. Highly motivated during the war, they were now sitting around wallowing in boredom. Rumors flew in every direction and in some camps troops became so restive there was serious talk among them about going on strike.

Understanding the role radio could play, Neal called on General Eisenhower to discuss with the General his using radio to speak to the troops. The troops were listening. According to the NAB, only one person in ten in the U. S. had a radio in those days. According to sales figures from the European post exchanges, one soldier in three had purchased a radio. Eisenhower agreed and stunned everyone with his talk to the troops. It turned out not to be the word from the top about troop rotation. Instead, it was Ike's farewell address and it was fed to the world through AFN's facilities.

Eisenhower's successor, General Joseph McNarney was then approached and bought Neal's next idea. It was basic: talk to people, communicate with them on their level and tell the truth. This leads to reason. The program was simple enough. Through spot announcements, troops were urged to send in letters or cards asking questions to which they felt they weren't getting answers. Then the staff would choose those most representative and fly the person to Frankfurt to ask, face to face, what he wanted to know and be answered by the General who had the answers. The idea worked and there were immediate results. According to Neal, "... *within about two to three weeks, the troops knew when they could plan to go home. They knew how the point system for replacements was working. They knew what kinds troops were coming in and in what numbers. They knew where these replacement troops were going. They knew because the Generals were answering their questions.*"

As the troops restiveness died down, McNarney and his staff sent letters of appreciation. But even better, the general noted that there wasn't much rank among the staff of broadcasters. Neal explained that was because everyone there was on temporary duty from some other outfit. AFN itself did not have what the Army calls a Table of Organization which authorizes spaces and appropriate ranks to fill the spaces. The General suggested Neal ask the new commander, Lieutenant Colonel Oren Swain, to prepare one and he promised to approve it.

AFN's credibility took a forward leap following the efforts it made to bring troops up to date on when they might be going home. A large measure of credibility had already been gained by the network's coverage of the Nuernberg war crimes trials. Most of the villains from the German side had been captured and a mass trial was being held in the Palace of Justice in Nuernberg. Judging them were representatives of the major allies.

The world wanted to know what was going to happen to this miserable band of humans and Nuernberg began filling up with representatives of the world's press. The men who had been responsible for their downfall, the troops in Europe, had more than a passing interest as well and it was up to AFN to satisfy that interest.

Word came down from General Paul Thompson, commander of the Army Information and Education Division, that he wanted full, complete and accurate coverage of the trial. He specifically asked the most experienced and mature newswriter be assigned. Corporal Roy Heatley was the acting news director when the word came down and he suggested to John Hayes that Corporal Harold Burson, a former reporter for the Memphis *Commercial Appeal*, was the man for the job. Burson met with both Thompson and Hayes and, having little or no choice in the matter, accepted the job. The job of covering the trial turned out to be a cinch compared to the

NUERNBERG—1945-1946, One of Germany's most delightful cities. Unfortunately Hitler felt the same way and Allied Bombers and Patton's Third Army showed Adolf a thing or two by leaving it in this condition. Enough was left standing to house the famed Nuernberg War Crime trials which AFN reporters covered with live and recorded reports during the months the trials dragged on.

Today the city has been restored to its former charm and is the home of an AFRTS station and numerous military units including the famed 2nd Armored Cavalry unit and Third Armored Division, both of which fought with distinction during the Gulf War.

Hitler would probably not have seen the irony, but for years the American High School football team played their home games on the field where the infamous Nuernberg Nazi Party Rallys had formerly been held. (Photo: George Coulter, AFRTS)

job of doing their work as enlisted men in what was now an essentially civilian situation. It was the first recorded instance of this problem which was to plague enlisted reporters at all AFRS outlets through the coming years.

Burson, the writer, and announcer Sy Bernhard who was assigned to read the reports, arrived in Nuernberg several days before the trial was to start. They found two major problems waiting for them. First, no arrangements had been made for a telephone land line from Nuernberg to Munich, the nearest point which could feed the network. The city was a huge mass of rubble and this was a problem that couldn't be fixed by a quick call to the local phone company. A line was finally installed by the Signal Corps just hours before the first broadcast.

The second problem was accreditation. The press camp was headed, for reasons known only to a long-forgotten personnel officer, by Colonel Clarence Lovejoy whose experience as the Yachting Editor of the *New York Times* was his only apparent qualification for the job. Under pressure for food and housing by the hordes of civilian journalists, Lovejoy decided that Burson and Bernhard should be domiciled in the enlisted men's quarters reserved for guards, truck drivers and mess personnel. He explained that meals for civilian correspondents were twenty-five cents each and they would be able to eat free in the enlisted mess hall.

They protested, knowing that isolation from the remainder of the press corps would inhibit their ability to cover the trial. They also knew the accommodations at the press camp were a lot better than in the enlisted barracks. They sought the help of Howard K. Smith who was covering the trial for CBS and was chairman of the Correspondent's Committee. He strongly supported their stand. Lovejoy did not. Finally Burson casually mentioned to Lovejoy that he was unable to carry out his orders and was going to telephone General Thompson and explain the problem to him. After some hemming and hawing, Lovejoy capitulated and the two men were both billeted and fed at the press camp. And they never paid their quarter a meal, either.

They broadcast each evening from 9 to 9:15. Part of the broadcasts covered the day's events, the remainder was devoted to interviews with trial personalities including Chief American Prosecutor, and late Supreme Court Justice, Robert Jackson.

Bernhard remained for a month or so and rotated home. He was replaced by news reader Ted Pearce. Pearce left and was replaced by Jack Hooper. Burson stayed on for months as the trial dragged on and the story was becoming repetitive as each of the accused sought to justify his wartime actions. The broadcasts were cut back to six to eight minutes each evening. Burson stayed through the testimony of Göring, Ribbentrop and Keitel, three of the principal defendants. By then he had accrued enough points for overseas service to go home, which he did. He was replaced by a young 18-year old draftee named Herb Kaplow who subsequently went on to a distinguished career with two American networks.

Chief Engineer Walt Cleary also spent time as the engineer on the trial coverage and still recalls clearly the last days when sentences were being passed. Announcers were now Saul Green and Grady Edney. Cleary put a microphone by the elevator which brought the prisoners up from the basement cells. In the hushed courtroom, the sound of the elevator rising, the door opening and the prisoner coming out would be heard. Then he opened a circuit tied into the English language portion of the four-language translation system and listeners heard sentences being pronounced. The two announcers would then describe the reactions and soon there would again be the ominous sound of the approaching elevator with another defendant inside.

Such detailed coverage helped the troops listening understand what the war was all about and why they were there as occupying troops.

"Troops" calls to mind a mental picture of spit and polish soldiers, splendid in their martial spirit. By the end of the trial, Lieutenant Colonel Oren Swain, a West Point graduate, was the new AFN commander. That picture was exactly what he had in mind for his AFN "troops." He was determined to shape up the "civilians

Why are these unidentified members of the AFN Berlin staff in 1948 laughing?. It's probably because of the equipment they are still using. Although AFN's chief engineer in Frankfurt, Walt Cleary, was instrumental in discovering German-invented audiotape recorders and making AFN the first network in the world to use them. They were in short supply and some stations still relied on wire recorders such as these. These models had no doubt seen better days. The inscriptions indicate they were used by the Allied Expeditionary Forces broadcast unit of the BBC which operated during the hostilities following the Normandy invasion. The author is indebted to AFN's Bob Harlan for indentifying these units. (Photo: Oscar Skocik)

in uniform" who were attempting to maintain a freewheeling attitude and who in no way fit his image of what soldiers should be.

His first attempt was to order all troops to fall out for training every morning. Roy Neal tried to explain that many of them worked all night because radio was an around-the clock operation. Not deterred, Swain insisted. Not deterred either, Neal played dirty pool. Knowing he couldn't sustain a late night schedule and make the troops get up for training, he merely canceled the major 7 p.m. newscast -- knowing that it was a regular listening habit of the European Theater commander. It was only canceled for one night. The next morning Swain got a rather heated telephone call from an irate four star general and morning training vanished as quickly as it had started.

Swain saw the light and under his command, many of the military staff members took overseas discharges and remained with the network for years to come.

♣ ♣ ♣ ♣ ♣

ANNOUNCER: DO I HEAR A WALTZ? PROBABLY, BECAUSE WE'RE OFF TO VIENNA FOR A VISIT TO THE *BLUE DANUBE NETWORK.* AT THE SOUND OF THE LAST TONE, IT WILL BE THE YEAR 1945, IN THREE-QUARTER TIME.

♣ ♣ ♣ ♣ ♣

Anyone who has ever seen the Blue Danube should have his or her eyes examined, because the famed river is, was and probably always will be brown. This deterred neither Strauss nor the AFRS staff who named their operation in Austria **The Blue Danube Network.** It began operations as an offshoot of the Fifth Army mobile units which had been chasing after the German armies from North Africa through Sicily and up the considerable length of Italy. At war's end, it found itself in Austria. It also found itself in rather sad shape at the beginning of its Austrian adventure

Reshaping it was Bob Cranston's next assignment. AFN stations in France were closing slowly as troops moved out and although Cranston had been in Europe a long time, he extended his tour to take his first command -- the moribund Blue Danube Network. The idea of having his very own network in the land of the Waltz struck a chord. He accepted the job, and then learned it was only open because the present commander had managed to take a shaky operation and turn it into a total mess. The local military commanders in Austria were totally disenchanted with the stations which were supposed to be there to help.

The Salzburg area commander had walked in on the station one day to find the staff lying around without shirts, drinking beer and failing to stand when he entered. "Yeah? What can I do fer ya?" asked one. They soon found themselves moved out of lovely quarters in a chateau and in to two tacky rooms elsewhere. Vienna was as bad. Cranston's first job was convincing the military commanders that he was there to straighten things out and obtain their cooperation.

The first problem was money. In Austria, unlike Germany, AFRS stations were under Special Services and scrambled for money with the folks who also needed money for ping-pong balls and fishing tackle. Cranston soon got that changed and the money problem was eased. He started making the staff at least look like soldiers and, to the disgust of some, decided that the former practice of allowing nightgown-clad ladies to share the barracks would stop. His decision to make such a far-reaching change was made for him one morning when he went into the station early and discovered a naughty nighty-clad nymphet helping the morning DJ select his records. The DJ was told to cease and desist by Cranston. The DJ replied that **HE** didn't have to obey because he was a star. When Cranston next saw him, some weeks later, the man was checking passes in the rain at a lonely checkpoint.

All this changed things imperceptibly for the better. He than pulled off a stunt that is perhaps unique in the history of broadcasting. Because the staff in Salzburg and the one in Vienna were both totally immersed in their own off duty pursuits in their own familiar haunts, to the detriment of their broadcasting duties, he decided to reverse them. Perhaps, he felt, they would shape up if taken away from their familiar surroundings.

Each night a train, named the *Mozart*, left each city and headed for the other, passing half way there. He called Salzburg and told them to be on the train, one and all. He then packed up the Vienna staff and put them aboard. The next morning, the Vienna staff opened up the Salzburg station and vice versa.

Commanders were now beginning to come around and appreciate the job the Blue Danube gang could do for them now that they had gotten their minds out of their pants and back to broadcasting. Then Cranston almost blew it. He had the audacity to cancel an hour of organ music which had been a part of the schedule every Sunday morning from the beginning. His rationale was that there was only a limited amount of organ music in the library and, besides, it was dull. He cut it back to a half hour. He was told this was the local commander's favorite show and, sure enough, the first week he was unsticking his ear from the phone following an angry call from an angry Colonel who wanted to know "what have you done to MY show?"

This was Cranston's first confrontation with the firm belief of many local area commanders that the local radio station is run for their exclusive benefit and enjoyment. The classic example is the Admiral at Guantanamo Bay who returned from a party to find a movie in progress. It looked interesting so he called the station and ordered them to rewind it to the beginning. Which they did, to the confusion of all.

Cranston handled it like subsequent network and station chiefs have been trained to do -- take a firm stand. First you try sweet reason and explain the difference between good and bad programming. Explain that YOU are the expert. Explain that the commander, by virtue of age and rank, is a minority listener and the station must serve the majority. All this is both logical and sensible. It also almost never works. Next you accept with all the grace possible the invective heaped on you as you pray that this guy doesn't remember all this when it's time for your efficiency report to be written. Today, station and network programmers can buck the problem up to a higher headquarters and avoid having their careers wounded beyond repair. Then, all that could be done was to do as Cranston did: refuse to back down and tell the man you were there to do a job and you were going to do it with or without his help.

The services do not give medals for this kind of bravery...but they should.

Meanwhile, the stations were improving physically although the land lines linking them between Vienna, Linz, Graz and Salzburg sometimes sounded like they were connected by tin cans and a tight string. The new and more powerful transmitter in Vienna had some strange characteristics. It was in the courtyard of the station and when anyone walked by, the RF made their hair stand straight up. Lights turned off at the switch inside the surrounding buildings continued to glow. To turn off the lights, it was necessary to unscrew the bulbs. The connecting lines between stations were not only terrible, they were pre-emptable by any high ranking officer who wanted to make a phone call. More than one General found his conversation being broadcast to thousands of listeners.

Some equipment was obtained from the Munich dump. Hearing that the Signal Corps was junking a number of items, the BDN people drove into Germany and raided the dump. They didn't know what was in many of the boxes but some of them contained teletype machines. One went to each of the stations and the rest of them went to the American Signal Corp group in exchange for permanent lines. Other cases contained gas masks. These were put into the Chemical Warfare Officer's office during the dark of night and if he should happen to read this, he now knows where those darn things came from. BDN's biggest coup was when Cranston heard the station in Trieste was closing. AFN in Germany also heard. Both wanted the equipment. Cranston borrowed a converted B-17 from a friendly General and was already back in Austria with the loot while the AFN group was on its way to Trieste by train.

Still, there was a certain spirit of cooperation between the two networks. Walt Cleary, the chief Engineer of AFN, was always ready to lend any assistance he could. And Cranston was able to arrange a landline connection between Germany and Austria -- a considerable feat in those days -- so the two networks could share the morning newscast. This worked fine until Germany went on daylight time and Austria didn't. This meant the Blue Danube Network had to record the news for later playback and, using antiquated equipment, the news frequently began: "Good Morning. Today in Washington today in Washington today in Washington today in....."

Cranston extended his tour of duty three times and during that time faced the same problems as AFRS and all the other networks and stations. Personnel was pouring out of the operation and returning home. Replacements were scarce and not always fully trained. All this was happening while AFRS was cutting back and readjusting its programming which meant that the local stations around the world had to temporarily provide more of their own locally produced programs.

A spot opened up in the office of the Army Chief of Information in Washington and Cranston decided it was time to go home. He was replaced by Major Frank Tourtellotte. Both men were destined to later command AFN, long after BDN closed its doors for good when the four power occupation of that country ended.

In the Pacific, in Northern and Southern Europe, things were tough for the Brass Button Broadcasters. Money was tight. Frequently, so was the staff. AFRS was struggling to adapt to the changing times. Someone joked that what we needed was a good war.

And damned if that isn't what happened.

In 1952, Germany was still putting itself back together after the destruction inflicted on it during the war. This is the site of the sight that greeted AFN staffers as they strolled out of their castle into the city of Hoechst. Over the years, the townspeople and the AFRTS military staff developed a number of lasting friendships and marriages—some of which have stayed both to this very day. (Photo courtesy of Jim O'Gorman)

STATION BREAK NUMBER TWO

We pause now for Station Glorification

Get a bunch of broadcasters together and the conversation inevitably turns to the weird, the wild and the wacky things that can and do happen at radio and television stations. The business, by its very nature, attracts people of great imagination and well developed egos. This frequently can result in somewhat skewed minds. Put this type of person in an environment demanding precision, timing and discipline -- often during times of stress and inflexible deadlines -- and strange things often happen. Sometimes they happen because of the tensions created by the medium itself. At other times, by the peculiar, and fascinating, personality which is frequently the mark of a broadcaster.

This personality type is the exact antithesis of the kind of person generally considered to be "military." In order to be a good military person, a certain respect for order, standardization and conformity is required. In order to be a good broadcaster, the exact opposite approach is often needed. The constant conflict within the individuals and between the individuals and their more military-oriented and orthodox superiors often resulted in bizarre episodes.

When sitting around the table with a group of present and former AFRTS staff members, what follows during this short break are a few of the kind of stories a person could expect to hear.

There was one announcer in Munich who produced a daily five minute show called "Hymns from Home" which ran on AFN every morning. It consisted of the usual short prayer and a hymn. Before he rotated back to the States for discharge, he had completed about 300 of the things. Number 20 or 30 was a disaster. Downright nasty in fact. Vulgar. And, of course, it got on the air to the dismay of all chaplains and other right-thinking people. He also left a note saying there was only **one more** similar program. Rather than throw out the whole, otherwise excellent, series, the production chief had to sit down for days on end until his end was dazed and listen to several hundred hymns and beatitudes. It wasn't until the job was completed that he discovered the insidiousness of the joke. There **was** no other such show in the batch.

☆

Former AFN newsman Milt Fullerton recalls the day when news announcer Bill DeArmond was on the air reading a newscast and finding to his horror that the story he was well into had never been finished by the news writer. The last line of his copy read, "The Navy attributes the sinking of the submarine **Thresher** off Cape Cod to a ruptured..." He thought frantically to think of a missing word.
He could have said "pipe." Or "valve." Or "hull." Not Bill. He picked the worst possible choice. "...to a ruptured nut," he said before both he and the engineer began rolling on the floor in hysterical laughter.

☆

Dave Johnson, a long-time Air Force member of the AFRTS group, has a special talent for producing spot announcements and today works for a television production company which does just that. It's unlikely that he will ever again be connected with announcements such as the ones he describes here:

"Thailand, the Land of Smiles, is famous for both its natural beauty and its natural beauties. At one point during the days of the American Forces Thailand Network, the station in Korat was asked to perform a service in the interest of public health. At the time, 'ladies of the evening' were issued numbers by local health authorities and received periodic medical checks. Those that failed the examinations were supposed to be taken out of circulation until they recovered. Our duty, of course, was to make sure the troops knew which numbers were good and which, at least temporarily, were not. We handled the assignment with the subtlety and sophistication that the task demanded. Without any fanfare, during certain portions of the broadcast day, the music would be gently faded down and a well-modulated voice would softly intone: 34, 23, 67, etc. Newly assigned personnel to the base always gave themselves away by their inability to determine whether our announcement was a recommendation or a warning."

Warning: Dick Hiner, who was also there, has more on this subject later.

☆

If Rodney Dangerfield don't get no respect, consider the case of Colonel Don Macaluso. As commanding officer of the Air Force's 7122d Broadcasting Squadron, responsible for all the AFRTS outlets in the Mediterranean, Middle East, North Africa and Atlantic, he was being taken on his first formal inspection of his new domain. Accompanied by

The five o'clock traffic jam is passing in front of the Lajes station in the Azores. It was perhaps one of these same critters who deposited donkey doo-doo on the highly polished shoes of LTC Don Macaluso as he arrived to inspect the station (AFRTS photo courtesy John Bradley)

Chief Master Sergeant Roger Maynard, he arrived at the station at Lajes in the Azores. On the front porch, with due ceremony, he was greeted by Chief Master Sergeant Jim Ennis, the station manager, and a donkey, which on that tiny island many used as local transportation. Both were very pleased to see him, especially the donkey who proceeded to deposit several pounds of fresh, steaming appreciation on the Colonel's feet, which is precisely what a person would expect to get from an ass.

☆

Hal Kelley, station manager in SHAPE, Belgium, one day near Christmas told one of his very young announcers to begin playing more traditional holiday music such as "Oh, Come All Ye Faithful."

"Sure," said the bright and eager young man. "What group put that out?"

☆

The nomination for the absolutely worst program ever presented on an AFRTS station goes to Berlin for a series produced there in 1946. Called "Know Your Lyrics," the program played the popular records of the day at half speed so listeners could copy down the words.

Second place nomination goes to Bremerhaven, Germany, for "Radio Bingo." Listeners could pick up bingo cards at the local service club and the station would read endless lists of numbers so the people at home could play.

...we return now to our regular programming.

CHAPTER TWELVE

In which we learn that even Five-Star Generals can learn something from radio. General MacArthur learns from AFRS that President Truman has fired him, thus proving that:
"War is too important to be left to the Generals."
--Georges Clemenceau

Here we Go Again

It was still an uneasy peace in 1948, although troops in Europe and in the Far East had been drawn down to the smallest levels since the end of World War II. Russia and the Western Allies had been at swords-point since the end of the war in 1945. Now things were beginning to heat up. The Russians were particularly incensed that the Marshall Plan, devised to help insure the economic recovery of Europe, seemed to be working. The so-called "Truman Doctrine," which offered help to nations "resisting attempted subjugation," had Bolshevik blood boiling. In March the **Treaty of Brussels**, later to be forged into the NATO alliance, was signed. The Western allies agreed to make the Deutschmark the legal tender Germany and in the four-power occupied city of Berlin.

This was too much for the Russians who wanted to keep Germany weak by forcing them to continue using the almost worthless Reichsmark. They began cutting off allied access to Berlin which lay deep within their Zone of Occupation. First it was for hours or days at a time. Finally, at 6 a.m. on June 24, the Soviet Government slammed the door to the West. Berlin, isolated more than one hundred miles away from the Western zones of Germany administered by the U. S., Britain and France, was totally cut off from the supplies of food, fuel and medicine needed to sustain it.

With land access blocked, the three Western nations began flying the biggest airlift in history. For the next 462 days, they flew in to Berlin the estimated 4,000 tons of material needed daily to provide at least minimum requirements. Around the clock, the small C-46 and

The AFKN control room in Seoul, Korea. At the switcher, upper right, is John Otterman; the video man is MSGT Richard E. A. Pirtle and in the background at the audio console is SSGT Raul Casteneda. (U.S. Army Photograph)

C-54 aircraft of those days arrived and departed at approximately one minute intervals.

The German population of Berlin was understandably a trifle nervous about all this. Many Berliners, to this day, remember AFN Berlin as a constant reminder of the American resolve to stay in Berlin. Other stations in the network were then broadcasting on an 18-hour schedule, but Berlin went to around-the-clock broadcasting. Weary pilots and crews used its signal as a homing device while flying into Berlin. They also became used to AFN news personnel flying back and forth, producing a steady stream of stories which, while this was not their intention, helped the morale of both the military and the civil population. In addition, according to many, it solidified in the minds of the Germans that the allies were sincere in their determination to make Germany a partner rather than remaining an occupied nation.

The confrontation with Russia during the period of the Berlin Airlift may well have saved the network. As the cold war grew more frigid, the military build-up throughout Europe became a reality and the need for the network by its ever-increasing audience became daily more necessary.

...Now, getting to the Seoul of the Matter

While the troop build-up began in Europe, the military began pulling Americans out of the Far East. The Far East Network had established stations in Korea after World War II at Seoul, Pusan, Chonju and Kwangju. Each was an independent outlet and none was interconnected. As troops withdrew in 1948, stations were closed down and by mid-1949 only the Seoul station remained in operation. FEN then transferred it to the Korea Military Advisory Group and it temporarily ceased to be a part of the Far East Network.

A year later, the cold war in Europe suddenly became a hot war in the Orient. In June, 1950, North Korean forces burst across the 38th Parallel and launched an attack toward Seoul. Once again, AFRTS was operating under very real and very dangerous wartime conditions. The Seoul station stayed on the air until the last moment, broadcasting emergency instructions to American personnel and proving once again that radio is the fastest, most efficient way to reach an audience during an emergency situation.

Shortly before the North Korean troops occupied the South Korean capitol of Seoul, the station hastily packed its gear and headed south to Taegu. Broadcasting briefly from there, it was again forced to move even further south to Pusan as the war flowed toward them.

Soon the perimeter around Pusan became so small the station was again forced to evacuate, this time back to Japan where broadcasts directed to Korea were transmitted from the FEN station in Fukuoka.

At this point, United Nations forces were pretty well squashed into the southern quadrant of the Korean peninsula. Plans were well underway at General Douglas MacArthur's headquarters to make an end run by sea and land at Inchon on the coast near Seoul, thus cutting off the invaders. At the same time, plans were being completed to reopen the Seoul station once the city was recaptured. Equipment was acquired and FEN's radio officer, Ed Tidwell, appointed Lieutenant Albert Jones as officer-in-charge, along with Francis Crosby, a civilian engineer, and seven enlisted men. At the moment, their station was a pile of crates on a Yokohama dock.

Seoul was recaptured on September 25 and on September 27, Crosby and two men, Sergeant Allen Larkin and Corporal Lawrence Butcher, flew into Kimpo Air Field in Seoul to find a home for the station and make the other necessary arrangements. On the same day, General MacArthur officially signed a General Order creating AFKN -- the American Forces Korea Network, which, to this day, continues to operate a network of stations. Today, however, it is somewhat more sophisticated.

The three original AFKN pilgrims found space in the basement of the American embassy. As the North Koreans had departed the city, they had exploded two bombs in the basement but the men figured it could be made useable after it had been fluffed up a bit. The rest of the staff arrived several days later and the nine men worked 48 straight hours getting the studios ready. Although they had brought most of the necessary equipment with them, a few small items had been forgotten. Among them were tools such as screwdrivers. Larkin recalls that getting the station installed involved such non-standard implements as fingernail files, nail clippers and the edge of mess kits. In 48-hours they turned on the power and AFKN Seoul was on the air. For a couple of hours, that is.

◆

Sign on went as scheduled. So did the newscast following. Then...silence.

Crosby had strung the antenna up the side of the eight story embassy, run it across the street and up a smokestack atop a building 30 feet away. Then a Korean, cleaning up the heavily damaged embassy building, dumped trash out of the top floor window, ripping the antenna down in the process.

Crosby climbed the eight flights of stairs and repaired the damage, putting the station back on the air.

Off it went again. This time someone had thrown a chair out of one of the windows and once more destroyed the antenna. Crosby fixed it again. It happened a third time.

Figuring the Army hadn't given him a carbine for nothing, he posted himself in the street and a few warning shots fired into the air when people tried to use the windows as a public dump soon convinced everyone to abide by more customary garbage disposal techniques. It worked, and AFKN Seoul stayed on the air.

◆

To cover the fighting, FEN equipped its correspondents with .45 automatics and tape recorders, both essential pieces of equipment when required to simultaneously talk and fight. One correspondent recalls being with the 1st Cavalry Regiment north of Taegu with whom he remained until he was wounded and evacuated back to Japan. One day, he says, he looked out of a courtyard and saw tanks approaching. Jubilantly he yelled, "Here come our tanks." The soldier next to him said matter-of-factly, "We ain't got no tanks." Oops. Bug out!

When the North Korean troops went back north of the 38th parallel, the AFKN stations which had been rolling around the countryside in mobile vans got a chance to hunker down in lavish quarters such as this quonset hut. It would be nice to identify this group by name, but they will know who they are when they see themselves. The Lieutenant is believed to be A. K. Szczerbowski, perhaps not the easiest name for a listener to remember, but one has to put up with a lot during wartime. (U.S Army Photo)

All networks have more wheels than they really need—except AFRTS when they are in a wartime situation. Then they need all the wheels they can get—real ones. Pictured is a Korean war style mobile unit on wheels which is slightly more advanced than the World War II version but still primitive compared with what is to come in the future. The careful reader will note the very latest in antenna configurations—a long wire stretched between wooden telephone poles. Units such as these racked up a lot of mileage as they chased the troops up and down the Korean peninsula. (U.S. Army Photo)

77

What with the Cold War in Europe and the Hot War in the Far East, AFRS was now back on a wartime footing. Like its predecessors in World War II, it was necessary for AFKN to keep moving along with the troops if the stations were to do their job. Although AFKN was no longer under the command of FEN in Japan, FEN continued to provide considerable help. They came up with the idea for a mobile station modeled after those used by AFN in Europe during World War II. Somewhere they found an old ordnance van and built a studio, complete with transmitter. Dubbed "The Monster," it was airlifted to Kimpo Air Base and from there dragged to Seoul where it replaced the studios in the embassy basement.

Meanwhile, the war turned fluid and troops chased each other up and down the peninsula. United Nations forces had driven well into North Korea. Now the North Koreans and their new Chinese allies turned around and were driving back toward Seoul, only a few miles from the border of the two Koreas. As they entered the northern suburbs of Seoul, the AFKN staff piled into the Monster and took off in a southerly direction.

Kilroy is STILL Here...

The last radio announcement asked listeners to be patient. "We don't know where we'll be next," the announcer said, "and nobody else does either. That's why we're the Kilroy station."

The Kilroy allusion was familiar to everyone. He was the imaginary character, drawn with his eyes, huge nose and bald head peering over a fence with the inscription "Kilroy was Here," who appeared in every unlikely place imaginable throughout World War II. Eventually ending their journey in the far south of Korea at Taegu, the Monster and its residents became known forevermore as Radio Kilroy.

By then, more mobile stations had arrived and AFKN became a network on wheels, although it was not a true network as the stations were not interconnected. Most of the station names were apt in that they all indicated movement. There was "Radio Gypsy" which was continually on the move. "Radio Rambler" covered the Chunchon area. "Nomad" was with I Corps and "Mercury" with the Air Force.

"Radio Vagabond" moved back into Seoul when the city was again recaptured. Deciding to be vagabonds no longer, the staff set up a permanent station in the Bando Hotel. Two hours later they were vagabonds once more as the area came under attack. It was back on the road again for Vagabond.

Programming was both produced locally and received from AFRS in Hollywood. Newscasts came via shortwave from AFRS and other sources as well as from AFKN correspondents who followed the troops into battle and flew missions with the Air Force.

Max Lash was one of the AFKN crew. If anyone could be called "an old Orient hand," Lash is the man. Starting with the FEN Manila station in 1949, he served in the Philippines, Japan and, in Korea, with "Homesteader" in Pusan as well as with Kilroy, Rambler and Vagabond. He left the army in 1963 and became both a print and broadcast correspondent in the Orient where he remained through 1971.

He points out that rolling around Korea in a perambulating radio station was not all fun and games.

While with "Rambler" in Chunchon, he vividly recalls the day two AFKN staffers hitchhiked a ride to Japan where they were due for Rest and Recreation leave. One, was PFC Robert Pipes. The other was the new executive officer from headquarters in Seoul, in Chunchon to conduct an investigation.

The plane was first to land in Seoul so they could get the necessary documentation for their leave and then continue on.

Waiting for them at the Seoul airport was Captain George Kennedy, then Seoul station commander and later to become famed for his motion picture work. The plane never arrived.

The small, unarmed six-place aircraft inadvertently strayed into North Korean airspace and, although unarmed, was shot down by anti-aircraft fire. The AFKN personnel did not survive.

One minor skirmish took place between the Army and Marines. In the midst of the grim realities of combat, **this** war had the troops laughing.

Radio Vagabond decided to put on a fund-raising marathon for the benefit of crippled Korean children. Contributions poured in from throughout the listening area. A group of Marines paid $400 to hear **The Marine Hymn**. Another group of Marines offered $100 to keep **The Artillery Song** off the air. The artillery-men fired back a $155 pledge to play it. Then the Army paid to have **Anything You Can Do, I Can Do Better** dedicated to the Marines. They followed this up with **Baby Face** and **Too Young**. Finally the Marines dedicated **I Surrender, Dear** to the Army and the musical war ended -- but not before they all as a group dedicated **I'll Be Glad When You're Dead, You Rascal You"** to Communist General Nam Il.

The rolling radio men often got close to the front. One Christmas program featuring a choir and a chaplain was produced with the distinct sounds of artillery in the background. This is perhaps the only time a chaplain wished his congregation "Peace on Earth, Goodwill to Men," followed by the blast of a 60mm shell.

THE OLD SOLDIER FADES AWAY

As the troops, and the radio stations, marched up and down Korea, a less publicized war was in progress. This one was between Supreme Commander of the Forces, General of the Armies Douglas MacArthur, and the feisty President, Harry S. Truman. As President, Truman was Commander in Chief. And he was ticked off more than somewhat that his Commanding General was becoming offensive toward him. After 50 years as a soldier, MacArthur gave the impression that only he understood the world situation. He publicly disagreed with his Commander in Chief, called for a war against China and, possibly, the Soviet Union. He wrote letters to influential congressmen and railed against Truman s policies in order to gain congressional approval of his very personal foreign policy.

This was too much for the short-tempered President. He, with the knowledge and approval of the Chiefs of Staff, prepared a presidential order relieving MacArthur of his command and sent it off by special courier to MacArthur.

While the courier was in the air, a press release with all the pertinent details was prepared and given to the news agencies with an iron-clad hold for release embargo until after the estimated time of arrival of the courier.

AFKN news editor Leo Cross did all the right things. The embargo was on until 10:15 p.m. Korea time. Knowing the importance of the story, Cross put an announcement on at 10 p.m. saying an important message would follow the regular 10 o clock newscast. At 10:15 there was a stay tuned announcement. Moments later, announcer George Tyson was handed the copy by Cross and read:

WE INTERRUPT THIS PROGRAM TO BRING YOU A SPECIAL NEWS BULLETIN RELEASED FROM THE WHITE HOUSE JUST MOMENTS AGO. PRESIDENT TRUMAN HAS RELIEVED GENERAL DOUGLAS MACARTHUR OF ALL HIS COMMANDS. MR. TRUMAN S EXTRAORDINARY ACTION WAS JUST ANNOUNCED AT A WHITE HOUSE NEWS CONFERENCE. MR. TRUMAN SAID HE HAD CONCLUDED THAT MACARTHUR IS UNABLE TO GIVE HIS WHOLEHEARTED SUPPORT TO UNITED STATES AND UNITED NATIONS POLICIES. THE PRESIDENT IMMEDIATELY DESIGNATED LIEUTENANT GENERAL MATTHEW RIDGEWAY AS MACARTHUR S SUCCESSOR.

The broadcast cliche You heard it here first held true for AFKN. Mac Arthur was listening to the newscast and discovered, from the very broadcast network he had formed, that after a lifetime of service, he had been relieved by AFRS.

It seems Truman s emissary had not yet arrived because his plane had been held up by bad weather after a refueling stop en route.

Although MacArthur faded away, the network did not and it continued to grow. "Radio Meteor" was added to serve the First Marine Air Wing and became the ninth AFKN station; seven broadcasting from mobile units, two from permanent studios.

Because the Korean "police action" was a United Nations campaign, AFKN was given special permission to broadcast in languages other than English which, except in these special circumstances, is a distinct AFRS no-no. Although not frequent, the network presented disk jockey shows and newscasts from time to time in French, Dutch, Korean, Flemish, Turkish, Greek, two dialects of Spanish and two versions of English -- American and that which can be understood by Commonwealth troops from Great Britain, Australia, Canada and New Zealand.

Finally the armistice came on July 27, 1953. The wheels of the stations stopped rolling and they found permanent quarters. Vagabond opened up on Vagabond Hill in Seoul. "Comet" moved to Osan Air Base. "Gypsy" moved to Camp St. Barbara in Kuma-ri. Others closed permanently, their job done.

The fighting troops went home. Fresh new faces arrived, full of tales about a new gadget back in the States.

It was sort of like radio with pictures and called Television...and they wanted it.

Things got better in Korea fairly soon after the "police action." By 1960, the headquarters station of AFKN in Seoul looked like this. Shown are PFC Robert Pyle on camera and floor manager PFC Ashley Hawken. By this time, about 25 percent of the stations' programming was done live. On Dage cameras. (U.S. Army)

Right, The Limestone station managed to pump out more than 100 film presentations each week through this primitive film chain. The picture was projected directly on to the tube of the videocon camera, unlike today's more sophisticated multiplexers which can handle several films and slide projectors. With this early model, when the reel of film ran out, everybody waited until the operator loaded the next one. Adjusting it (C) are engineer A/1C Arthur J. Hanlin being assisted by (L) SSGT James Ranne and (R) SSGT Robert Dean. Below, The original "Elevator Shaft Gang," the group that put the first AFRTS television station on the air at Limestone, Maine, from studios high atop the base hospital in the elevator shaft housing. All duded up here for their Christmas party are, back row L to R, Bill Birchfield, Bob Slezak, Charles R. Hughes, John Bradley, Ray Proffit, James A. Bob Dean, Homer F. (Rick) Welsh and David Freeman. Front row, L to R, Bill Lynn, Peter O. E. Bekker, A.J. Hanlin, Bob Miles, David Ward. (Photos courtesy of John Bradley)

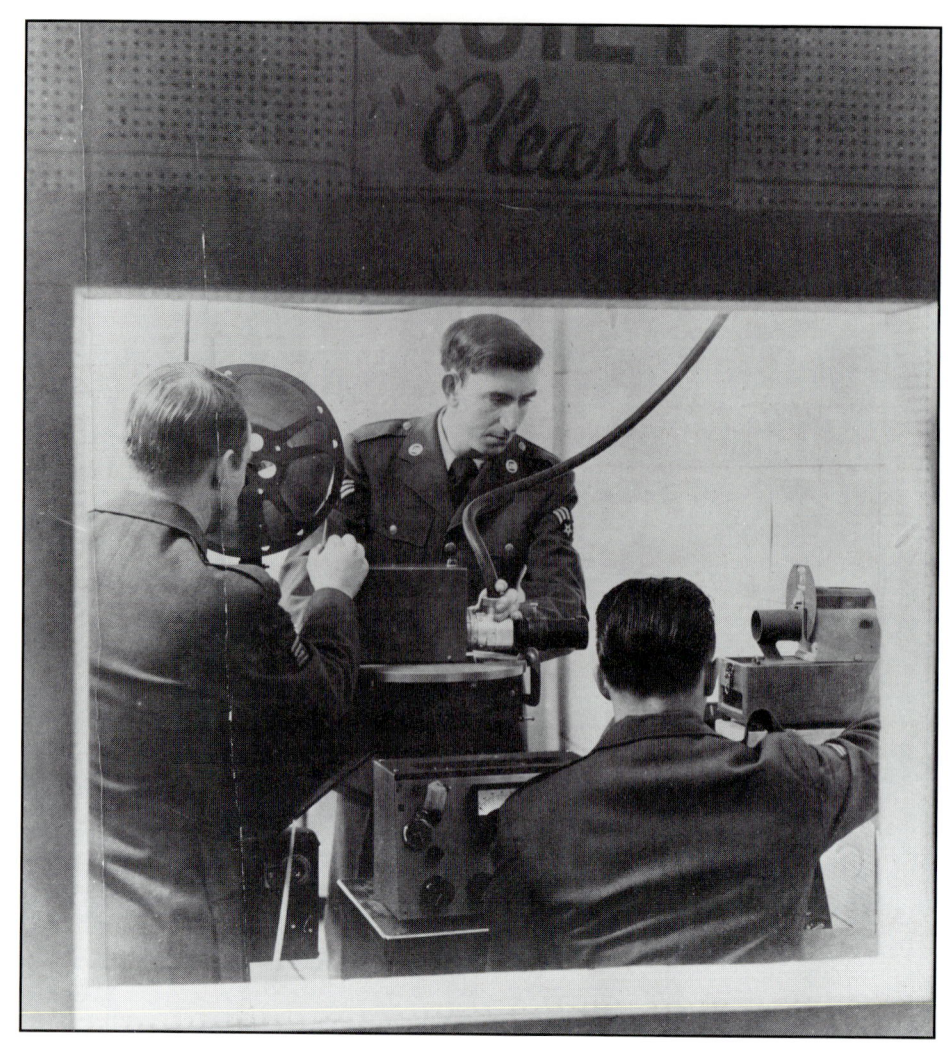

CHAPTER THIRTEEN

*Just think! If it weren't for Television,
people would be sitting around eating
frozen radio dinners.
-- Johnny Carson*

"Well, Picture That. AFRTS Gets Television"

The Korean conflict and the Cold War insured several things. One was that the world would be tense for awhile. Another was that AFRS would be needed for the foreseeable future, although the changing world and the changing patterns of broadcasting dictated that it would be a different AFRS.

Radio, the rock upon which broadcasting had built, was turning to sand. Now, in the 50s, the broadcasters and the audiences were turning their attention to television. Talent who had made their careers on radio were anxious to move into television. Nor did it take an anvil falling on their heads to make the writers, directors and technicians realize that television offered infinitely larger playing fields for their particular skills. Almost overnight, the radio in the corner was replaced by a 10 or 12 inch television receiver, flickering away in colorless black and white.

The radio was moved into the bedroom or the kitchen and soon local disk jockeys replaced the kings and queens of the airwaves.

This presented a challenge to the programming people at AFRS. The most popular part of the radio package sent to stations each week was disappearing as the radio stars moved to television. Using the theory, "If you can't lick 'em, join 'em," AFRS joined 'em. Several plans were undertaken. One was to record local disk jockey programs, generally from the Los Angeles area, and remove the commercial spot announcements. These were replaced by public service announcements and sent to the affiliates on disk. This technique worked, but had the unfortunate side effect of making AFRS stations everywhere sound like they were in Los Angeles because of the normal localized mentions of LA-unique events and activities.

The better, and more lasting, solution was to increase the number of programs produced in the AFRS studios by top-notch DJs, auditioned and contracted to produce programs exclusively for AFRS audiences. This remains the system in use today and the DJs, many of whom have never been heard outside a limited area before, are known world-wide.

AFRS also increased the size of its record package and began sending more individual music cuts for use by local overseas DJs. As time passed, stations built up large music libraries and utilized more and more of their broadcast hours producing local shows directed specifically to their individual audiences.

The worries over what to do about television continued to plague the staff at AFRS but, perhaps hoping it would go away, little or nothing was done to initiate television programming overseas. The very thought of the complexities involved boggled minds throughout the system. It took a very strong mind, indeed, to unboggle the problem.

His name was Curtis LeMay, the cigar-chomping, never-take-no-for-an-answer boss of the Strategic Air Command. Because the cold war had forced new strategic thinking on the military, services were deploying their forces to remote locations in order to avoid the hordes of Russian bombers many expected to appear on the horizon momentarily. SAC, in those days before reliable intercontinental ballistic missiles, was America's primary retaliatory force. Its remote bases were the remotest of all, and, because of that, SAC had a morale problem.

LeMay and his buddy Arthur Godfrey often got together, and, with base troops, would race automobiles on SAC runways. One day LeMay told Godfrey about a particularly bad problem with AWOLs and a low re-enlistment rate at the newest and most remote SAC base at Limestone, Maine. Located in Aroostock County, Maine, Limestone Air Base had a population of 15,000 while the townsfolk numbered only 864. The nearest bright lights were in the big city of Caribou, ten miles away, with a population of 4,500. To misquote Fred Allen, it was a nice place to live if you were a potato. In fact, it was so far in the wilderness, you could throw a potato into Canada if you had only a fair pitching arm.

With no war going on to occupy the service members, they found less urgency in their work and more spare time which needed filling. LeMay was concerned that boredom was driving many of them out of the service and SAC, requiring highly trained personnel, didn't want those losses. He asked Godfrey, who was at that time one of the most popular personalities on TV, if perhaps a television station on base might not help fill some of the empty hours. Godfrey thought it was a great idea and LeMay put his staff to work to develop plans for a SAC television station.

Because there were no applications for a civilian

TV station in the area, the FCC agreed to let the Air Force proceed, providing it agreed to close down should a commercial station apply for a license. It also limited the power of the proposed station to a mighty 10 watts. LeMay gave the go ahead and said to get the thing operating by Christmas. Lieutenant Colonel P. L. Moen was selected as the project officer and after a number of unsuccessful inquiries to manufacturers of television equipment, brought his problem to General LeMay. As one who believed in the direct approach, LeMay called David Sarnoff, president of RCA, and soon two RCA representatives appeared to do an on-the-spot survey at Limestone Air Force Base.

The planners chose space on top of the base hospital as the best area for the studios and antenna. They then returned to the RCA headquarters in New Jersey to draw up plans and select equipment. The plans were driven by limitations of both space and money, so they came up with three: $18,000, $24,000 and a big-budget $28,000. None provided for live studio productions. The Air Force went all out and took the big package, with modifications that would allow for live cameras at some future date. The money came from SAC's welfare funds.

The term "bare bones operation" must have been invented to describe what they got for their money. There was a one videcon camera film chain, a 16mm projector, monitoring equipment, a synchronizing generator and an eight-watt transmitter. They also threw in a turntable to play fill music and an audio tape recorder.

Even this elemental station required programming, and SAC decided to ignore AFRS and provide its own. Even though AFRS had years of experience dealing with program suppliers, SAC charted its own course and sent a team to New York to talk to the major networks and obtain clearances to use network programming. It turned out to be more difficult than was first imagined. Contacts and permissions had to be obtained from sponsors, agen-

The early studio at Limestone TV measured 8 X 8 feet which put a certain limitation on the type of studio productions that could be done from there. They couldn't present the Rockettes in performance, but they were able to do the news, weather and programs such as this. Here is Chaplain (1Lt) William Petrich getting the go-ahead cue from the station's program director. You'll note the camera is on a revolving platform. When the program is over, it will be turned around and once again assume its rightful role as the film-chain camera. (Photo courtesy John Bradley)

cies, networks, the talent and everyone but the janitor before anything could be done. The Department of Defense helped by preparing a blizzard of memorandums assigning responsibilities and laying out reasons why television was required.

Coordination for future television expansion was put under the supervision of the Office of Armed Forces Information and Education which, in effect, meant that AFRS would be the major player in the future. The directive also missed the essential point. It stated in its best bureaucratic baloney, that the mission of military television would be to foster in the serviceperson (called, in those days, "serviceman") attitudes conducive to military efficiency; the mission of the Armed Forces; American democratic principles; and increasing knowledge of national and international affairs.

It is no wonder the AFRS people backed away with that directive guiding them. They knew, if the Pentagon did not, that however noble the stated goals were, the primary mission of television would forever be to meet the leisure-time needs of service people and, if it did so, improve morale.

MEANWHILE, BACK ON THE RUNWAY

Returning now to LeMay and Arthur Godfrey, at the time the idea of using television as a morale tool first surfaced. The two men frequently met at what was called within SAC, "Racing on the Runways." LeMay was an avid sports car racer and Godfrey would go anyplace that was close to airplanes.

Assigned as a film cameraman to provide radio and television coverage of the SAC "Racing on the Runways" program was a young Air Force enlisted man named John Bradley. Bradley was eventually to become an AFRTS legend, but at this time he made his living peering through the viewfinder of a Bolex. LeMay knew him from his work with the racing program. Congress one day in 1953 suddenly awoke to the fact that SAC was using its runways as racetracks and could find no earthly excuse for them to do so. No matter how fast they got those cars moving, Congress reasoned, they would never fly. The expensive runways were meant for bombers, not souped-up Chevys, they told LeMay. Kindly and forthwith, get back to defending our great land and making it safe for democracy rather than for four-barrel carburetors. End of racing program, end of Bradley's job.

Later, flying back to SAC headquarters aboard LeMay's private C-97, LeMay called Bradley aside and told him he realized there was no longer a job for either him or his Bolex. He said that because Bradley had done a lot of filming for television, he might be interested in a job with the new television operation at Limestone Air Force Base. Later in his career, Bradley became famous for his tact in disagreeing with officers and his luck in getting away with it. That was later. This time he agreed that it sounded like a fine idea.

Perhaps he was overwhelmed by LeMay's plane which featured genuine engraved silver, real china service with the SAC crest and -- always -- steaks cooked to order followed by imported cigars and choice of after-dinner drinks. In any event, filled with the General's words, steak, booze and cigar smoke, he was soon on his way to

ATC William D. Ward adjusts the sync generator at Limestone. In those early days it took this whole rack of vacuum tubes, to synchronize the audio and video signal. Today it's done in a little box that only takes up about an inch and a half of rack space. You don't find too many vacuum tubes around any more either. (Photo courtesy of John Bradley)

the upper U.S.

The station had met LeMay's deadline of a Christmas sign-on; a second Christmas miracle. Half the equipment arrived on the 21st. The transmitter arrived on the 23rd. An RCA engineer accompanied it and everyone worked until 4 p.m. the next afternoon getting it installed and operating, if somewhat haphazardly. It turned out the antenna was not only the wrong one but the duty operator had to go out on the roof and pound ice off the thing several times each shift. Still, no one could say it didn't send out a signal. One day the station got a reception report from someone in Pittsburgh, Pennsylvania. Either the airwaves weren't busy that day or the signal was just trying to get out of Aroostook County.

Film for the first day of broadcasting didn't arrive until the 24th. On Christmas day, as promised, military television broadcasting began. It began with what may well be one of the dullest program schedules in the annals of television history. The good folks at Limestone Air Force Base and those living eight watts out into the surrounding countryside were treated that Christmas day to the excitement of **The Littlest Angel, Rootie Kazootie, Child of Bethlehem, Guiding Star, You Can Change the World, Jamie, U. S. Steel Show, Comeback Story** and sign off at 10 p.m. Because programming was so limited, the station initially broadcast only on Friday, Saturday and Sunday for four hours each day.

Still, it was better than sitting around, listening to the mating call of the bull potato.

Imperfect it was, but LeMay and Godfrey turned out to be right in their assessment that television added a positive dimension to isolated living. When the broadcast operation began, the AWOL rate was 20 per 1,000 men. It soon fell to 2.5 per 1,000. Courts-martial figures went from an average of 22 per month to 4 or 5 and, later, dropped to 1 or 2 monthly. Re-enlistments had been zero before television. They soon jumped to 18 a month, surprising in view of what there was to view.

About this time the base was renamed Loring Air Force Base and the station slowly expanded its schedule of programming. Within a month or so, it was producing "live" shows including **The Chaplain's Hour** which ran fifteen minutes but was so named, the staff explained, because it "sounded better." The Wing intelligence officer did the newscasts and **LAFB Personalities** was a smash hit, presenting such sure-fire entertainers as an amateur fire-eater, a hypnotist and an airman who did three dimensional paintings.

Bradley arrived and to his dismay, it turned out that the station was indeed on top of the hospital -- in a structure made of steel I-beams supporting corrugated sheet metal walls with no insulation. It was an extremely small space, so small in fact that while in it, a person didn't have room to eat a foot-long hot dog. Worse, the limited space, was shared by the elevator machinery. This machinery was on the same electrical circuit as the station. Every time someone used the elevator, the picture shrunk to half its normal size. Each night the staff looked forward to the end of hospital visiting hours.

In the heat of summer, the six man staff worked in the minimum amount of clothing the law allows. In the midst of the Maine winter, they piled on parkas and worked in gloves. This made changing film difficult, particularly as it had to be done every half hour because there was only one projector. It was necessary to take gloves off for that task and there was jubilation all around when they finally learned to thread the projector in five seconds by using a board with nails simulating the film path which pre-formed the film to fit on the sprockets quickly.

Bradley's reputation is solidly based on the fact that one way or another -- frequently another -- he gets the job done. This trait became obvious when he and the group built their own multiplexer, a gadget for which RCA charged many thousands of dollars and which essentially consists of mirrors at 90 degree angles, allowing two projectors to project film at right angles off the mirrors, bouncing the pictures into a single camera.

His trusty Bolex also soon became a part of the operation and film of local events, developed through the courtesy of the Base Photo Lab, became a regular part of the news.

Now that it was possible to project slides by use of the homemade multiplexer, the group wrote the Disney Corporation and asked them to design a logo for the station. They explained who, what and why they were. Soon, back came their logo: A giant thumb holding up what could only be an old fashioned outhouse with a giant antenna sprouting from the roof. Under it were the words: **The Tom Thumb Station at the Top of the Nation.** Even Disney has off days.

The next improvement was a studio of sorts. Base engineering cut a hole in the wall behind the camera and built a giant studio, all of eight feet square. They then built a swivel so the station's single camera, permanently fixed to the film chain, could be turned to the hole in the wall and shoot into the studio area. Here is how Bradley describes it:

Live operations were simple. At six o clock we would do a wipe to black -- that is, we would put a hand over the camera lens, swivel the camera, rotate the turret and wipe to a newscaster sitting behind a desk. Later we got a special service troop to be a sports man. We also added a weather segment with a forecaster from the local weather unit. We wiped to black at the end of the sports. The weather man and the sports man lifted a prepared transparent map on to the desk and exchanged places. The whole change took less than 3 seconds. The whole show was done in the eight foot square studio.

At times relations between the local populace and the air base were strained. When it was first announced that a television station was to be built, it was emphasized that it was to be directed specifically at the military. This started a rumor that the Air Force intended to build a huge fence around the base so the locals could not receive a picture. This being North of Northern Maine where the winter goes to spend the summer, there is always great interest in something to do other than discuss the price of potatoes. The interest in TV was considerable and created even more excitement than the time someone found a potato that in dim light looked a little like the late, great Calvin Coolidge. When the rumor that television would be restricted to the base took hold, riots were threatened.

Although interest was high, knowledge about TV was somewhat limited. Again, John Bradley:

A few days before the station signed on, the station got a call from a lady with an accent that can only be described as Aroostookian, for northern Aroostook County. She was very concerned and wanted specific information about the size of the picture we would be transmitting. We tried to explain that we didn t control the size of the picture as the limitation was the size of the receiver. She was not convinced and forcefully explained that she planned to buy her father a television set for Christmas and didn t want to buy one of those expensive 19 inch ones if we were only going to broadcast a 13 inch picture.

To the unmitigated horror of AFRS, the films being received through Air Force channels from the networks came complete with commercials. The policy then was -- Run 'em. Although a disclaimer was put on at the beginning and end of each broadcast day, the running of commercials had a certain interesting aspect. When the Eddie Fisher **Coke Time** program started, a Coke machine magically appeared, was always full and refused to accept your nickel. Yeah, nickel. This was 1954. Local merchants, through the goodness of their hearts, and the fact the station was running commercials for the products they offered, always seemed to want to supply food and beverages for station picnics. No one objected and station personnel made certain everyone shared alike in the largess.

For more than a year the hospital used the elevator and the station got the shaft. The Air Force knows a shaft when it gets one and realized that if any real improvements were to be made, they would have to move and equipment would have to be added. As additional funds became available and space was selected, Bradley and Bob Dean, another staff member, were sent to Camden, New Jersey, home of all the RCA goodies, to attend a special course in the latest equipment. Just like in the real world of TV, they got new treasures including a second film projector, a television slide projector and a real honest-to-goodness multiplexer. The real treasures, however, were a PAIR of the latest image orthicon cameras with a full complement of lenses, studio lighting, a switcher with a fader so no one had to hold their hand in front of the lens anymore. Also with the package came a truckload of audio equipment.

The plum, on the peak of the pudding: a brand new FIFTY watt transmitter.

Life suddenly changed. The station moved into a wing of a two story barracks with space for a larger studio, a master control room, projection room, film storage room and office space. There was only one problem. Directly below was the base nursery, populated by tiny sadists who could sense when the microphone was turned on and would pick that micro-second to yowl. Finally, someone at the station acquired a couple of cases of acoustic tile while the supply room was out to lunch and quickly glued them to the nursery ceiling. As noted, Bradley knew how to get things done.

For example, he knew how to get the General's attention. In the summer, the studio got unbearably hot with no air conditioning and lots of TV lights. It was the custom of the news anchor to put on bermuda shorts which couldn't be seen behind the desk, although on top he was impeccably uniformed. Bradley had just slipped out of his pants one evening when the film broke. He dashed in his underwear shorts to fix the problem. Then he dashed back to slip into his bermudas. The film broke again. Same thing. Time to go on the air. He grabbed the uniform jacket and sat down. Over the intercom there was a great deal of distracting chatter. Then the cameramen began making strange and non-standard hand signals.

When the newscast was over, he got up and started to chew out the crew for unprofessional conduct. "Well," one explained, "we were just trying to tell you to stay in your chair because the General, his wife and three visiting Canadian Generals are watching you from the control room."

The General told him later he could never again watch the news without wondering whether Bradley had his pants on.

About this time, SAC made the transition from B-36 bombers to the then new B-52s. Bradley and his Bolex were sent to California to shoot transition training at Castle AFB and then returned to Maine. His documentation of the transition was used on the major U.S. networks. The big story was the successful attempt of the B-52s to fly around the world non-stop with mid-air refueling. His film of the takeoff was needed NOW by ABC in New York. Knowing it would be a tight schedule, an old C-45 was modified with film editing equipment in the back. Grabbing the film out of the developer, Bradley spent the flight to New York editing negative film while lying on the floor, calling out the length and subject of each edit to writers Percy Brown and Bob Spearman so they could write copy to fit. It was a full team effort of the "Elevator Shaft Gang." The pilot on the flight was the Information Officer of the Base, and thus the Officer in Charge of the station, Major Peter O. E. Bekker. He was later to show up assigned to AFRTS in Los Angeles.

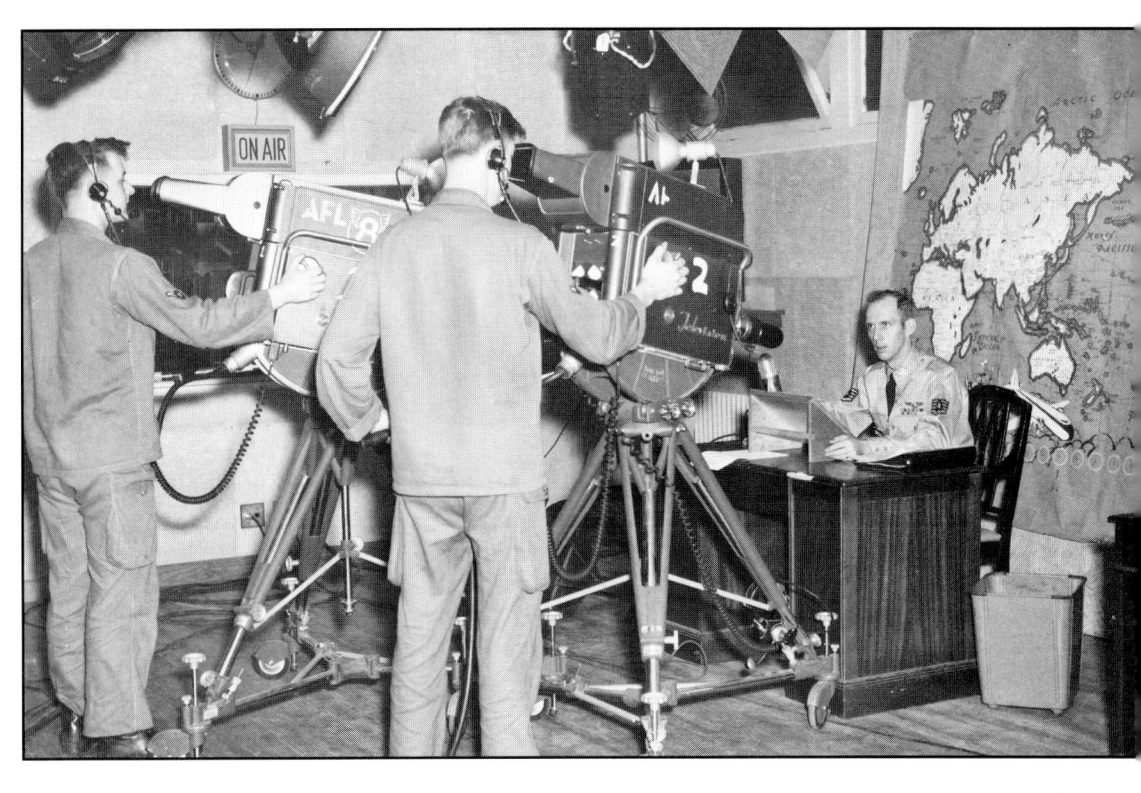

They made it to the ABC studios with minutes to spare and John Charles Daly read it without a hitch.

In early 1956 someone applied for a commercial television license for the area and that was the end of AFTV. Why anyone would want to start a TV station in a town so small the local MacDonalds had only one Golden Arch is beyond comprehension. In October, the "elevator gang" signed off for the last time and threw the going-away party of the century. The new commercial station, with no network affiliation, put television back several years. The first Sunday after AFTV signed off and the new station signed on, Aroostookians and Air Force alike were stunned to find a new show where Ed Sullivan normally was seen. It was a production of the University of Maine and bore the exciting title, **The Artificial Insemination of Cattle.** That's no bull.

The staff was scattered and most of them headed for a new station at Ramey AFB in Puerto Rico. Bradley was the exception. He was sent to the Azores. We will meet him there subsequently but leave him now on his very last day in Maine. A B-52 has crashed in the Canadian woods nearby and he is tramping through the snow, filming and documenting the wreckage.

Now, though, he is wearing his pants.

Later the studio space was expanded and RCA TK-47 cameras, then the very latest type, were added to the station at Limestone. John Bradley is pretending here to be about to go on the air with the six o'clock news. At least we think he is pretending because the clock says eight minutes to one. Either he is very early or the clock is very late. Bradley had the "modesty panel" in the knee-hole of the desk added because the studio got so hot in warm weather that he often worked pantless. (Photo courtesy of John Bradley)

THE STARS AND STRIPES

Daily Newspaper of U.S. Armed Forces — **in the European Theater of Operations**

Vol. 3 No. 207 — New York, N.Y.—London, England — Saturday, July 3, 1943

1D — 1D

Jap New Georgia Headquarters Struck

ETO Radio Network on Air Sunday

American-Type Programs Scheduled Daily for U.S. Forces

A "back home" radio program, designed to provide U.S. forces in the ETO with American-slanted broadcasts, will get under way tomorrow at 5.45 PM. Shows recorded in the United States for troop broadcasts, Stars and Stripes news bulletins and special features will be presented.

Offered by the American Forces Network and administered by the Special Service Section, SOS, ETO, the program will operate with the cooperation of the British Broadcasting Corporation. The new service marks the first time Britain has granted broadcasting facilities on its home territory to an allied nation. There will be no interference with reception of BBC programs in areas covered by the American Forces Network.

An all-American radio staff, made up of soldiers who have had radio experience back in the States, will prepare part of the program. Two daily features will be "Sports News," gathered and presented by The Stars and Stripes, at 7 PM, and "Final Edition," the S and S round-up of world and sports events at 10 PM.

Top-Flight Radio Stars

Recorded shows of Bob Hope, Bing Crosby, Jack Benny, Dinah Shore, Fibber McGee and Molly, Red Skelton, Fred Waring and other top-flight radio entertainers will be given regularly.

The schedule calls for BBC's news broadcasts, as well as some of its musical and comedy shows.

On the first day of operation tomorrow, the "back home" program will be on the air from 5.45 to 10.30 PM. The weekly schedule thereafter will be from 5.45 to 10.30 PM, Monday through Friday; 5.45 to 11 PM Saturday; and 8 AM to 10.30 PM Sunday.

The American Forces Network, through a complicated system of land lines and regional transmitters, will reach a limited number of U.S. troops at first. Eventually, it is expected that every American station and camp in the United Kingdom will be able to hear the programs, as the service will be extended further each month.

Clubs to 'Pipe' Shows

Broadcasts cannot be heard in the London area, except as "piped" by direct wire into various Red Cross clubs. Every Red Cross club in London eventually will carry the programs.

Much of the equipment being used in the American Forces Network has been "lease-lent" by the BBC. The rest has been brought direct from the United States for the new network.

The opening program tomorrow:
5.45 PM—Program summary and network opening.
6 PM—News (BBC).
6.15 PM—Harry James and his orchestra.
6.30 PM—Transatlantic Call—People to People.
7 PM—Sports summary, prepared and presented by The Stars and Stripes.
7.5 PM—Crosby Music Hall.
7.30 PM—Front Line Theater, Robert Young and Joan Bennett in "Mr. and Mrs. Smith," with guest stars Ralph Bellamy, Gene Krupa and A. Q. Bryan.
8 PM—Anglo-American services from St. Paul's Cathedral and Washington Cathedral.
8.40 PM—Charlie McCarthy and Edgar Bergen, Kay Noble and his orchestra, Don Ameche and Bert Lahr.
9 PM—News (BBC).
9.15 PM—Dinah Shore show.
9.30 PM—Independence Day program.
10.15 PM—"Final Edition"—late world sports and army news, presented by The Stars and Stripes.
10.20 PM—Artie Shaw and his orchestra.
10.30 PM—Sign off until Monday, 5.45 PM.

Next Stop—Jap Territory?

ANNOUNCEMENT OF AFN AIR DATE

New York Times Photo

These fresh U.S. troops disembarked recently at an undisclosed South Pacific port as reinforcements for Lt. Gen. Walter Kreuger's Sixth Army, according to the caption on this picture just received in London. The Sixth Army, last reports show, was based in Australia as part of the Allied force commanded by Gen. Douglas MacArthur, who is directing the present island offensive in the South Seas.

Allied Air Attacks Crippling Nazis' War Transport System

RAF heavy bombers have destroyed two locks of the important Rhine-Heren barge canal and paralysed Germany's principal artery for transporting Ruhr coal to Amsterdam, dispatches from Stockholm said yesterday.

This information, disclosed a day after news of the smashing of Germany's key U-

U.S. Calls Advances Merely Preparatory To Major Offensive

These Outfits May Be In New Pacific Drive

Here are identification numbers of several Army and Air Force units and their commanders which already have closed in previous Stars and news stories as based in the Southwest Pacific. Unfortunately these organizations are in the U.S. forces now taking the Solomons and New Guinea.

___ Division—Maj. Gen. Joseph Collins, Guadalcanal.
32nd Division—Commander not given, New Guinea.
Sixth Army—Lt. Gen. Walter Krueger, Australia.
5th Air Force—Maj. Gen. Ennis C. Whitehead, Australia.
13th Air Force, Maj. Gen. Nathan F. Twining, Guadalcanal, South Pacific.

Foe's Munda Airport Menaced by Planes, Artillery, Troops

WASHINGTON, July 2—American dive-bombers hammered today at what was believed to be the headquarters of Japanese forces on New Georgia Island, four miles from the big air field at Munda, as ground forces continued mopping-up operations on nearby Rendova Island and other scattered points of resistance.

Authoritative sources here, meanwhile, emphasized that the sudden American sweep northward through the Solomons was not a major offensive in itself, but rather a prelude to such an offensive.

Rejecting the much-criticized and necessarily slow "island-to-island" policy, the United States forces have attempted to gain bases—particularly air fields—from which a real offensive can be launched, it was said.

Air Score: 123 to 25

The cost of the advance so far has been surprisingly small, according to figures reported by Adm. William F. Halsey Jr., commander of South Pacific naval forces. His headquarters announced today that 123 Japanese planes were shot down over the New Georgia islands Wednesday and Thursday, against American losses of 25 planes, ten of whose pilots were saved. Of the total Jap losses, 101 were scored the first day and 22 the second.

Munda, the principal Jap air base on New Georgia Island, was menaced from three directions today. From Rendova

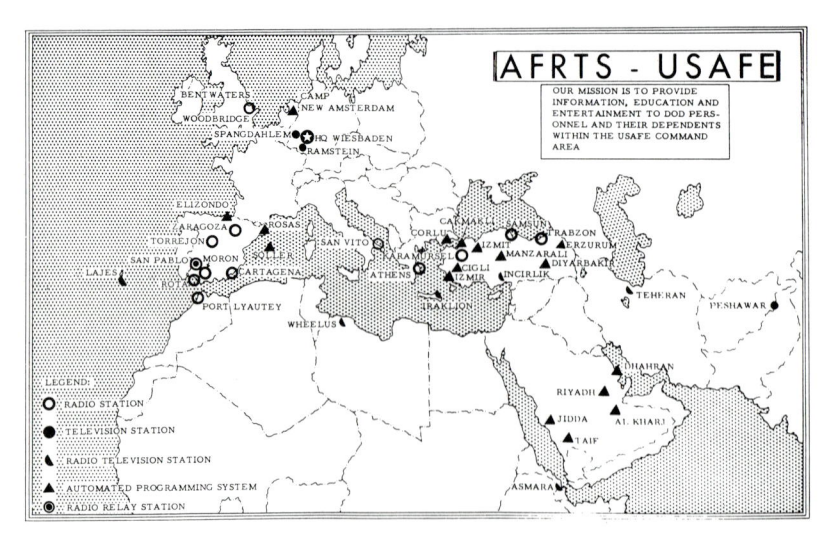

Germans Shoot Greek Leaders Of Crete Riots

Disorders Follow USAAF Raids Near Salonika;

The Air Force was the quickest off the starting blocks in bringing television to its people overseas. In the early 1960s, they put out this map showing the location off their radio and television outlets in Europe and the Middle East. As bases closed or moved for operational reason, the stations did the same thing.

Soon after this map was prepared, Morocco closed down. Libya tossed Americans out of the Wheelus AFB. Teheran was soon out of business. So, too, was Asmara in Eritrea as were most of the outlets in Saudi Arabia. The latter were to return in a big way some years hence. (Courtesy U.S. Air Force)

*It was front page news in **Stars and Stripes** when AFN signed on the air July 4, 1943. Not only did they help publicize the baby network, the **Stripes** staffers pitched in to augment the radio staff by reading the news and sports. A couple of them, including a reporter named Andy Rooney, got pretty good at this broadcasting thing. (Courtesy **European Stars and Stripes**)*

CHAPTER FOURTEEN

*Do you realize that if it weren't for
Thomas Edison, we would all be sitting
around watching television by candlelight?
-- Al Boliska*

Television Goes Overseas

It has been "AFRS" up until this point. In February 1954, the Department of Defense established the Television Section of the office of Armed Forces Information and Education and AFRS was no more. The acronym now became AFRTS as military broadcasters let the television monster into their tent. No one dreamed how quickly the monster would grow nor how much of the organizational energy would be expended on its care and feeding. By September the Television Section had become the Television Branch, an arcane difference of importance only to those employees who had to change their business cards because being a "branch" is more important than a "section" in the military scheme of things.

Preparing for AFRTS to enter the video age was required because the Commander of the Military Air Transport Service (MATS) decreed that all his isolated bases would have television, following the enthusiastic acceptance by viewers at Limestone Air Base. Approval was obtained from the various commercial networks to extend the somewhat tenuous approval to broadcast at Limestone and negotiations were begun with the Portuguese government to open a station at Lajes in the Azores. Lajes fit the category of "isolated" perfectly. It's on an island in the mid-Atlantic some 500 miles off the Portuguese mainland for those readers who missed school the day they taught geography.

Lajes was followed by a station at Wheelus Field, Tripoli. These stations required a headquarters so the 7122nd Broadcasting Squadron was formed to provide support and policy guidance. The squadron, located in Wiesbaden, Germany, later added Keflavik, Iceland and Bermuda to its roster and kept growing as stations continued to be opened in future years at Air Force bases in half the world. Eventually it would control stations stretching from Iceland to Pakistan.

It took the Army somewhat longer to begin moving toward movies, but by 1957 they were operating in Korea and two years later in Japan.

All of these new stations required programming and, just as it had done with radio, AFRTS in Hollywood supplied it. To do so required an entirely new mind-set and created a completely new set of problems. The commonly understood AFRS radio rules, regulations and requirements no longer held true. Some personnel, nurtured in radio studios, wanted nothing to do with the television upstart. Costs of equipment soared. Trained personnel were in short supply. Everyone, from the commander on down, found themselves learning on the job.

From the very earliest wartime days, it had been an article of faith at AFRS that there would be no commercials. The first major problem to be solved was how to handle this situation on television. Limestone had solved it by doing nothing at all. It merely got the programs from the networks and ran them, commercials and all. Had Colonel Tom Lewis been dead at this time, which he wasn't, he would have turned over in his grave.

Early day television arrived on film which was hard to work with, hard to store, expensive to ship and easy to break. This is what a corner of a film library looked like in those days as a large part of the staff was tied up checking, repairing and rewinding film. This particular library was at Ramstein Air Base in Germany. (U.S. Air Force Photo)

This view of the video recording room at AFRTS in the early days of television when both videotape and kinescope recordings were made. Normally there would be the rear end of an engineer protruding from one of the units as he tried to fix the balky devil. In the foreground are the hated kinescope recorders. In the rear, early model 2" videotape recorders. (Photo courtesy of John Bradley)

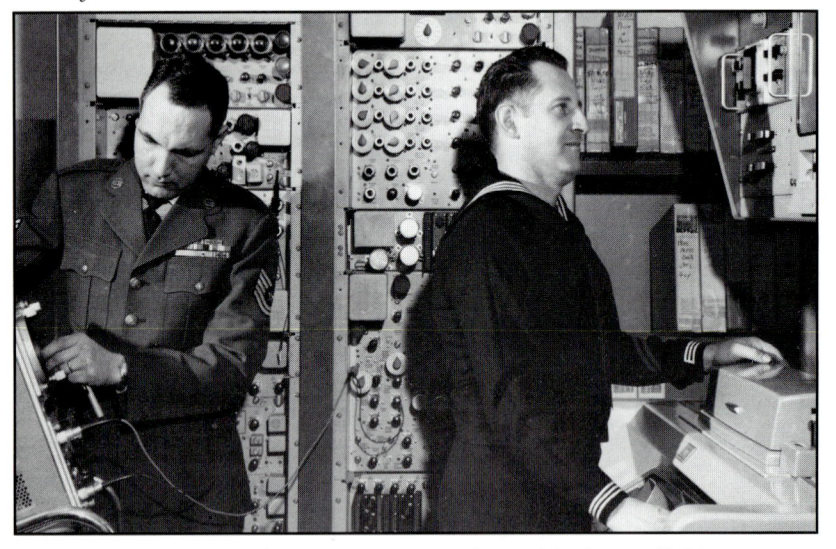

Today it seems impossible that a video recorder could take up all this space. Today a recorder can hang from the shoulders or fit in a pocket. This one, among the first to arrive at AFRTS, took all the rack-mounted equipment on the wall to make it work. It also took a couple of trained technicians such as TSGT Harold Smeder (L) and PM1 Curtis Luhman (R). (Photo courtesy of John Bradley)

Colonel Harold L. Smith, who retired later as a Major General, inaugurates AFRTS's first overseas television at Lajes in the Azores. If described as "austere," people would think you were bragging. It was housed in two bedrooms of a Bachelors Officer's Quarters but it was the start of what was to become the most widespread network on earth. from this tiny oak, a number of nuts grew and multiplied. (U.S. Air Force Photo)

The Office of Information and Education in Washington which supervised AFRTS, decreed through its director, General Harland Hartness, that "it would be inescapable" that AFRTS stations run programs with the commercials still in them. Navy Commander E. F. Hutchins, the officer in charge at AFRTS Hollywood, found himself in a difficult situation. He understood his boss's reasoning in that sponsors of the programs would be more likely to furnish them at little or no cost if the commercials were left in. He also understood that removing them would be difficult, labor-intensive and costly.

Conversely he understood that Tom Lewis' insistence on removing such material had served the system well and that leaving commercials in would probably, in the long run, create more problems that it would solve. He expressed his concerns to General Hartness who repeated his arguments in favor of retaining commercials and further explained he was concerned that the guilds and unions would consider decommercialized programs as new material and, as such, require additional fees for their use.

By judicious and subtle probing on the part of AFRTS, it was soon learned that once more the Hollywood entertainment community was willing and anxious to help provide its product to the men and women serving overseas. Agreements were reached with all interested parties and, just as they had done during the war, the guilds and unions continued to exhibit total cooperation and understanding of this unique situation.

Finding methods of physically removing the commercials was quite another problem. At this early stage, the overseas stations were equipped to run programming only from film. Videotape was still in its infancy and was ruinously expensive. Nor was it easy to edit, as it is today. By default, AFRTS and its stations were stuck with the kinescope process, the long dead and little lamented system of photographing the program off a tiny, high-intensity picture tube and recording the sound on an accompanying optical track at the edge of the film.

No, it isn't possible to snip the film and remove the commercial in all cases. Because the picture is 28 frames away from the matching sound, a snip causes either the sound to be wrong if the snip is made to edit the picture or the picture to be wrong if it is snipped at the sound. This not being an engineering manual, enough to say that it was often necessary to compromise and editors did the best they could. Viewers overseas were generally so grateful to have television that an occasional missing word was a small price to pay for the chance to see Jack Benny or Bob Hope. Or even Crusader Rabbit.

AFRTS was slow off the mark in moving to videotape as a means of supplying programming to its stations overseas and the continuing use of kinescope film caused numerous problems. Editing it was the least of them. The weight of packages sent weekly to overseas stations containing many hours of film caused excessively high postage costs. Storage of the large reels was difficult and outside contractors had to be hired to inventory and stock it. Overseas stations often worked from inadequate space and maintaining a library of film caused constant problems.

The Los Angeles Video Operations Branch was presided over by engineer John Scales and it was impressive even though the technology was a rapidly dying art. The kinescope recording machines were fed from miles of coaxial cable originating at the various networks and terminating in the AFRTS kinescope recording section. Along with them was other circuitry including a "hot key" which allowed the AFRTS operators to listen to control room commands at the network origination points. Operators could then hear when a commercial was coming up and could edit it out on the fly.

There were no cool transistors and integrated circuits here. These film devouring monsters were driven by vacuum tubes and the heat produced was hellish. As videotape became more and more common, less and less kinescoping equipment was used by the industry and keeping the five machines in operation was more than a full time job. It was essential that at least three be awake and ready to work at any one time as often recordings were being made from each of the three networks.

As the machines aged, it became nearly impossible to keep them on line. The picture tube from which the film camera photographed the picture was a type used only by kinescope recorders. It used extremely high voltages and consequently burned out frequently. There came the day when they were no longer manufactured due to lack of customers. AFRTS found a company which could rebuild them. Sometimes it was even necessary to remove and replace the burned phosphors on the inside face of the tube. Unfortunately this was done by hand and did not lend itself to consistency. Each tube, and each machine, developed its own set of engineering parameters and the operators had to learn the idiosyncrasies of each.

Once learned, it could be expected that a consistent kinescope could be turned out each time. Not so. By the time the kinescope process was about to pass into a well deserved graveyard of ideas which had died of old age and inefficiency, Eastman Kodak decided that it was corporately irresponsible to continue making the special film required when AFRTS was practically the only customer left on earth. Although millions of feet of film were required by AFRTS each year, Eastman was not anxious to tool up to produce what to them was a relatively small amount. Normally film was ordered quarterly because of limited storage space at the AFRTS program center building at 1016 North McCadden. Eastman didn't want to turn out such small batches. Okay, said AFRTS, how about turning out large batches and storing it for us. Sorry, said Eastman, we make it, not store it. Eventually it was agreed that Eastman would run off a six months supply at a time and AFRTS would somehow find a place for it. It was a poor solution because film deteriorates in storage and each batch changed characteristics as it aged.

The recording process was under the supervision of the Engineering Branch. The decision on whether the finished product was acceptable was placed under a new Quality Control Section which reported to Graf Boepple, now a Lieutenant Colonel home from the far Pacific and chief of the Program Branch. The quality control man, Bob Trumble, was equipped with an eagle eye and had no compunctions about turning back kinescopes with the tiniest of imperfections. This, in the time honored tradition of broadcasters everywhere, caused undeleted expletives to be continually hurled between those stupid

It took awhile for the station at Lajes to get built but, when it was finished, they had a pretty nice view—this one. More than one pilot, making his way across the Atlantic, has breathed a sigh of relief to see that . (Photo Trent Christman)

Traffic passing in front of the Azores AFRTS station. Milkmen there really get off their ass to deliver the milk on time. (Photo Trent Christman)

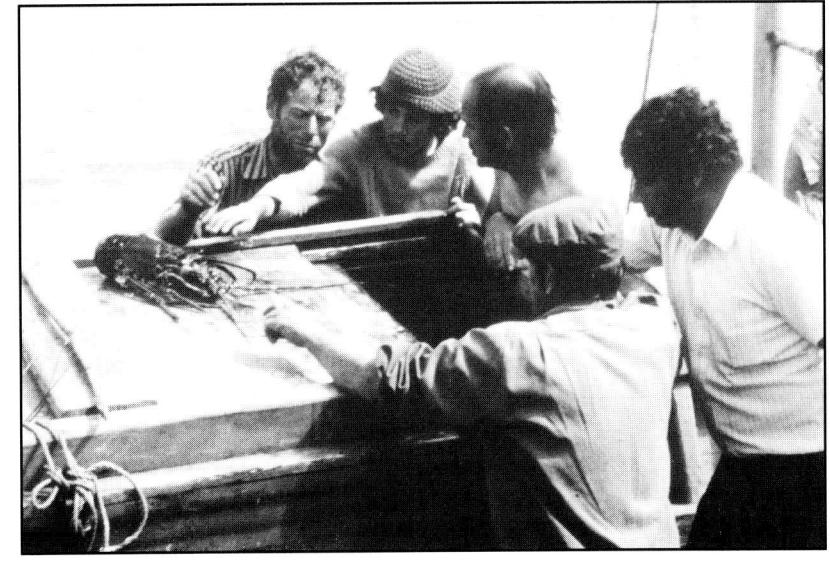

In the village near the Lajes air station, the lobster fleet goes out and comes back with monsters like these. This one went nine pounds and, served with crusty bread and a quart of melted butter, plus wine, set the author back five bucks. Easy come, easy go. (Photo Trent Christman)

Long before the days of the handy-dandy, hand-held videocamera, this is what a news cameraman had to lug around in order to film some guy saying "no comment." Still, it was better than no picture at all of someone saying "no comment" so the trusty single-system Auricon was lugged on strong backs to the scene of the action. Shown here is ace cameraman SSGT Juan Vela with the Lajes, Azores, camera. (U.S. Air Force Photo)

Because the staff at Lajes was less than impressed by the quality of the kinescope they were getting from AFRTS in Hollywood, they built their own, shown here. It took a lot of ingenuity but eventually it worked so well Manager John Bradley could brag to Los Angeles about his superior product. Los Angeles must have been impressed. They dragged him back there and put him to work in their kinescope department. (Photo courtesy of John Bradley)

engineers and those fancy-pants program people.

As the kinescope era drew toward its inevitable close, it got its greatest challenge. The date was November 22, 1963. John Bradley, taking a short hiatus from his endless overseas assignments, was in charge of the kinescope room. Waiting for a feed from CBS, he got nothing on the line. He called CBS Master Control on the intercom system and the CBS control operator was practically out of control. "Start Rolling. Start Rolling," he screamed. John did -- and continued to do so from all three networks for the next 120 consecutive hours. AFRTS ended up with the most comprehensive coverage of the John Kennedy assassination, funeral and national mourning in existence. Although the crew slept in brief snatches throughout the days following the events in Dallas, so good was their product that one of the networks came to them for copies after the network videotape machines broke down.

The kinescope days were about over although the budget problems in funding for the hideously expensive videotape machines for the Broadcast Center in Hollywood and at all the overseas outlets were not. The AFRTS commander, Lieutenant Colonel Luther B. Siebert, a gentleman from the deepest of the deep south, one day was overheard explaining to his boss back in Washington, John Broger, how he felt about absorbing another budget cut. Shooting from the lip, he said, "Wal, sir, that's gonna be mighty hard to handle. It's kinda like tryin' to fertilize a forty acre field with a faaart. It spreads it mighty thin."

...and on this agricultural note, we move on to see what is happening in the field.

IN THE AZORES The clock now rolls back to 1957 and once again, there is John Bradley, arriving at Lajes Field in the Azores. It was to be the first of several assignments at the station serving the troops on this isolated speck of rock in mid-Atlantic. The island group is administered by Portugal, some five-hundred miles East across open ocean. The U. S. Air Force and various associated components, including weather and search and rescue teams, worked out of the base on Terceira Island. For many years the Azores had been the welcome refueling stop for aircraft using the southern route to Europe, Africa and the Middle East.

On arrival, Bradley found that although the Commander of the U. S. Forces, Azores, Brigadier General Harold L. Smith, was a supporter of radio and TV for his personnel, not much had yet been done about furnishing adequate space for the operation. The AM radio was housed in a small cubbyhole in the Service Club and TV, such as it was, occupied a former day room and barracks room. General Smith promised a building for the stations would soon be built on the highest point of the Northeast part of the island.

The Portuguese government mandated that only U. S. and Portuguese military could own television receivers, as they had not as yet brought in Portuguese TV to the islands. Receivers were closely controlled and were registered much like firearms.

Because television, under the auspices of the Air Force, was proliferating rapidly at this point, the Air Force, through a sense of misguided economics, decided that cheap was the way to go. Much of the equipment supplied their stations, including the Azores, was industrial-grade DAGE brand -- fine for industrial use watching doorways and assembly lines, but totally inadequate for on-air production. Endless hours, days and weeks were spent keeping it operational and the technicians soon learned why DAGE spelled backwards is EGAD. In the long term, the Air Force and other services learned that top grade equipment is actually cheaper because of decreased maintenance, longer life, increased reliability and the unmeasurable value of fewer viewer complaints. That blessed day was still far in the future, however, and in the mean time Bradley realized if the station was to put out a nice, crisp picture that didn't look like a surveillance shot of a robbery in a convenience store, he would have to use film. And he just happened to have his film camera with him. What he didn't have was a way to develop the film. He tried sloshing it around in a light-tight container but got mixed results ranging from merely miserable to totally terrible. Anyone who has ever served in the armed forces will have no trouble believing how he solved the problem. He found a friend and, in the age-old tradition of the non-commissioned officer's corps, the friend just happened to be in charge of stockpiling materials for war readiness and, further, just happened to have a gen-u-wine film processing machine sitting in the corner of his warehouse. Using methods known only to wizards and NCOs, the friend caused the machine to appear on the station's doorstep and the station went into the film business in a big way. A deal was worked out with the Military Airlift Command to return much of the output to the States and much of it was used on various U. S. outlets as part of the then current "People to People" program.

In mid-1958, the new studios were completed on the hilltop overlooking the Atlantic and, directly below, the Air Base. In these days before the common use of videotape recorders within the still growing AFRTS television system, it was only natural that the station staff wanted a way to retain their programming or to record it for later use. That meant using the dreaded kinescope process of recording on film. It also meant that if they wanted a kinescope recorder, they would have to build it because the Air Force was not about to lay out tens of thousands of dollars for one of the monsters.
Build one they did by the simple expedient of making every mistake possible and learning as they went. They learned that their new 1200-foot Auricon film camera shot film at the normal 24 frames a second rather than the 30 frames per second on the picture tube it was supposed to photograph. They learned that a normal picture tube doesn't work well and that special, high-voltage tubes with special phosphors are required. They learned that normal motion picture film doesn't have the proper contrast ratio and special film is needed. What they didn't learn was that it was impossible to build a workable system on an island in the middle of nowhere so they went ahead and did it.

After endless errors, they ended up with a system good enough so in their arrogance they could call the Program Center in Hollywood and criticize the kinescopes being sent out from there. This example of transmitting when he should have been receiving was to backfire on Bradley later when, as we have seen, he was sent to Hollywood to work in their kinescope branch. Now, though, with their new toy, the station was busy producing documentary programs on everything from Portuguese cultural activities to the Strategic Air Command refueling operation.

The Lajes television operation is being used here as an example of the type of TV installed and operated by the Air Force during the early and mid-1950s. It was the first of the stations which the Military Airlift Command had decreed would be put in place at all its isolated sites. Following soon was Wheelus Field, Tripoli, Dhahran, Saudi Arabia, Keflevik, Iceland, and Kindley Air Force Base, Bermuda. Each was supervised by the newly formed 7122nd Support Squadron (AFRS-TV), with headquarters in Wiesbaden, Germany. The squadron was immediately christened the "Seventy one double doo doo" by its members and so it was to remain for years to come. Although it was located directly in the center of AFN-Land, AFN, busy with radio, showed little or no interest in getting into the television business. The Air Force and the 7122d lost little time in seeing that their people were able to gain the cultural benefits of flickering black and white

It was a matter of pride for the early day overseas television stations to try to look as much like their stateside counterparts as possible. It, of course, wasn't possible, but it wasn't for lack of trying. They did with cardboard, spit and glue what the big boys did with platoons of graphic artists and set builders. Here they are set for election night, 1960. (Photo courtesy AFRTS)

The Air Force station at Spangdahlem AFB had guts if nothing else. And they had little else in the way of equipment when they decided to broadcast what turned out to be the first live football game done by a local AFRTS station. It meant ripping out all the studio camera equipment and jury-rigging transmission lines back to the transmitter—but it worked. (U.S. Air Force Photo)

television and put stations into American air bases in Germany at Ramstein, Spangdahlem and, later, Bitburg. The Air Force system of selecting state-of-the-lost-art, bottom-of-the-line equipment remained in force and the stations by today's standards were primitive beyond measure. DAGE remained the standard and studios were generally built in existing structures with normal height ceilings and minimal sound insulation. Still, it WAS television, and because it was impossible to keep the fact television was around a secret, the Army folks made no secret of the fact they were discouraged, disgusted and disenchanted that their Air Force counterparts could see

The AFTV station at Spangdahlem presenting one of the daily newscasts with anchorman TSGT Jerry Huddleston. Air Force outlets insisted that their personnel wear civilian clothes on the theory that the audience looked at uniforms all day and would prefer not to see them in their living rooms or billets at night.

When the Army took over European television in 1972, AFN tried to continue this policy but higher headquarters insisted that seeing their contemporaries wearing the same uniform on camera that they wore during the day would instill pride in service. AFN resisted by pointing out it was possible that the viewer would feel a man or woman with PFC stripes simply didn't have the credibility of Walter Cronkite in Armani suits.

This argument failed to fly and, except when wearing the uniform on the air was inappropriate (such as scuba diving or interviewing a rock star wearing a rhinestone jock strap) uniforms were now a permanent part of the television picture. Initially there were numerous complaints from viewers, but they soon got used to it and complaints stopped. (Photo courtesy AFRTS)

Japanese television is technically marvelous but runs a bit heavy toward Samurai swords and Sumo wrestling. For the Americans stationed in Okinawa, this is their link with the Cosby Show, Cheers and Tom Brokaw. (Photo courtesy AFRTS)

Gilligan's Island and they couldn't. For years, the Letters to the Editor column in the unofficial military newspaper *Stars and Stripes* carried angry "why doesn't the damn Army get off its lazy butt and get us TV" letters.

It finally did, two years later. The only problem was that where they did it was half-way around the world in Korea.

IN KOREA the Army controlled AFKN, the American Forces Korea Network, which had spent the recent Korean "police action" marching up and down the peninsula with mobile radio vans, but had now settled down into more permanent quarters. It began to broadcast film programming over television in September 1957 in Seoul but was to wait more than a year before live programming became available. In the meantime, additional film capability was added to a station at Camp Kaiser. Later Kunsan Air Base and Taegu joined the television network.

By 1963 AFKN was beginning to look like a real television operation. Radio and television facilities were now combined in the headquarters compound at Yongsan and the picture was looking good, thanks to the installation of modern Image Orthicon cameras and a new studio lighting system. Plans were also well underway to install a microwave system linking all the stations to Seoul. This came about in 1965 and Seoul could now originate all programming in real network style through the distribution system which reached as far south as Taegu. Later a link was installed to Pusan.

Station personnel built a mobile van which was put to good use covering the Bob Hope Christmas Show, sports events, military ceremonies, USO shows and chapel services. And wonder of wonders, AFKN became the first AFRTS outlet to receive a videotape recorder for its studio operation. Take that!!! you Air Force guys.

The van, covering a meeting of the Military Armistice Commission at Panmunjom captured through a window some unforgettable footage of an angry exchange between the two sides. When President and Mrs. Lyndon Johnson made a 44 hour visit to Korea, both radio and television stayed on the air for 34 hours of the presidential visit. They covered everything but the sleepy time and potty breaks of the couple. Their efforts paid off when radio Vagabond, the Seoul station, won the American Heritage Foundation award as the military

In 1959, Airman Wade Hawkins gets ready for his weekly "Cowboy Caravan" show from AFRTS Taiwan, People's Republic of China. In anyone doubts that this was an integrated organization, Harkins was in the Air Force. The station was Navy controlled. The picture was taken by Army Specialist Harold Slate. (U.S. Army Photo)

radio station of the year and AFKN Television won the same award for military TV.

...remember the PUEBLO

Almost from the moment the North Koreans captured the USS *Pueblo*, the U.S. surveillance ship, along with its entire crew, it became certain that AFKN would play a key role in coverage of the story. Because Panmunjom was the only official access between the two Koreas and because it was deep in the demilitarized zone with its controlled access routes, it was obvious that AFKN would provide pool coverage to the world. Planning began long before there was any real progress in the negotiations to gain return of the *Pueblo*'s crew.

The network worked directly with the military's public affairs officers in planning for coverage of the return of the crew, whenever that might happen. Broadcast circuits from key locations within Korea back to Washington were arranged. Plans were made to have videotapes of the release flown by special aircraft to Japan where they could be uplinked by satellite to the U.S. and on to the rest of the world. AFKN also arranged to broadcast all the proceedings live on radio and use duplicate videotapes on television. In this way they beat the rest of the world by hours in breaking the story.

The mobile van was rebuilt and new equipment was installed in it. Again it was used to cover a Bob Hope Christmas Show but the real story had to wait while delicate negotiations continued month after month. Military authorities would not even let the network make any test runs from Panmunjom for fear of starting unfounded rumors that the release was about to take place.

Finally, on Saturday morning December 21, the network commander, Lieutenant Colonel Frank Tennant, Jr., called in the staff and under strict security told them the military expected the release to be made Monday morning. Timetables, assignments and changes in ground rules were gone over and by noon the TV van, radio van, lighting truck and a bus with the 27 man crew were headed north over roads made icy-slick following a two inch snowfall. The group arrived in the early evening and by working all night everything was in readiness the next morning.

AFKN began its five and a half hour broadcast at 9 a.m. with the signing of the formal document which contained the so-called "apology" which the North Koreans insisted upon as a condition of the release. The North Koreans purposely held up the actual release for a half hour past the agreed upon 11 a.m. time and then the chief negotiator, Major General Gilbert Woodward, explained to the AFKN microphone -- and the world -- about the contents of the document and shortly thereafter returned and officially repudiated the "apology."

Both radio and television broadcast the return of the *Pueblo* crew as they trudged over the "Bridge of No Return" and followed this with coverage of the first hours of freedom of the 82 men. Later in the day they broadcast the press conference in which Commander Lloyd Bucher described their ordeal.

Videotapes were brought by military jet to Japan and AFKN sent their copy to Seoul by helicopter where it was edited and used on the late news. By the next day, AFKN crews had set up to cover the departure of the *Pueblo* group from Seoul's Kimpo airport.

All the planning and hard work had paid off. Every scheduled event was covered and there were no breakdowns, no misses, no foul-ups. It was a model of how to handle a story which other AFRTS stations would benefit from for years to come.

During those years, unfortunately, equipment became ever more antiquated and physical facilities deteriorated. As time passed, President Carter proposed withdrawing all American forces from Korea and immediately AFKN's budget was reduced to the point where not even basic repairs could be maintained. Only when President Reagan decided to retain troops in Korea was the budget restored and was the network able to replace and modernize facilities.

The FAR EAST NETWORK.

FEN, operating in its early days under the direction of the Army, had its own set of problems. One of them was geography. Like its predecessors, the Jungle and Mosquito networks, its area of responsibility stretched at various times from Japan to Guam to Taiwan to the Philippines. Nor is it possible to escape the conclusion that the staff spent more time dealing with orders from on high making changes in organizational structures, jurisdictional controls and opening and closing outlets than it did on providing a high quality radio and television service to its far-flung audience. On the other hand, perhaps the organizational changes will make sense to some. Just follow this twisted path:

The radio network reached a peak of 20 outlets in 1954 and that year the FEN station at Clark Air Base in the Philippines became the first FEN television outlet. The next month Okinawa became the second. From here on it gets a bit complicated, so hang on.

In 1956, troops began being withdrawn from the Japan area and a number of stations were closed down, relocated or modified to become relay sites. Sendai was shut down and the equipment moved to Misawa Air Base which then opened up as a new station. FEN lost Okinawa when the Air Force commander on the island took it over and made it an independent station.

Then the whole of FEN was placed under Air Force control with authority delegated to the 5th Air Force

MSgt Bill Mansfield interviews actor Jimmy Stewart at the Imperial Hotel in Tokyo for the Far East Network. The time: 1955. (U.S. Army Photo)

93

headquarters in Japan. Television reached the Japanese mainland on Christmas Eve 1960. Misawa was the first station, followed by Wakkanai. The later was closed circuit, Misawa was UHF, both systems used in order to get required Japanese government approval and to insure American TV did not interfere with the Japanese VHF system.

Now began a series of reorganizations that kept the staff wondering who they were working for. The Air Force formed the 6120th Broadcasting Squadron and put FEN under its jurisdiction. Then they took the Clark Air Base station away from FEN and gave it to the local base commander. That station became headquarters for a brand new network called AFPS, the Armed Forces Philippine Network which had stations at Clark, Subic Bay and San Miguel. Then the Air Force established another layer of control and called it the 6204th Broadcasting Squadron which assumed command of the Philippine stations. The U.S. now returned control of Okinawa to Japan and AFRTS Okinawa was given back to FEN.

(Perhaps you better take notes. There may be a quiz.)

In 1974, to the surprise of no one, the Air Force decided to make some changes. The 6001st Aerospace Support Squadron (AEROSS) in Thailand merged with the 6204th Broadcasting Squadron and by combining the names became the 6204th AEROSS with responsibility for all AFRTS stations in the Philippines, Thailand and Taiwan. This lasted about a year when Thailand and Taiwan suddenly were no more and the 6204th was left with only the Philippines. "Hey, gang," somebody said. "Let's reorganize!" This seemed logical so the 6204th AEROSS was combined with the 6120th Broadcasting Squadron. But, what to call it? How about the 6204th Broadcasting Squadron? Great idea! And so it came about that the 6204th BS ended up with supervision of the Philippines, Japan and Okinawa for AFRTS stations. Today its headquarters are in a large and efficient building designed especially for it at Yokota Air Base from which it feeds AFRTS stations throughout Japan and supervises those in the far distant Philippines and Okinawa.

It is hoped this explanation has made everything crystal clear.

❖ ❖ ❖ ❖ ❖

ANNOUNCER: EXCITEMENT CONTINUES TO RUN HIGH AS WE JOURNEY TO MORE FAR AWAY PLACES WITH STRANGE SOUNDING NAMES. COME WITH US TO SUNNY ITALY, CLOUDY GERMANY, EXOTIC LEBANON AND TURKEY. AND OUR TRAVELS WILL ALSO TAKE US TO WASHINGTON, D.C. READY?
LETS GO!

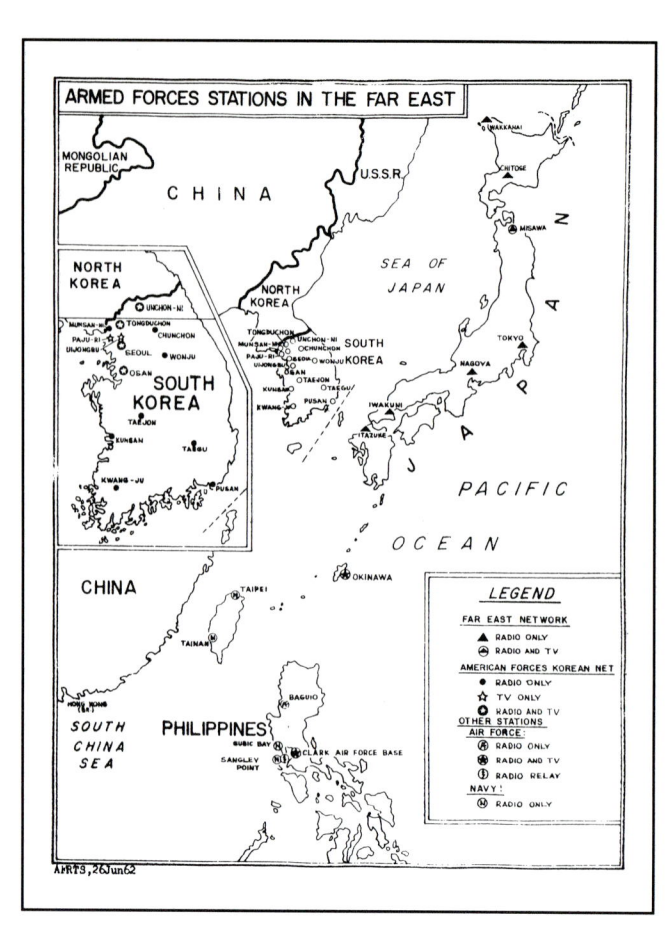

❖ ❖ ❖ ❖ ❖

In 1962, these were the stations located throughout the Far East. The Army managed the Korean stations, the Navy controlled Taiwan and the Air Force had the rest of the area including Japan and the Philippines. Just as in Europe, the vagaries of politics and military necessity caused the opening and closing of stations, but there was sure to be an AFRTS outlet if there were sufficient troops in the area to warrant it. This map probably indicates the high water mark of facilities in the Pacific since the end of World War II. While many of the stations still remain, many are now mere memories. Taiwan has long since closed down. Mount Pinatubo near Clark AFB in the Philippines did in that station, helped by the political situation in the islands.

At Clark Air Force Base in the Philippines, the staff of the first AFRTS TV station in the Far East decided to go all out. So they all went out to do this remote broadcast to raise money for the Combined United Overseas Fund Drive. (U.S. Air Force Photo)

CHAPTER FIFTEEN

If you can keep your head while those about you are losing theirs, chances are you either belong in an overseas AFRTS TV station...or should have it examined.

TELEVISION GOES EVEN FARTHER OVERSEAS

By the time the Korean conflict had ended, television had become as much a fixture in most American homes as a telephone, refrigerator or flush toilet. To the generation now reaching the age at which the military services would take them, and with the draft doing just that, the television was perhaps the most important appliance in the house they were about to leave. They could accept being stationed in some strange far-away land. They could accept the fact they couldn't understand the local language. They could understand that the local food was concocted of mysterious ingredients. What they couldn't understand was why they couldn't see *All in the Family* or *Bonanza* in English on the local television. Dammit! What kind of a (choose one) Army, Navy, Air Force, Marines was it that didn't look after its people any better than that?

The Air Force, especially on SAC and MATS isolated bases, was doing the best job of filling this absolutely essential requirement for happiness although the pictures frequently left something serious to be desired in the way of quality. Still, they moved and talked and that was better than nothing. This group found plenty to complain about including the fact that news and sports were generally several weeks late due to the necessity to ship them to their destination by the vagaries of the U.S. and military postal system. Because radio carried the major sports events live, the final score was no secret by the time the kinescope arrived a week or two later, thereby spoiling the tension about who won the game. One base commander with a tenuous grasp of reality and with his head inserted snugly into his hemorrhoidal infrastructure suggested strongly that the local radio station only broadcast the first part of all the games so the final score would be a surprise to the TV audience weeks later. The station, to its credit, refused.

While the relatively small percentage of service people who had access to American TV were relatively happy with the relatively adequate service being transmitted over the relatively efficient facilities by the relatively well trained AFRTS personnel, the great majority of the overseas military population could only complain and write letters to their congressmen.

A number of these letters ended up at AFRTS in Hollywood or at the office of John Broger, chief of the Office of Information and Education for the Armed Forces. Neither could do much about increasing service. Their charter was to provide programming and policy guidance to existing overseas outlets. The request to open such outlets and to provide logistical support, space, personnel and other support lay with the various overseas commands and military services. Once a command or service obtained approval, AFRTS gladly added them to its roster of stations and began supplying the program material needed, engineering support, host country liaison, and various other technical services for both radio and television.

That program material was the key to the whole system. Without programs people wanted to hear and see, the entire operation was pointless. Obtaining it was and is the job of the Industry Liaison Branch of the AFRTS Program Center, which, in true military fashion, is now renamed the Broadcast Center.

The function of the Center will be looked at in greater detail later, but it should be explained prior to visiting the overseas outlets springing up around the world how the programming material is obtained.

The Programming Division is responsible for the selection of the programs sent out in a weekly shipment and for providing a balance of programs to appeal to the widest possible spectrum of listeners and viewers. Because AFRTS outlets are most often the only game in town, this means a single channel must serve viewers from every state, educational level, age, gender, personal prejudice, political persuasion and economic background. Commercial broadcasters can use a rifle to target a specific segment of the audience. AFRTS stations must use a shotgun and hope to hit as many people as possible.

To obtain clearances and necessary rights to broadcast this programming is primarily the job of the Industry Liaison people at the Broadcast Center in Los

Angeles. It is their responsibility to deal with the owners or syndicators of the programs and get permission to reproduce it and ship it to destinations around the world. Once cleared for AFRTS use by the owners of this highly protected material, the contracting and legal people take over. Bob Vinson was the first to assume this near-impossible task. On his retirement, the redoubtable Vincent Harris, Vinnie, took over. Both men had been with AFRTS since its earliest days and had functioned in every nook and cranny of the program arena. Not only did they understand AFRTS, they knew intimately the ins and outs of the broadcast industry. More important still, they both possessed an innate sense of diplomacy; a critical requirement for dealing with the sometimes volatile egos of those in the entertainment business.

A fact of life in that business is that the owners of a program seldom if ever make back their production costs on the first network run. To get rich, live in Beverly Hills and drive a Rolls Royce, a producer has to make any profit from his show by selling repeat showings on the networks, putting the program into syndication by leasing it to independent stations and by sales to overseas markets. Today, the home video market is another huge source of revenue for certain types of programs. The job of Vinson and Harris has been to obtain this invaluable material while it is still new and fresh and get it at laughably low prices — generally just enough to cover the owners' administrative costs.

True, it has never been possible for AFRTS to get EVERY show. If an owner feels he has a shot at selling the program to an overseas civilian network, chances are he will temporarily hold back release to AFRTS. However, because most AFRTS outlets broadcast on low power and often on different transmission standards than those of the host country, owners and distributors have been exceptionally generous in supplying their product. Year after year AFRTS is able to bring its viewers in excess of 90 per cent of the top rated programs. Probably no other network on earth can match that. Much of the credit must go to Bob Vinson and to Vinnie Harris who, through their reasoned approach and diplomatic explanations, plus knowing the major players around the globe, have managed to provide military viewers and listeners the finest single channel broadcast schedule anywhere. Today that job is made even more difficult with the satellite distribution of real-time television to much of the world, adding live sports, news and a wide variety of special features to the normal schedule of videotaped programs distributed each week.

Most amazing of all, this is done in Hollywood where deals are made at lunch. An example of Harris' genius is that he is able to do the job without an expense account and with the knowledge that federal law forbids accepting favors, including lunch, from suppliers. In spite of the fact lunches are Dutch Treat by order of the Conflict of Interest laws, to the great amusement of the producers, a recent estimate shows that AFRTS furnishes its stations about $120 million worth of programming each year (figured on the basis of costs for comparable sized audiences) and for this it pays approximately $8 million. Figures like this make taxpayers and congressmen happy. Not happy enough, though, to let Harris pick up a lunch tab once in awhile. Now, when he says to a potential supplier of programming, "Let's do lunch," he means, "You do yours and I'll do mine." It gets embarrassing when trying to get a few hundred thousand dollars worth of programming for very little money and you can't even seal the deal by popping for a pepperoni pizza.

While Harris is wrestling with this problem, come along overseas for a look at how a few of the outlets came into being and developed an insatiable appetite for the programs AFRTS sends them each week.

READY OR NOT, AFN GETS TELEVISION

CENTRAL EUROPE. The American Forces Network, AFN, continued to operate under Army control as a radio network, just as it always had, while the Air Force began bringing television to its areas. Home continued to be the Von Bruening castle on the banks of the Main, in the small town of Hoechst halfway between Frankfurt and Wiesbaden. It had excellent relations with the Air Force 7122nd Broadcast Squadron and it was from them that it received Air Force staff members. Because about 20 percent of the military population of Germany and the Benelux countries was Air Force, it was mutually agreed that the Air Force would furnish that percentage of the AFN staff. This integrated Army/Air Force group of broadcasters enjoyed life in what must certainly be the world's only radio operation housed in buildings constructed between the 14th and 16th centuries. And certainly no other radio station can claim that Napoleon once stayed there, which he did after the inhospitable Russians tossed him out and told him never to darken their steppes again. Does any other station boast a moat or an entry to the courtyard through a portcullis?

Life in the castle was good and the staff thoroughly enjoyed the Knight life. Gardens overlooked the placid Main River and the castle's genuine dungeon was only used for Halloween parties and never used by the commander for its original purpose.

By 1962 it became obvious that the romance of living and working in a real castle was about to end. The *Farbewerke Hoechst*, the nearby giant chemical complex, bought the castle from the Von Bruening family and politely asked if the Army would kindly move AFN elsewhere. The European Command headquarters, the Army headquarters and AFN began negotiations to secure a new home. The German government in Bonn selected a location just north of downtown Frankfurt. It stood near the majority of the American installations such as the PX and V Corps Headquarters. It was also directly next door to the First German radio and television network and as time passed this turned out to be a blessing. Strong ties were established with the German counterparts who invariably went out of their way to take care of "their little brother."

Construction of the new building began in 1964 and AFN moved out of the romantic but drafty castle in 1966. What they moved into was a sparkling new building, faced with blue tile, and such unique new ideas as adequate bathrooms and air conditioning. Well, air conditioning for the equipment at least. It had a large audience studio, lots of smaller studios, editing rooms, warehouse and office space galore. The top floor contained rooms for the unmarried staff and the basement, along with the warehouse, had a club and a bar complete with slot machines. (These were to disappear in future years when militant mothers convinced the congress that the services were teaching their innocent sons and daughters to gamble and that might lead to heaven-knows-what.)

Getting moved in and settled was not without its obstacles. German building regulations and military engineering specifications kept bumping heads. The military insisted buildings needed outside fire escapes. The more practical Germans pointed out that the building was concrete and brick and absolutely fireproof as well as having fire doors in all stairwells and hallways. This held up construction for months. Eventually the German view prevailed concerning the exterior. The interior was another story.

Tom Collins, the Fire Marshal for all of Europe, was practically a member of the AFN staff in that he did a program on fire prevention each week and was a close personal friend of the AFNers. However, he was adamant about the type of sound treatment to be used on the studio walls. The German contractor had hired an acoustical expert who selected a panelling both aesthetically pleasing and acoustically perfect. Collins insisted it was not fireproof and demanded the wall panelling in studios be perforated metal strips. After endless negotiations, his view won out by the simple expedient of refusing to let Americans occupy the building until it met long-outdated specifications which were still on the books. When the metal panels went up, the acoustical experts tested the sound quality by having a trumpeter play some high notes. It was like being inside a tympani as the metal walls rattled with sympathetic vibrations. Down came the walls. Rubber baffles were added. Up went the walls. Trumpet, please. Awful. Down came the walls. More baffles.

This went on and on until eventually an acceptable compromise was reached.

In a reversal of the norm, AFN radio was revelling in space. The Air Force, which was running television at four locations in Germany, was strangling for it. Its major studio was in Kaiserslautern, just outside of Ramstein Air Base 90 miles from Frankfurt. And, yes, they were still using DAGE equipment and playing scratchy old kinescopes. The live newscasts required all the anchors to remain seated so they wouldn't bump their heads on the lights should they stand up.

The Air Force, and the Army personnel lucky enough to be stationed in range of the station, loved it. When the Air Force began to expand to other air bases, word naturally spread through the entire military community that the Air Force had television and, dammit all, why don't we? Each month the volume of complaints grew. Because AFN was generally synonymous in the public's mind with broadcasting in Germany, many of the letters came there. Stars and Stripes devoted columns to the issue. No amount of reason in replies to the public did any good. They wanted television and they wanted it NOW and that was that. So serious did the issue become with the Army that some officials were concerned that the lack of television was affecting the re-enlistment rate.

After twenty-one years of spending their days living like knights in a castle, the AFN staff had to make do with this new building in Frankfurt. Built by the Germans, and to be returned to them should AFN ever leave Germany, the building core rests on giant springs to dampen vibrations. The first floor contains studios and equipment; the second is primarily offices and the unmarried staff lives on the third. (Photo Trent Christman)

Some unknown and unsung electronic genius in the Army community of Bad Kreuznach discovered that the Air Force microwave signal feeding from Ramstein to Weisbaden passed right through his area. Somehow capturing the signal, he and a lot of willing helpers wired the entire community and fed the picture into the housing area. This was the crack in the dam which soon burst. Every Army community raised a howl and wanted TV.

They were still howling when Secretary of the Army Robert Froehlke made an inspection trip through the area. Talking to the troops and their families, he was bombarded with complaints about the lack of television. In his report on returning, he wrote, "...the one biggest boost to the morale of our troops in Europe would be to give them American television." This got the Army in motion and soon the Army, the Air Force, the 5th Signal Command, the U. S. Army Communications Command, the Television Audio Support Activity, the 7122d Broadcast Squadron and AFN were spending endless days and weeks in meetings planning for expansion.

It was agreed that the expansion would be done in three phases which were cleverly named phase one, phase two and phase three, under the over all name of Scope Picture. The Air Force was elected to complete the first two phases. The third, and by far the largest, was to be an Army project and involved interconnecting by microwave the hundreds of far flung Army areas throughout Germany and, later, the Benelux countries. Each microwave site, which had to be within line of sight of the next one, required purchasing real estate for the tower and negotiating with the local and federal German governments to insure it met aesthetic standards and did not interfere with the German National television systems. The entire system, as envisioned, was budgeted at $31 million.

The television tentacles began reaching farther and farther into the various military communities and the Fifth Signal Command, which had the responsibility of planning the microwave paths was somewhat less than enthusiastic when General Michael Davison, the Army commander in Europe, announced that he wanted the most isolated and most distant troops to receive it first before the signal got into the larger cities and headquarters areas. This perhaps made good sense from the point of view of troop morale but was a technological disaster. The entire time table and endless hours of negotiations and planning were tossed into the round file. It might be asked why an Army general could do this to an Air Force project. He could because AFRTS works on the logical assumption that the predominant military service in an area should run the broadcast system. At this point, the Germany system had expanded into so many Army areas that the Army was the predominant user.

And AFN was in the television business. On AFN's twenty-ninth birthday, July 4, 1972, the network took over the partially completed television system and the studios in Ramstein. Staff members were recruited to move from Frankfurt and other of the eight radio affiliates to try their hand at the new medium. Many of the former Air Force staffers stayed in place and, in general,

Bob Vinson, former head of Industry Relations for AFRTS, left, in Europe on one of his interminable overseas excursions to meet foreign broadcasters and program distributors in order to obtain broadcast rights. The author, right, and Vinson are waiting for the next meeting to begin. It is obvious from the high-tech equipment in the background that both men are preparing themselves to discuss the latest in television technology.

Should anyone doubt that Vinson's trips were successful, he managed to bring the number of programs restricted for use on AFN from about 30 percent of the total AFRTS program package down to zero in just a few years.

Proper preparation is obviously the answer. (Photo courtesy of Robert Vinson)

the transition went very smoothly. The most obvious change was in the newscasts. The Air Force newsreaders had worked in blazers. The Army insisted everyone on camera wear a uniform.

Equipment was upgraded. DAGE had by now disappeared. Picture quality improved although it was still in black and white and on kinescope. New live programming was added, including a program directed at the women in the audience and an unsuccessful attempt to do an early version of David Letterman. That one just never made it, especially with all those uniforms on camera which somehow took away from the spontaneity of the project.

In spite of the ever enlarging audience and the improved program schedule, the volume of mail continued high although the complaint now changed from "why can't we get television?" to "why can't we get COLOR television?"

Sure. Why not, indeed? When the Army took over control and AFN took over operations, Phases one and two had been completed and there were now 46 transmitters feeding that many communities over a system of 64 microwave links. The system was not designed to carry a color signal and all transmissions were still in black and white. The Army began work on the final Phase three and this was designed from the beginning for color. Of course, there was no color studio equipment nor color studios, for that matter. While the installation of the final expansion of the system continued, the renovation started on the Frankfurt studios which had been designed and built with only radio in mind.

What followed was a miserable year. When the Germans build something, they build it to last. Adding television meant tearing out walls, lowering floors, raising ceilings, widening doors and other tasks such as digging cable trenches. For almost a year, the jack hammers pounded away. A stage in the large audience studio which was destined to become the main television studio turned out to be a solid block of poured, reinforced concrete 40 feet wide, 20 feet deep and five feet high. The studio floor, through which dozens of ducts for air conditioning and cabling had to be cut was found to be 24 inches of solid concrete. The dust raised was so bad that the record library had to be sealed to keep powdered cement out of the record grooves. The noise level of the jack hammers was so intense that listeners all over Europe heard them every time an announcer opened a microphone. A studio was finally built in the garage so hourly newscasts could be read without the accompaniment of ear splitting air hammers.

After rebuilding the building, the new color equipment arrived and installation began. Everything was the latest including RCA color cameras, Grass Valley switchers, Ampex videotapes and top of the line lighting. Finally everything was ready. Plans were made to sign off the black and white facility at Ramstein at midnight and during the night reverse all the necessary microwave dishes so Frankfurt could begin feeding the system in color beginning at noon the next day, October 27, 1976. Dignitaries had been invited to the color sign-on and everything was in readiness.

At midnight the last piece of ancient black and white film flapped its way through the film chain at Ramstein. The remaining staff threw everything moveable into their cars and a few borrowed trucks and started up the road to Frankfurt. Crews climbed towers and realigned microwave dishes.

The AFN project manager throughout had been Major Larry Pollack and he had worked limitless miracles getting the massive job done. When morning came he checked once more to see that everything was ready — and it was.

Then the phone rang. Bonn calling. Would AFN mind putting off going on the air with the new system until further tests could be run to insure there was no interference with the German television? If Pollack ever earned his money, he did it that morning. Having worked closely with the Germans throughout the project, he was friends with many of the decision makers. He got on the phone and called in every favor he had earned. No one ever knew whether he blackmailed someone, whether he decided you can get more with a kind word and a gun than you can with a kind word or whether he truly is a miracle worker. In any case, by 11:30 permission to sign on had been received.

At noon, the entire staff was gathered in the studio. Up came a black and white picture of the Commander, Lieutenant Colonel Floyd McBride and his guest, Major General Dean Tice, deputy chief of staff for Europe. Tice went to one of the new color cameras and removed the lens cap — and there was the new AFN logo in glorious color.

Pollack went out for a well deserved martini.

SOUTHWARD NOW TO SUNNY ITALY

ITALY may well win the prize as one of the most peculiar areas on earth from which to broadcast. U. S. broadcasters, used to the sometimes onerous but generally understandable FCC regulations can seldom absorb the differences in the systems used by the two countries. The AFRTS facilities in Italy, still all radio in 1975, were known as SEN, Southern European Network.

The Italians had their own broadcasting organization known as RAI and, as in most European countries, it was government controlled. That meant monopoly no matter how you looked at it. RAI was the only network in the country and planned to stay that way. Then, in 1976, the Italian Supreme Court ruled that the government monopoly on broadcasting was illegal although it did agree that RAI would remain the only true network. This meant that SEN's name was illegal according to the Italian government which asked them to change it forthwith. They resisted as long as possible, which when making changes in Italy is quite a long time, but the U. S. Embassy finally put pressure on the station and the staff which included Al Edick, Mike Mullen and Dan Lawler, was asked by the Officer in Charge, LTC Mitch Marovitz, to draw up a list of names. One of them was Southern European Broadcasting Service and this is the one the local commanding general selected after taking a pen and crossing out "service." SEN became SEB and so remains today.

Renaming the organization was as nothing compared to the chaos on the airwaves following the Supreme Court Decision to outlaw the RAI monopoly. The decision, in effect, allowed anyone with the inclination, a few Lira for parts and just enough knowledge to be dangerous, to start a radio or television station. And they did. FM transmitter towers suddenly sprung up on top of pig pens, wine presses and chicken coops. Because there was no central agency, such as the FCC, to regulate anything, these stations picked any frequency available, pumped in as much power as they could afford, bought a couple of recordings and went on the air. Like Italian drivers on the Autostrada, it became a test of wills. Uncle Guiseppe's station on which he played old Caruso records from his living room would go on the air and as soon as he signed off, some guy would sign on the vacated frequency. Obviously the only way to keep your frequency was to not sign off. At first many of the amateur entrepreneurs put a tone on the air while they grabbed a little sleep or stomped a few grapes. Scrolling through the dial sounded like you had inadvertently tuned into an electronic madhouse.

Then someone discovered SEB which was on the air 24 hours a day with music, news and information of all types. Suddenly, although it didn't plan to be, it was indeed a network. When a nearby station operator wanted to take a break, he would tune in SEB and broadcast that to preserve his place on the dial. Then the next station would hear that and he, too, could become a repeater. Further away another station would pick it up. Before long SEB was being heard through most of Italy.

Many of the repeaters were jerry-built rigs and had a tendency to wander all over the dial. One that continually interfered with the headquarters station by broadcasting a tone around the clock turned out to be coming from behind the stone walls of a monastery nearby. It continued for weeks and it was suspected the monk in charge just wanted visitors. He got a lot of those in the person of Al Edick and other staff members at SEB who would knock on his door and piously pray that the good friar would knock off that crap. Finally he changed frequencies, perhaps to get a better type of visitor calling on him.

Frequently the Italian transmitters pirating the SEB signal wandered into the aircraft control channels and pilots would receive Wolfman Jack rather than instructions from the tower. This had a tendency to cause complaints to the U. S. Embassy. The Italians figured the way to solve the problem was to have SEB stop broadcasting altogether, rather than simply trying to make their own people quit playing radio. It took dozens of meetings and many years but eventually the situation calmed down and SEB addressed itself to a new problem — getting television.

For years, the Services had conducted polls to find out what was on the minds of the service members stationed in Italy and for years the most common reply was, "television." Edick convinced the command to support SEB's entry into TV and they agreed. This meant developing a piece of paper known and hated throughout the military as a "Schedule X" which is a document of infinite complexity listing in great detail how many people it takes to operate a military unit and justifying each and every one of them.

Edick and the new SEB engineer, Bob Tomko, worked it out in detail. Tomko had a solid background in commercial TV and was prepared to defend their personnel needs. The Command was in shock after reading it. A full colonel came by to discuss it and said he couldn't understand how SEB could possibly need more than a dozen people to operate TV seven days a week. He ticked them off on his fingers: "Let's see, you have one person reading the news, one person to run the camera, one person to run the switcher..." After considerable discussion, he agreed to submit the list but warned that "they are going to laugh at it."

He submitted it. They didn't laugh. SEB got its approval.

On the list was an Officer-in-Charge and the first one turned out to be Major Pete Satterlee who had been with AFN in Germany. Later in his career he would serve as OIC in Panama and with the Army Broadcast Service before retiring to sail his boat and grow a beard. He was followed by then Major Mitch Marovitz. Marovitz was in large part responsible for getting the money to allow SEB to begin bringing live satellite events to Italy on a per use basis.

Closed circuit television in Vicenza and at Aviano became a reality in September 1976, thanks in great part to Ron Manning, an engineer from the Television-Audio Support Agency in Sacramento who came to Italy to design the facility and, later, to install it. This was the first AFRTS color system in Europe and it operated out of the smallest studios. Like other AFRTS outlets, the building had never been designed for television which not only needs more manpower, it needs space.

Unlike today when there is around the clock satellite television over a dedicated circuit from AFRTS in Los Angeles from which to pick and choose, each satellite feed from the States in those days had to be booked long in advance. Costs were high — somewhere in the $60 a minute range — and live sports events were a rarity. Each command had to fund for as many events as they could afford and the in-fighting at high levels over which events to choose became ferocious. One could only hope that the commanding general wasn't a lacrosse fan or an aficionado of female Jello wrestling.

In Italy, because it narrowcast on closed circuit, people who did not live in government housing had to come on to the base and watch TV in dayrooms or clubs. The high spot of any season was the whole command getting together to watch Super Bowl or the World Series. None of them realized what SEB, or the other broadcasters, went through. Negotiations had to be carried out with the host government to use their satellite downlink and their equipment to connect with the signal from the States. The satellite had to be booked long in advance and was subject to preemption for any number of reasons and frequently had to be vacated at a specific time for regular users whether the game was over or not. Every event was chancy and the broadcast crew sweat buckets over each one.

So, came the Michigan-Ohio State game to determine the winner of the Big 10 Conference Title. Biggest game of the college season. The clubs were packed to the rafters with fans waiting for the game. People had been talking about little else for days.

Edick was in the station when the phone rang. It was Ed Bove, the chief engineer of AFRTS, saying that

President Carter was about to sign an historic agreement with Anwar Sadat of Egypt and Menachem Begin of Israel and everyone involved with the signing wanted the event to be seen in their respective countries. They were therefore going to appropriate one channel on the satellite — SEBs. The game start time was just moments away, just time enough to put up a disclaimer that the game wouldn't be seen, sorry about that.

Edick walked over to the club to exercise a little diplomacy himself and explain the situation in a reasoned way. No way. He was met with an angry mob ready to lynch Edick, the President and every other damn politician on Earth who failed to understand that a championship football game was more important than some guys signing a piece of paper. Edick didn't lose his life, although it was close. What he lost was a place to eat lunch for the next few weeks. Nobody at the club would sit with him.

Let fire come out of the bramble and devour the cedars of Lebanon.
— Judges 9:15

LEBANON. Tokyo...Italy...Germany...they all seem like romantic and exciting assignments. And they are. But not every AFRTS assignment is quite so glamorous. Quite often the less glamorous assignments seem to fall to the Navy. One such was setting up and operating stations in Lebanon.

In 1982, the fire truly came out of the bramble and threatened to devour the entire country. Beirut was split in half, part controlled by Christian militia, part by the PLO. Israeli jets were bombing Moslem West Beirut. Syrian ground forces were moving in. Terrorist bombings were occurring almost daily, one of which killed the Lebanese president, Bashir Gemayel. Once one of the loveliest and liveliest cities in the middle East, Beirut had become a battleground. President Reagan ordered in the U. S. Marines as part of a peacekeeping force which also included contingents from France and Italy.

Naturally they wanted radio — and the Navy got the assignment to provide it. The Navy was the first to develop a service-oriented broadcasting system. It's instigator was Jordon (Buzz) Rizer. Rizer had a long career in military broadcasting as an Air Force Officer, including tours at AFN and as commander of the 7122d Broadcast Squadron as well as assignments in the Far East. After retirement, he organized NBS, the Navy Broadcasting Service and developed so-called SITE systems which provided television service to the deployed fleets around the world. He faced awesome problems in that Navy ships are not known for their excess space, crews are on

When AFN assumed control of television in Germany and the Benelux from the Air Force, the color studios at the AFN Frankfurt headquarters were far from completed. AFN-TV became a resident of the former Air Force studios at Ramstein Air Base and the television staff moved there. Transmissions continued in black and white with the German PAL system until the move to Frankfurt.

Celebrity interviews were done as often as possible but somehow Ramstein didn't attract them in great numbers. Shown here is one of the earliest. L to R, SFC Dave Stewart, NCOIC of the station, Dan Allen of the AFN news staff and John Hart of Stars and Stripes prepare to interview James Schlessinger, Secretary of Defense during the Nixon administration.

Note the polite distance between them, unlike today's confrontational interviews in which reporters go nose to nose with the interviewee. (Photo courtesy of Dave Stewart)

duty around the clock while at sea and it's difficult to decide what is "prime time" when a third of your audience is working at any one time. And another third is sleeping. In addition, ships keep moving around and getting videotapes to them in sequence and on time presents difficult logistical problems. He solved them all, including the assignment to set up stations in the midst of a small war.

At first it was decided that a mini-SITE system would be brought ashore, to be manned by Navy/Marine broadcasters who would serve short terms, or waves, with the station. When President Reagan announced in September that he was extending the Marine's presence indefinitely, it was decided that a more permanent system would be used. Rizer and his staff had developed a prototype mobile station which had been put together at the request of the Armed Forces Information Service to support the Rapid Deployment Force. Unlike the cumbersome mobile vans used during World War II and in Korea, the new system took advantage of smaller equipment. The size was pared down so the whole unit would fit into a C-130 transport, could be moved by a forklift and could be up and running within hours. That's exactly what happened.

Navy Lieutenant Ray Nash of NBS with a five-man detachment became the leader of the first wave. They loaded the unit from the NBS parking lot in Arlington, Virginia, on to a truck. At Norfolk it was loaded on to a C-141 and flown to Italy where it was moved to a C-130 for the flight to Beirut. The last part of the journey was by fork lift to the hill overlooking the Beirut airport where the Marine headquarters was located. Ten hours later it was on the air with a complete schedule of programs from AFRTS.

News and sports programming presented a problem at first because shortwave reception from the AFRTS shortwave service was marginal. The Navy, which quickly learns to improvise when they are thousands of miles at sea and an equal distance from the nearest store, solved that simply by going downtown to the local Associated Press office and getting news copy. This system worked fine until permanent satellite, land line and cable feeds could be set up.

Radio from the mobile unit was on the air from 6:00 a.m. until midnight. Television, using a SITE system, fed at very low power covering the Marine perimeter. Both systems performed an important function. Normally Marines move in, finish their job quickly and move out. Here they found themselves acting in a totally passive role, sitting on a sand dune listening to their beards grow. Boredom was the biggest enemy and being able to stay in touch with their world through radio and television was a Godsend.

After four months, the first wave was replaced.

VINCE HARRIS

Today when anyone at the Los Angeles headquarters wants to know, "What do we do now?" chances are Vince Harris will be the focal point of the discussion. He, and his assistant Dorothy McAdam, were both sitting there when the building was put up around them and their breadth of experience encompasses the entire history of the organization. If a flap like it ever flapped before, chances are they will remember how it was handled and know what to do to keep it from flapping again.

Harris' job as Director of Industry Relations is tailor-made for him. He knows everybody in the industry who controls programming, knows what AFRTS needs, knows where to go to get it and somehow weaves a magic spell over suppliers who suddenly become anxious for AFRTS to take their programs. (Photo courtesy of AFRTS)

By July 1983, the fourth wave arrived, this one commanded by John Burlage who had served at Navy Broadcasting as Rizer's Command Master Chief. In April, during the tour of the third wave, the American embassy had been bombed, killing eight Americans and leaving eight more missing, presumed dead. Then things calmed down and Burlage expected a fairly calm tour. He was wrong. All hell broke loose. Fighting escalated and the Marine headquarters, including the station, came under mortar fire. One round landed only yards away from the station, which continued to broadcast.

The value of the station was brought home to Burlage when he stood outside watching a group of walking wounded, who had been hit by shrapnel, walk down the road. One man came up to the fence, bandaged and being led by a buddy. "I just want to thank you," he said. "You were really here when we needed you."

Burlage's group was relieved in September by the fifth wave, let by Joe Ciokin who had served just about every place the navy could think of to send him. This group was made up of volunteers because Lebanon had now been declared a high risk area. The volunteers, however, considered that the Marines were not the target and that the fighting was between the various factions within the country. As it turned out, four days before their arrival the pace of the fighting increased and hostile fire into their area continued throughout their entire tour.

By now the station had become somewhat more sophisticated. It had minicam equipment now which allowed it to put as many Marine faces on the screen as possible as a morale building factor. Most often it was simple production and often went directly from the camera on to the air. However, they did cover the visits of Vice President George Bush and Marine Commandant P. X. Kelley.

Ciokin worried about the security of the station and of his men. Their sleeping quarters were in an open air tent. The station was a natural target sitting in a tree line, wide open to the highway from which anyone with a hand-held rocket launcher could easily hit it. There were only sand bags for protection. Ciokin began looking for a more secure location because, it then seemed, they were going to be around for a long time.

He found a spot. It was in the Lebanese Civil Aviation building and, with the help of the marines, the van and all the other gear was moved in on October 22. The van was set up outside the building, a studio was created inside and the men slept under a roof for the first time.

On the morning of October 25, Bob Rucker signed on the station at 6:00 a.m. and introduced an AFRTS recorded program. As it began, he went to shave. At that moment, a suicide terrorist bomber drove his truck up to the front of the Marine barracks nearby and set off an explosion. Two hundred forty- one Marines and Navy personnel perished in the blast.

Rucker found himself trapped temporarily under a collapsed ceiling which also jammed the door. Ciokin was blown completely across his room and he and his roommate at first thought they had been hit by artillery fire.

The station was knocked off the air by the blast and the television equipment sustained heavy damage. This problem was secondary to the staff. Although most of them had suffered at least minor wounds, they immediately pitched in to help with the rescue efforts. When Ciokin later determined that enough other rescuers had arrived and his group was no longer needed, he collected them and began trying to put the station back together.

Radio was back on the air in five hours. Television, which had taken more damage, required a week during which parts were flown in from the States. Ciokin recalls that the radio programming had an enormous impact: "When the Marines were able to get out of the perimeter the next day, the first thing they told us was what it meant to them to hear radio come back on the air. We were their beacon of hope. It told them that we were surviving, that we were still alive."

The station concentrated on keeping the survivors of the attack informed of developments with constant updates so everyone would know what was happening in the wake of the tragedy. This, common sense said, was what was needed at this time. Then, five or six days after the explosion, a Marine Sergeant Major buttonholed Ciokin and said that everyone was depressed enough. "Do something," he said. "Everybody's down. We have to do something to bring them back up, build up their morale again and get them over this hump and revive their spirits."

Ciokin quickly realized that his staff, depressed as the rest of the audience, were reflecting their depression on the air. They were playing dark, depressive heavy metal music which added to the sense of gloom and melancholy. He called them together, explained the problem, and they began to change their downbeat approach. One was to begin kidding the Marines and they went so far as to run a contest to find the Ugliest Marine. They saluted people on their birthdays and changed the musical approach.

This had an immediate impact and soon the Marines started joking with and razzing the staff. To Ciokin it was a perfect example of why AFRTS exists: "To be aware of the audience, respond to its needs and create within the audience certain responses."

Right, During the Golden Days of Radio in the 60s, radio stations were able to send correspondents wherever there was a story that would be of interest. Following the traumatic events of November 1963 when President Kennedy was assaassinated, AFN dispatched Specialist Gary Bautell to Remirimont, France to dispatch dispatches back to network headquarters as the city renamed its town square John F. Kennedy Place.
Specialist Bautell later became Mr. Bautell and remains a mainstay of the AFN news operation. (Photo courtesy Pete Bissman) Below, the AFN staff gathers in the garden of the castle for a group portrait in 1964. Some had been working the overnight shift and weren't too happy to be immortalized. A glance at the picture, however, will show immediately why this book is named "The Brass Button Broadcasters." This particular group not only is loaded down with brass buttons, it represents perhaps the high-water mark in broadcast talent at AFN. Many other extremely talented individuals preceded and followed—but this gang was as good as it ever got. (Photo courtesy of John Liwski)

104

CHAPTER SIXTEEN

In which we learn how AFRTS reaches out and touches its listeners...and also learn that Edward R. Morrow was right when he said, "Just because your voice reaches half way around the world doesn't mean you are wiser than when it reached only to the end of the bar."

THE GOLDEN DAYS OF RADIO, PART ONE

Although television overseas, as at home, was the glamour girl of the airwaves, radio, during the period between the start of the first AFRTS TV stations and the 1970s, remained the fundamental broadcast voice. This was the case primarily because television was in a growth period and budget and host country restrictions did not allow for the wide geographic coverage radio was able to provide.

Radio was universal. Television was limited in many places to geographically constricted areas. In the U.S., television almost overnight had pushed radio down a different path. In the AFRTS world, radio reached its creative zenith while television only slowly found its niche.

Many commanders overseas felt more comfortable with radio. Its ability to react quickly and reach large segments of the audience provided a medium which television could not at that time match. Nor could television provide the instant news coverage which, in the pre-satellite days, gave radio its unequalled credibility.

Added to all this was an existing cadre of professional broadcasters who still remembered radio as it had been. Their mind-set was to provide the more elaborate style of programming which Stateside audiences were no longer getting. Rather than a steady diet of music directed exclusively to a narrow segment of the audience, as was common in the highly competitive U.S. radio marketplace, AFRTS stations presented a wide variety of programming with something for everyone at some portion of the day. Variety shows, talk shows, specialized musical shows, interviews, news, sports -- all were represented on the majority of AFRTS radio outlets.

To the delight of the "shadow audience", the foreign nationals of the countries in which AFRTS broadcast, this was a totally different style of radio than anything to which they had been accustomed. AFRTS broadcasters, typically young, brash, open, honest and controversial, were like nothing the foreign listeners had ever heard on their staid national radio systems.

It was, and is, AFRTS policy that their stations do not specifically address this audience. It would be foolish, however, to pretend that they are not there. Each new staff member has it pounded into his head to talk only to the American audience when on the air, but to remember you are a guest while broadcasting in another country. Remember, too, **they** own the airwaves you are merely borrowing.

The impact the AFRTS stations made on the various host countries was enormous. Although intangible, the very fact that a radio station could broadcast news potentially embarrassing to its government and assiduously avoid propaganda had an immense influence on the foreign post-World War II generation. Americans talked a great deal about nebulous "freedoms," but listening to an American station broadcast, for example, the entire Watergate hearings made that vague concept crystal clear to many foreign nationals.

...a Few Examples

So popular is AFRTS in Japan, Sony produces an inexpensive portable radio permanently tuned to the Far East Network.

A number of Japanese magazines devote themselves exclusively to AFRTS programs and include lengthy interviews with the local disk jockeys.

In 1964, the United States Information Agency took a poll in the United Kingdom to see how their Voice of America propaganda broadcasts were doing. What they discovered was that rather than listen to the staid BBC, 18 percent of the random sample were AFN listeners.

A poll taken in the 60s by the Bavarian State Radio in Munich was never released. In spite of the fact it had more than three thousand employees, as opposed to only several hundred AFN broadcasters scattered throughout Europe, it showed the German audience preferred AFN. Into a desk drawer went the poll.

German radio stations early-on decided if they couldn't lick 'em, join 'em. They wanted to do disk jockey shows that would draw a big audience like the Americans. Teams of Teutonic talkers descended on AFN for advice. The upshot was, never the twain would meet. The German unions, and work ethic, determined it required a minimum of 12 people to do a DJ show. That included three people to pick the music, a couple of writers to create the DJ's ad libs, a producer to point to the DJ before he was permitted

Bob Matthes was chief of "command information" (for which read "Spot Announcements") at AFN for more than a quarter of a century. During that time he taught hundreds of young broadcasters the arcane art of grabbing the listener's attention, convincing them to do something they did not necessarily want to do and making them grin when they went ahead and did it. He also was a genius at avoiding non sequitors such as the one in the example shown. Words flowed from his beat-up typewriter in a crystal-clear stream and one whole wall of the AFN building is devoted to awards he won for excellence.

Now retired, he spends half the year travelling the globe and the other half living in Las Vegas trying to figure out why every horse he picks decides to take a nap on the backstretch. (Photo Trent Christman)

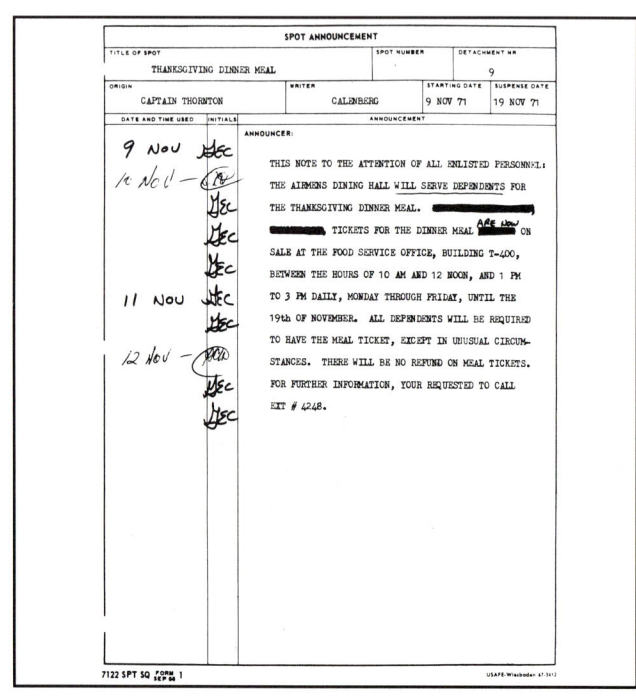

This spot announcement ran for four days with 10 airings before somebody noticed from listening carefully to the first paragraph that the writer very likely did not mean what he wrote when describing the Thanksgiving menu.

to open his mouth, plus an engineer on each turntable or tape deck. They were never able to understand how a single military enlisted man could pick his own music, write his own copy, run his own equipment, ad lib his introductions to the music and do it without a supervisor sitting behind him directing his every move.

The impact of American music around the world has been dramatic -- and in most locations, it was AFRTS that first introduced that sound. As far back as 1954, Variety observed, *It s no exaggeration to say that AFN stations are mostly to blame for the Germans strong predilection for American music.* Blame? Perhaps, but it created a whole new market. Four years later, Billboard commented: *It is AFN s tremendous European audience that is credited with creating the trans-Atlantic market for American music. There is scant doubt on that score.*

In the 70 s, a major European broadcasting magazine ran a reader s poll to pick their favorite radio personality. The winner was Wolfman Jack whose syndicated program from America was heard only on AFRTS.

The French never did understand what all this equal time business or freedom of expression stuff was all about. During the nine year period from 1959 until 1967, AFN maintained a network of three studio locations and some 60 small transmitters there. So terrified were the French that American comedians would poke fun of their cherished institutions that a French broadcaster was assigned to monitor every word. Charles DeGaulle, who himself admitted no one could govern a country which produced 469 varieties of cheese, was not about to let his beloved country become a joke on barbarian airwaves. Meanwhile, AFRTS broadcasters were amused by DeGaulle s concept of equal time. During the French elections, DeGaulle broadcast at 8:00 p.m. His opponent got equal time -- at 4:00 a.m.

One AFN staff member, while being interviewed on the French radio network s English language program, borrowed a line from Billy Wilder and commented that France was the only country he had ever seen where the paper money dissolved in your hand, but you couldn t tear the toilet paper. After an agonizing silence, the French host introduced the next guest and the offending, big mouth Ami was heard no more. Not even a *bon soir*.

These were the glory days of AFN radio. The military draft guaranteed a constant stream of experienced broadcasters who were not always the military's idea of what a soldier or airman should be, but who were responsible for an endless stream of memorable programs. This happy state of affairs held true throughout the entire AFRTS system. In addition, military funding was available in reasonably large quantities due to the Vietnam fracas then in progress. This allowed for AFN news bureaus to be set up in Paris, London, Bonn and for extra news personnel at many of the affiliate stations in Germany. Most locations had both an American and a local national newsman who knew the territory and who could deal with sources in his own language.

The AFN commander during these 1960s glory days was Lieutenant Colonel Bob Cranston whom the reader has already met. He appeared earlier as a young company-grade officer with both AFN and the Blue

Danube network. He had worked in the interim on Eisenhower's public affairs staff and by now had built a considerable reputation in military broadcasting and public affairs. He will appear later, as a Colonel, as commander of AFRTS in Hollywood and as all-high potentate of the American Forces Information Service in Washington.

His great strength lay in his willingness to allow his people to do the job for which they were best suited. Another strength was his understanding that he commanded a unique organization, complete with a crew of khaki-clad kooks who certainly could communicate but with whom you might think twice about sharing a foxhole.

The reader will get a better idea of what type of commander he was as Cranston tells it in his own words:

"The annual Inspector General inspection fanned out through the network and they inspected everything we had. Our people thought of themselves as civilian broadcasters, not military people. The military high command thought of them as military people and not broadcasters. It was up to me as commander to keep the military people off their backs so they could do their job as broadcasters, and that's what I tried to do.

"After the inspectors arrived in Frankfurt I thought I'd get them off to a good start by inviting them to dinner at my quarters. At that time the Commander in Chief in Europe was General Bruce C. Clark. He had decided that everybody in Europe should go to bed early so 'they would be fresh and ready to fight' in the morning. He declared an 11 o'clock curfew. This applied to my troops at AFN and, because we didn't go off the air until 1 a.m., and there were no food facilities at our castle studios, this created a lot of problems.

"When the IG inspectors came to dinner, we had a few wines and some cognac afterwards and the time got away from us. It was already 11:15, but I told them I would take them back to their hotel. Just as we pulled up, so did the MPs, who collared us all for being out after curfew. The next morning I was ordered by Colonel Meyer, the district commander, to report to him along with my fellow culprits. He was more than a little taken aback when he saw the IG insignia on a full colonel, especially when the IG told him that in a few months he would be inspecting Colonel Meyer's organization. The interview ended quickly after that.

"AFN ended up with an outstanding rating in the IG report.

"The curfew remained a big problem but I believe I solved it. One of the regulations was that there would be a bed check at 11 o'clock each night. I gave the order to Sergeant Major Clay Brooks and told him to carry it out. The next morning he appeared with a list of people who were not in their beds the night before.

"I said, 'The order specifically states there will be a bed check. Were all the beds there?'

"'Yes, sir, every bed was there.'

Nolan Kenner represents the hundreds of top-notch broadcasters who during the 60s made AFRTS stations around the world so creative. Kenner and his buddy, Jim Davis, created a program featuring an imaginary pirate captain and his sidekick. A number of real captains complained that it made the rank a laughing stock and in spite of arguing that pirate captains were in short supply in the army, the program was concelled by orders from on high. Kenner is still active in broadcast management and listeners in Michigan can enjoy Davis' wild sense of humor. (Photo courtesy of Nolan Kenner)

If you think rank doesn't have any privileges, think again. While the AFN staff is busy working, the boss, Bob Cranston, gets to goof off with his arm around Caterina Valente, for many years Europe's most popular entertainer. As a very young girl right after World War II, Caterina's career was given a big boost by frequent appearances on AFN. From then on, she was always gracious about appearing at special occasions such as anniversary parties and to this day remains friendly to the network. (Photo courtesy of AFRTS)

"'Okay, you've done your job. Don't bring me any more lists unless some beds are missing.'

"I passed the word no one had better get caught after curfew. Then I poured drinks for the mayor and chief of police who forever after returned our people to us rather than turning them over to the MPs. It all worked out nicely."

During the Cranston era, the radio network reached its production peak. For several years, the network churned out 75-hours of live programming each week, including live drama, play-by-play sports, extended local newscasts and three programs still remembered today by people who were then serving in Europe. One was called _Tempo._ Another was _Weekend World_, which may have been the most elaborate long-running show ever done by an AFRTS affiliate. The third was _Nightside._ _Tempo_ was a sort of radio version of _Good Morning, America_ with a potpourri of music, household tips, commentary, humorous bits contributed from all over the network, interviews with celebrities passing through, guest experts on just about anything and movie and concert reviews.

Weekend World was the network's showcase. The highest accolade a staff member could get was to host the program which ran four hours on Saturday and four more on Sunday. If it could be compared to any other program, it would be the old NBC _Monitor_ show, except everyone at AFN agreed _Weekend World_ was better. Anything went, and the only rule was to be exciting, different and interesting. The only sin was to be dull. Troops who couldn't travel halfway across Europe to hear their favorite musicians in concert could be pretty sure it would be on _Weekend World._ Celebrities didn't stand a chance. An AFN microphone was invariably poked in their face at some point in their European travels.

Van Cliburn gave his interview stark naked while sitting in a sauna after winning the Tchaikovsky Piano Competition in Moscow. Peggy Fleming did hers while stuffing German sausages down a starving stomach after having watched her weight for months prior to winning the world's figure skating championship. When French composer Michel LeGrande was asked what he was working on, he proceeded to play the major theme from his latest movie score.

The first American interview with the Beatles, just prior to the then unknown group's first trip to the States, was done by Paris correspondent, Hal Kelley.

As often as time and budget would allow, usually six to eight times a year, the program would travel. A team would be selected and a handful of lucky broadcasters would shoulder their tape recorders and take off for London, or Paris, or Holland, or Denmark, or Austria, or Munich, or Italy, or Berlin or the Alps. Working with the official travel authorities of the city or country involved, they would swarm over everything interesting like a herd of inquisitive locusts. They would also record the best of the local musicians, put the eight hours of material together, and do the show from the local national broadcasting studios by land line back to network headquarters in Frankfurt. Usually an English speaking local co-host was utilized as well.

Welcome to _Weekend World_

A few of the more memorable events on _Weekend World_ travel shows include the official witch of the British Royal family putting a curse on the interviewer because 'I don't like you.'

One interview that never got on the air was done in a liqueur factory in Holland. The idea was to sample the many different flavors. The reason it didn't get on the air is that the first question was, 'Tell me exactly where we are,' and the Dutch owner proudly explained, 'This is the Focking Liqueur Company.' The interviewer continued the interview on tape but didn't have the heart to tell Mr. Focking that chances were American listeners would misinterpret his unfortunate name. (The next year, the same interviewer tried again at the Bols Liqueur Company, only to discover that in Holland it's pronounced 'balls.' This didn't get on either.)

A wonderfully named Major Courage described the entire morning inspection of the colorful Horseguards in London and when asked, shamefacedly admitted that the reason for the inspection was that once one of the guards had been discovered drunk by Queen Victoria, who proceeded to order a daily inspection for the next one-hundred years.

A radio tour of the famed Lowenbrau Brewery conducted by a Bavarian braumeister ended with a simple question: 'Where did you learn to make beer?' In a juicy, rolling Cherman ackzent, he said, 'At Rheingold, in New York.'

One of the more memorable interviews was held in London in a ballroom filled with jewelers from Texas, in town to promote a solid silver memorial plate just issued in honor of the late Winston Churchill. The AFN interviewer was doing his thing with son Randolph Churchill on one side and the Duke of Marlborough on the other. As the interview got rolling nicely, a tall Texan with a five ounce brain in a 20 gallon hat, wearing an embroidered suit and $800 boots, pushed into the group and asked His Grace, 'Are you Marlborough?' 'Yes, I am,' said the startled Duke.

'Wallll,' drawled the intruder, 'I just wanted to tell ya you make the best f--king cigarettes I ever tasted.'

Nightside broke with long-standing tradition and used a commissioned officer -- Captain Chris Davala -- on the air, although he was never identified as an officer on the program. The reason was simple. He was, an is, an absolutely outstanding communicator. The program started out running five nights a week from 6:30 p.m. until 11. Later it cut back to a more modest two-hours, five nights a week.

In-studio the program was produced by Bob Myers and, later, Jim Kirby. The Special Events staff which supplied much of the material was headed by Herb Glover and included Gary Webb and John Dedakis. Practically the whole AFN staff contributed at one time or another. So did outsiders. The program normally opened with a different comedy vignette. One night the military boss of all Europe, General Mike Davison, shed his dignity and listeners heard, "It's 6:30 in Central Europe. I'm

General Mike Davison, and this is Nightside."

The basic yardstick was to be different and entertaining. Listeners never knew what was going to happen. The guest might be a top Hollywood star or, as happened one night, two scruffy characters who eked out a living playing guitar and singing in the Frankfurt subway. Another night, it was a well edited recording of the birth of a baby; perhaps a radio first. Reporters from bureaus around Europe who normally furnished reports running about a minute were brought in and "debriefed", getting to expand on their coverage of stories of particular interest to the audience. If anything happened during the program, it was often possible to switch to the scene for instant updates. Such was the case when the local officer's club was bombed by the Baader-Meinhof terrorist group. Herb Glover and reporter Dan Allen were on the scene in minutes and continued coverage for hours.

NOW, HEADING SOUTH...

Northern Europe was AFN territory, but the troops stationed in Southern Europe needed radio as well and they listened -- if they could -- to what is now SEB, (Southern European Broadcasters) but was originally called SEN (Southern European Network.) AFN staffers, working out of fancy studios to large audiences, sometimes had a tendency to feel they were the cream of the crop and it was not unknown for them to say that SEB stood for Spaghetti Eating Broadcasters. While it started small, and for a time grew even smaller, today's SEB is one of the most important and efficient networks in the AFRTS system. It is something of a miracle that this is so because, when God created the earth, he created Italy as long and narrow rather than squarish, as countries should be. Just getting around the SEB stations from its present headquarters in Vicenza in Northern Italy to outlets in the sole of the Italian boot and in Sicily and Sardinia requires days of travel time. And lots of spaghetti.

It would be fair to say that Al Edick, who spent twenty-three years with the network, is its father. Who the mother is, he will not say. Perhaps there was none, because Edick nursed the baby network from infancy to puberty before returning to Washington and other duties at AFRTS. Since then others, Mitch Merovitz and Steve Mason among them, brought it to full adulthood. It is now seen and heard throughout the Mediterranean area from its own satellite transponder, covering areas as distant as Spain.

Shortly after World War II, the Blue Danube Network was opened for the occupation troops in Austria. When the occupation ended in 1955, much of the equipment was shipped to Leghorn in Italy. Leghorn, so called by the British for reasons known only to them, is now Livorno. It is a port city close enough to Pisa so that if the tower ever falls, stones will hit Livorno. No one is sure why this location was chosen but it was, by order of the American military command known as SETAF -- the Southern European Task Force which was then headquartered in Verona, the home town of Romeo and Juliet.

SETAF, while setting up the station using the equipment from Austria, began negotiations with the Italian government for permission to broadcast. Their target date was early 1957, but they neglected to allow for the way negotiations are done in Italy.

Singer/songwriter Paul Anka drops by the AFN castle while on tour in Germany and stops to chat with AFN commander, LTC Bob Cranston. (Photo courtesy of AFRTS)

109

Slowly.

Italian radio is government controlled and, at that time, had only one network, known as RAI, *Radio Eduzione Italiana.* No English language broadcasts were available in Italy so SETAF started putting a staff together for their one-station network. A number of former AFN staff members, including Edick, were hired and moved to Italy on the promise that "we'll be on the air any day now." And so they were.

Ten years later.

The ill-fated Livorno station was moved to SETAF headquarters in Verona at Caserma Passalaqua, which sounds like an Italian dessert but is actually a military installation. What the staff found was that SEN was "broadcasting" by wire lines to loud speakers in troop barracks across town. Because no one could hear the station unless they lived in the barracks, the broadcast schedule was designed to fit the audience. It was on from 6 to 8 in the morning while the troops were dressing. Again from 11 a.m. to 2 p.m. so troops could listen during lunch. Then back on at 5 p.m. and off again for beddy-bye time at 11 p.m.

Getting current news was a constant problem but was solved by the common Army technique of "moonlight requisitioning." Or, not to put too fine a point on it, they stole it out of the air. News services in that part of the world used short wave morse code to transmit their stories. SEN, the owner of an ancient gadget called a Hammerlund Sp600, could feed the dots and dashes into their one and only teletype and out would come the news copy. Sometimes. Frequencies drifted and out came garbage. Sunspots. Out came garbage. Local interference. Out came garbage. Electrical storms. More garbage. The end result was that sometimes there was a newscast, sometimes garbage and sometimes neither. At times the same cast was read three or four times.

Knowing they would be actually transmitting over the air "any day now" SETAF furnished SEN new studios in the Spring of 1958. It even had a dedicated direct circuit to the teletype machine and darned if the thing didn't consistently print copy. The only problem was that it was copy designed for newspapers which meant the newsman had to rewrite every cast to put it into radio style.

With the new studios, the broadcast day was extended to cover the entire period between 5 a.m. and midnight -- and with a total of eight newscasts daily. By now, Edick was the boss, which was something less than a high honor, considering the station was only feeding loudspeakers in a barracks.

Then, AFN to the rescue! They had a batch of small 5 watt carrier current transmitters, doohickeys which were fed by a telephone line from the station and in turn put the signal into the electrical wiring of a building. By plugging your radio anywhere into that electrical circuit, wonder of wonders, out came music and news.

Now people who lived in government housing areas could get SEN. Let Al Edick explain how that happened in his own words:

"By this time SETAF was comprised of three military installations: Caserma Passalaqua in Verona, Caserma Ederle in Vicenza and Camp Darby in Leghorn (Livorno).

"We placed the little transmitters in buildings at all three bases but unfortunately there was no government housing in Verona or Livorno. There was one housing area in Vicenza, however, in which more than 300 U.S. families lived. We installed our little transmitters in the power vaults throughout the housing facility, and then ran wires over the tops of the houses. One power vault served three or four houses and we put transmitters in all of them. Our genius Italian engineer, Frank Paris, and I spent more than two weeks crawling over the tops of houses, hauling ladders and usually working in the rain.

"Technically it was a nightmare. When the radios were plugged in, they not only picked up SEN, they also amplified the sounds of mixmasters, blenders, shavers and other appliances. Everyone complained. Soon we discovered people were doing things to the wire on the roofs...splicing it, cutting it and, in one case, running it directly to the back of their radio "to make it better." Of course it didn't and we had to keep making a 35 mile drive over back roads from Verona to fix things."

The questions remaining are: did SEN ever actually get on the air; did it ever radiate a signal like real radio stations do; did it ever hack its way through the thicket of Italian bureaucracy?

The answer will be found "any day now" by patient readers. If impatient, try Chapter Twenty-two.

CHAPTER SEVENTEEN

In which we visit stations everywhere and learn why the people at AFRTS headquarters sometimes suspected the people running the overseas outlets must be Baptists. They knew they were out there raising hell... but they could never catch them at it.

BIG THINGS FROM SMALL PACKAGES

While AFRTS in both Washington and Los Angeles dealt with the more cosmic matters ranging from budgets to what programs to send overseas to where-shall-we-have-lunch, the hundreds of outlets around the world continued to do the job for which they were designed. That job was to act as a sort of electronic campfire around which the community could gather and exchange news and views, express their opinions, hear what was happening around them as they were both entertained and informed.

Stations and networks each developed distinct personalities depending on their management, past policies, geographical location, the cooperation from the local commanders, physical facilities and the type of audience they served. These things constituted the core of the overall station personality. The shell, which was most obvious to the listener or viewer, was a direct result of the men and women assigned to the station. A station's staff was usually assigned by the next higher headquarters and in most cases those doing the assigning tried to keep a balance between more experienced broadcasters and those just entering the field. For them, it was on-the-job training and many who have now gone on to bigger careers in broadcasting will quickly admit, it was a priceless experience.

Entry level broadcasters at most commercial stations are doomed to occupy a small niche at low level for a reasonably lengthy period. This was not the case at AFRTS. Newcomers there had to quickly fill the shoes of the person they replaced. Non-commissioned officers frequently pushed them to their limits. Mistakes were tolerated. Once. Most importantly, young broadcasters who considered themselves hot- shot heavy metal disk jockeys might well find themselves doing a country western show. Or interviewing the conductor of the local symphony. Or lying face down in the mud, cuddling a tape recorder and describing a night-firing exercise.

Television people quickly learned to do it all. A person could well be the reporter on the story one day, the camera operator the next, the editor of the piece the

This is a rare picture in which a broadcaster gets a lesson on how to throw the bull. In the center is John Bradley, then managing the Lajes station, about to get the shaft from the world's foremost cavallhiero, Simon Da Vega (L) and further information on bull from internationally famed bullfighter Armond Soares (R). Portugese bullfighters fight on horseback unlike their Spanish contemporaries. They both have better looking uniforms than the Air Force. (Photo courtesy of John Bradley)

following day and, no matter what the job, be responsible for washing down the mobile unit and filling it with gas.

No matter what the job, from anchoring the news to writing a spot announcement to covering a Presidential visit, the young staffer soon learned to do it all; an opportunity to expand his professional horizons to limits his civilian counterpart probably wouldn't reach for years. While this tended to bring bubbling egos close to the surface at times, it was kept under control by the NCOs who could, and did, hand a young staffer a paint brush or a broom as the rising star left the studio after an on-the-air ego trip. It was a quick and effective lesson in humility.

The views from the Ivory Towers in Hollywood and Washington were quite different from that in the trenches where the real action was taking place. What follows are snapshots from stations and networks around the globe to illustrate the point.

DOWN IN SUNNY ITALY Al Edick continued the good fight in the effort to get the Italian government off its fanny and get SEN on-air frequencies at a power that could be heard for a reasonable distance. At that moment the Italians were about as interested in this as they were in ugly women. SEN was being heard, sometimes, over carrier current signals which radiated from the interior electrical wiring in military housing areas and facilities such as clubs, commissaries and other gathering places. Obviously this limited the potential audience. Most people lived in rental housing off the bases and could not receive the wired-in signal.

The absolute necessity for a radiated signal was brought home graphically the early evening of November 22, 1963. Edick got a call from the station that President Kennedy had been shot. He was there in minutes and already word was spreading through the community from person to person. Crowds, made up of those who could not receive the signal, gathered at SEN's front door. It was the only place they could think of to go in order to get the latest news. The staff did everything it could to get the latest developments out including posting bulletins ripped from the teletype machine. A speaker in the lobby carried the news as well, but this was obviously not the way to get news of such importance to your audience, many of whom were standing stunned outside the station with tears streaming down their faces.

Distasteful as it was to the SEN broadcasters, they published mimeographed bulletins starting the next morning and gave them to callers. They also had the MPs at the main gate hand them to arriving drivers. It was not the preferred method for a radio station to get out the news but it was better than not furnishing the news at all.

A station manager of any AFRTS outlet soon learns that it is absolutely essential to the successful operation of his station to make, maintain and constantly renew good relationships with the power structure within his area. Often a little extra cooperation can work wonders, as it did for Edick when the whole SEN operation looked like it might be heading for the dumpster because of lack of coverage and lack of funds. Some felt that there was no point in paying for something that reached such a small group. In 1964, there was a meeting of the non-appropriated fund group. These are locally generated funds which are not appropriated by congress and at that time were used to pay salaries and buy much of the equipment.

The prevailing opinion on the base was that the money could be better spent elsewhere. Edick knew that Howard Berger, the ranking civilian on the base and a man with whom Edick had maintained close relations, was a stock market buff. Edick and the staff had been glad to furnish him the latest market reports off the news wire as well as report on developments during newscasts. Prior to the meeting at which the station's fate would be decided, Edick just happened to mention to Berger that it would be nice if stock market reports could continue to be easily available. Berger controlled the meeting and announced that SEN would get the funds it needed. This gave it the impetus to begin serious negotiations to actually get on the air rather than continue to be a wired service.

At least one announcer was delighted few people could hear him one morning. Following a rather serious, and losing, bout with a bottle, he stumbled in to the station for the 5 a.m. sign on. He put a newscast together. He pulled his records for the morning DJ show. He turned on the equipment, played the National Anthem, read the headline news and went into his first record. "This is Sergeant Mort on SEN's Daybreak show," he said, "and the time is now...2 a.m.???? Oh, my God!!!" He shut off the equipment and went back to the barracks for another three hours of sleep.

Another announcer was even more cool. Playing an AFRTS program one evening, he heeded the call of nature. The program ran a half-hour with a cross-over from one disk to the other at the mid-way point. Edick sitting at home suddenly heard the click, click, click, click of the needle in the last cutoff groove. He called the station. No answer. Again. Still no answer. The man must have died, he thought. He called the MP station which was near the station. They pounded on the door. No answer. Finally, there was the program, back after thirteen minutes of clicks. Turned out that the man had turned down the speaker and gotten interested in a magazine story while sitting on the throne. Edick could only be thankful that the man wasn't reading War and Peace.

Listeners, too, caused occasional concerns. One night a new female broadcaster was handling the overnight shift. Very late, and all alone, she heard the crash of glass and found a man trying to crawl into the building. She quickly called the MPs who arrived in time to pull the struggling intruder out of the window. It turned out he wasn't interested in the young lady. After a few snorts at the club, he had decided to come by and meet Wolfman Jack. The Wolfman was there all right, but only on an AFRTS transcription.

At last negotiations started to get SEN on the air. This was high level stuff with a high level negotiating team from the U. S. Embassy meeting with the Italians in Rome. The sessions dragged on and on for days. One sticking point was that the Italians said they simply could not agree to allow the Americans to broadcast on the AM band but they certainly would consider allowing FM broadcasting. In the way of negotiators, the U. S. team, with a knee-jerk reaction, insisted that it MUST be AM. The Italians then said that in that case, there was no need for further discussion as the negotiations seemed to be stalemated. Agreeing, the U.S. team departed. As they left the building, the Chief U.S. negotiator was heard asking one of the team members, "What the hell is FM, anyway?"

On which note we bid a reluctant farewell to Italy, still not knowing whether or not they ever get on the air like a proper radio and television station. All will finally be revealed later but first we must move on to:

BERLIN, BEHIND THE IRON CURTAIN

Berlin, until the reunion of East and West Germany, lay 110 miles behind the Iron Curtain, a city administered by four occupying powers and bisected by a wall. Access was difficult and in most ways troops stationed in Berlin were living on an island. An island luxurious in many ways, true, but always with the feeling that you were somehow detached from the rest of the free world.

AFN Berlin served its American audience from the days immediately following World War II and if anyone could represent the continuity of the station through those years it would be Mark White. White joined AFN in Munich in 1946 as an announcer. Later he served as a newsman at AFN Nuernberg and, in 1952, as a civilian, went to Berlin where he remains to this day. Now retired, he makes his home there after having been with the Berlin station for more than 35 years. The list of celebrities who have passed in front of his microphone goes on and on — Marlene Dietrich, Bob Hope, Jimmy Stewart, Sylvester Stallone, Kirk Douglas, William Holden, George Hamilton, Alfred Hitchcock, Liza Minnelli, Sally Field as well as every major musical star who ever played Berlin. He also covered presidential visits to the divided city by Presidents Kennedy (Ich bin ein Berliner), Nixon, (Ich ain't bin no crook), Carter and Reagan.

Berlin remained an occupied city until reunification and was in the enviable position of having many of its operating funds paid by the Germany government. It was always the showplace of the AFN network and White spent a great deal of his time showing visitors around the city. Originally AFN Berlin, as did the rest of the network, ignored television. Its radio station occupied a rather lavish mansion which had been the former home of a Berlin banker.

The Air Force opened a television station as it had done in other areas of Germany and this station, along with the others, passed to Army and AFN management when the Army assumed the television mission in the early 1970s. The German government funded for an expanded TV station and poured millions of Deutschmarks into the construction of a state-of-the-art color studio. Because of the difficulty in feeding a television signal across the 110 miles of East Germany, AFN Berlin TV for many years operated independently from the headquarters station in Frankfurt. More recently, with satellite technology, it is able to receive AFRTS programming from Frankfurt and free its staff to produce more locally oriented programs.

Because of its semi-isolation, Berlin was forced to provide more locally created programming through the years and its staff perhaps had more opportunity than others to expand their broadcasting experiences. One such is Steve Reichl, now a top film and television publicist. Tragedy opened the door for him while he was a reporter at AFN Berlin in the 1960s. After interviewing film director Anthony Mann, he went back to the studio and Mann went back to his hotel where he suffered a fatal heart attack. Reichl was taken in by the cast and crew of the film Mann had been shooting and decided then and there to have a career in film publicity.

Dick Rosse, now an NBC newsman, was in the news section in Berlin in 1960. Director Billy Wilder was in town shooting the film *One, Two Three* and Rosse became something of a drinking buddy with Jimmy Cagney, the film's star. The East-West crisis was heating up and the morning the unit was ready to shoot a scene of Horst Bucholtz driving a motorcycle through the Brandenburg Gate, the East Germans closed the border. While Wilder and crew were trying to figure out what to do, Rosse took Cagney on a short tour. The movie tough guy who had gunned down dozens of innocent victims in his films, pointed nervously at the East German guards. "Hey," he said, "those guys have GUNS!"

The whole film unit then moved to Munich where the scene was shot using miniatures and real guns were not in evidence.

Although directed at the Americans, AFN Berlin was voted the favorite station in Berlin on a number of occasions. So popular was it that frequently East Germans

Berlin Station Manager Mark White holds the European record for the longest continuous term of service with AFRTS—43 years. Starting as a GI announcer in Munich, he served in Nuernberg and soon moved to Berlin where he spent most of his career. Now retired, he lives, (where else) in Berlin. He watched the wall go up; he watched it come down. Probably no American today is more of an expert on the fascinating city of Berlin than Mark White. (Photo Pete Bissman)

In Berlin, SSG Herb Olsen and reporter Billy Shaw describe Berlin's Allied Forces Day Parade in front of the famed Brandenburg Gate. (Photo courtesy of Mark White)

Formerly the home of a prominent Nazi banker, AFN put his lovely home to better use as the AFN radio studios and billets for the staff in Berlin. Located in the posh Dahlem section of the city, it was perhaps the most elegant of all the AFN stations.

To the dismay of the staff who quickly got used to gracious living, the radio studios were combined with the TV studios which were constructed with occupation funds and located near the military headquarters several miles away. This building was turned into a fancy BOQ for a few lucky officers.

And no longer could the designated flag raiser send his lady friend out in the cold in her nighty to hold the world's most informal reveille while he grabbed a few extra winks. That's what one new commander witnessed his first day on the job. (Photo courtesy of Pete Bissman)

The AFN Berlin waiting room proudly shows off photos of some of the stars who have appeared before their microphones. Every performer who visited Germany seemed to want to come to Berlin—and the station managed to interview almost all of them. This kind of radio also made the station a favorite with German listeners—both East and West. One East German was so anxious to get a request played that he threw it, wrapped around a rock, into the guard post at Checkpoint Charlie. (Photo courtesy U.S. Army)

A corner of the AFN Berlin record library. Berlin's library was second only to Frankfurt's and the station was designated to be the backup to Frankfurt in case of war. For years people kept asking why this should be, seeing as how Berlin would undoubtedly be the first to go should Warsaw pact troops invade. For years no one was able to provide an answer—and they still haven't.

Seen here L to R are Army Specialists Dan Eads, Milt Fullerton and Dick Chapin standing in front of the 16" transcriptions which were still commonly used back in 1965. (Photo courtesy of Milt Fullerton)

would send in letters and musical requests. One staff member remembers a request coming over the Berlin Wall, tied to a rock.

Many of the friendships made by the staff with the German population have survived to this day. Staff Sergeant Oscar Skocik was a particular friend of the Berlin children. He and his wife would decorate a tree in their front yard with candies and chocolate bars during holidays — items many young Germans had never seen during the dark days of the Berlin blockade in 1948. Neighborhood kids could always get a candy bar or dish of rare ice cream at the Skocik door. When he was reassigned and returned to the States, the neighborhood kids all signed a going-away card. He returned for the first time exactly forty years later and discovered he was still remembered. The German press treated him like a celebrity and a number of his "kids", now in their mid-40s, gathered together to recall their first taste of chocolate and ice cream — courtesy of their "Onkle Oscar."

IN NUERNBERG the station for many years was located on the top floor of an Army operated hotel directly across the street from the main train station. Prior to that it had been in a German hotel next door. Still earlier, it had been in a castle-like chateau owned by the Faber-Castell family of pencil fame who had gotten extremely rich because an early family member had invented the eraser on the top of the pencil. It now occupies a new and modern building on the grounds of the military reservation in the adjoining city of Fuerth, which prides itself on being the birthplace of Henry Kissinger.

If any one person represented continuity at Nuernberg through the years, it would have to be Inge Geier. She began her career as a translator during the Nuernberg War Crime trials and later became the German secretary at the station where she worked until her retirement after forty years of service with the Americans. Frau Geier characterizes the invaluable group of local national employees who make life bearable for the young American troops serving around the world. Not only do they act as secretaries, librarians or technicians, they explain the ways of the country to the young American troops who are often away from home for the first time. They translate their telephone and utility bills, they pay their parking tickets, they show them how to use the bus or subway, they introduce them to the culinary delights of the country and sometimes to a local companion. In short, they are irreplaceable. Years after returning home and leaving the service, former AFRTS people say they often can't remember the names of the other staff members but they NEVER forget the name of the local national employees who made their stay enjoyable.

Inge Geier remembers a number of the odd-balls who made AFN Nuernberg somewhat different from the other stations. There was the sergeant who wouldn't look in a mirror because he was afraid his soul would get sucked from his body. There was another who craved privacy and would sit on the sloping roof, legs dangling six stories over the street while he read poetry to himself. One, suffering from emotional problems caused by a recent tour in Viet Nam, although no one realized it at the time, locked himself in his room and tried to drink himself to death. He failed on that try but, upon discharge, succeeded.

At times the entire staff seemed to go slightly bonkers. The studios, on the sixth floor, looked out on an airshaft in the center of the hotel. The station manager one night discovered they had put every piece of microphone cable in the station together and were lowering a microphone on a pole to dangle outside of hotel windows, eavesdropping on conversations, or other activities, going on inside the room.

The same manager one morning heard the morning DJ sign on and sounding extremely ill. His teeth were chattering and much of what he said was garbled. Rushing to the station he soon found the trouble. The DJ had been interrupted in the midst of a tryst with a married lady whose husband, he knew, was a pistol instructor. When he heard the sound of the front door of the lady's house opening at 3 a.m., the brave DJ grabbed what clothing he could and departed through the window. Unfortunately, he grabbed everything except his shoes. It being the middle of March, and 3 a.m., the streets were both covered with ice and snow and were deserted. This meant a four mile walk to open up the station. He made it in time to sign on but not to warm up.

Almost every station in the AFRTS family has a Swap Shop program of one sort or another on which listeners are given the opportunity to sell unneeded items. Used cars pass down from one soldier to another until they go for $25 or twenty-five miles, whichever comes first. Generally these programs border on the boring but Nuernberg disk jockey Paul Dunn decided to do something with it. He assumed the role of a man with a degree in salesmanship from the University of Death Valley and announced that he was the only soldier in the Army with the occupational code 97Z40 — Official United States Army Used Car Salesman. Listeners quickly began adding to the festivities by sending in such bizarre items for sale as a genuine Turkish coffin, never used, and a seven foot rubber tree plant named Albert which was "good with kids." One man who was allergic to his wedding ring kept trying to sell it but insisted that the wife did not go with the deal. A doctor who had tried in vain to sell his car offered to throw in a free nose job. Another medic offered a free hernia operation because, he explained, the car needed pushing from time to time. The program quickly became one of the most listened to features on the station.

AFN MUNICH had continued to broadcast from the mansion commandeered by Major Bob Light back in 1945. The former home of the famed German artist Kaulbach, it featured such amenities as a giant ball room, concert grand piano, ceilings of museum quality and an underground labyrinth of tunnels and rooms. In the early years, it functioned as headquarters for AFN Germany while Orleans, France was headquarters of AFN France and Frankfurt was the over all AFN headquarters. When Charles DeGaulle pulled out of NATO's military arm in the l960s, stations in France closed down and Frankfurt became the sole headquarters.

Munich, however, remained just about every AFN staffer's favorite station to visit. Station manager Neil Fontaine became AFN's Conrad Hilton and suffered through more "inspections" than any other manager in history. By some odd coincidence, these generally happened during the German Fasching season or while the

AFN Nurenberg began life in the Grand Hotel at the far end of the building complex shown here. It later moved to the attic of the Bavarian American Hotel, foreground, where it occupied the entire attic area. Directly across the street from the main train station and a short walk to the inner city, still surrounded by centuries old protective city walls, Nurenburg was a choice assignment—especially in that the staff lived in their own hotel room, could order room service meals and, unless they got caught, have a martini delivered while they were on the air. (Photo Trent Christman)

AFN Headquarters in Frankfurt expected all affiliated stations to provide a reasonable amount of programming for use on the entire network. One program orginated and produced by AFN Nurenburg was a clone of "College Bowl". Called "Knowledge Bowl", it pitted teams from service clubs throughout Northern Europe against each other in answering questions mainly on military subjects. The author, center, is shown producing one episode with Sergeant Paul Hartwig engineering. Due to the limited budget ($0.00 per program), the author's wife Pat (L) was given the entire amount to act as scorekeeper. (Photo courtesy of Trent Christman)

beery Oktoberfest was in progress. Fasching, the German Carnival or Mardi Gras, is a seemingly endless series of non-stop balls, parties and general whoop-do-do. Fontaine and his wife Erika hosted an AFN Fasching Ball each year which became famous throughout Bavaria. The station offered the perfect setting and people fought for invitations. For years it was THE big social event until tragedy struck in the form of the Army Engineers. One of them noted that the dance floor of the ball room looked suspiciously like a trampoline as several hundred guests did the bunny-hop on it. In the interest of saving lives, limbs, and the building, he was forced to ask that such activities cease. The parties continued in future years in other locations but somehow they were never quite the same.

Station NCO Dick Dail recalls how newly assigned personnel would be awed when they got their first glimpse of their new home. One said, "Dear God...they've given me a mansion." He also recalls Fontaine's distress when one of his engineers, getting ready for a forthcoming inspection, had his troops use Brasso on the gorgeous brass chandelier in the ballroom, ruining a hundred years of patina. The same man apparently had a fetish for cleanliness. Not long after, he decided to clean the dust out of a tape editing unit and did so by dunking it in a sink full of soapy water. Even though he dried it with a hair dryer, it never worked again.

OTHER AFN STATIONS have their share of anecdotes and each has developed its own personality over the years. Dave Stewart, now a civilian in the Army's public affairs office in Heidelberg, was a top NCO at three of them during his military career. He somehow managed to avoid duty at headquarters and was able to pretty much run his own show during his various tours with AFN.

Cleverly, he managed to find assignments at stations in cities not thought particularly noteworthy in the tourist guides. At AFN Bremerhaven, on the North Sea, he was generally spared the headquarters inspections which would bring platoons of inspectors to Munich during Fasching, Nuernberg during Christkindlesmarkt or Berlin during Green Week. Unless, that is, the commander suddenly developed a craving for a good fish dinner. Sometimes, too, a few members of an inspection team would develop a different sort of craving and decide they wished to inspect the famous Reeperbahn in nearby Hamburg. Luckily for Stewart, his German chief engineer, Werner Kretschmar, was always willing to play tour guide on these excursions.

Stewart is an outstanding broadcaster and a be-

AFN Stuttgart was captained by Bud Miller for many years and was unique in that it was located in the Stuttgart American elementary school. Engineer Rudy Strobel got so tired of announcers playing music too loudly over his precious equipment that he rigged the studio lights to flash on and off when the signal became over-modulated. This was a tough station to manage in that the headquarters for the United States European Command is located nearby and every General and high ranking public affairs officer, of whom there are many, felt duty bound to offer advice. Shown here are Pete Bissman at the control board and announcer/newsman Gary Bautell at the microphone getting ready to try to please everyone. (Photo courtesy Pete Bissman)

liever in doing something in moments of crisis rather than remaining mute and stupid. He instilled this in his staff. One rainy Springtime, floods knocked out the network line and Bremerhaven was forced to fill its air time locally for thirty-six hours. One young DJ picked a copy of *Boston Blackie* out of the library and started it on the air, only to discover that the library only contained part one. As the announcer on the program invited the audience to stay tuned for the exciting conclusion, Stewart's man opened the microphone and explained the situation. He then recounted an entire second half which he proceeded to make up.

TURKEY is Air Force territory and the stations there are controlled by the European Broadcasting Squadron located at Ramstein Air Base in Germany. The squadron was formerly known as the 7122nd Broadcast Squadron and at one time was commanded by Buzz Rizer, former Air Force Detachment Commander at AFN and future capo di capo of the whole shootin' match as Director of the Armed Forces Information Service in Washington.

In 1968 Rizer selected Air Force sergeant Roger Maynard to manage the station at Karamursel Air Station in Turkey. He briefed him before departure and said that the station was one of the "best equipped." After looking over the extensive equipment list, Maynard was impressed and completely sold on his new assignment.

When he arrived at the station, located on the Bay of Izmit off the sea of Marmara between the Bosphorus and the Dardenelles, he discovered that all the promised equipment was there. The only problem was that more than eighty percent of it was out of order. Worse, two weeks before his arrival, the control room had caught fire and the station was receiving its signal from another Air Force station located in Incirlik. The signal arrived by "Troposcan," a type of tactical transmission noted for using extremely high power to transmit extremely low quality signals.

Maynard also discovered: 1) The previous manager had departed in disgust two weeks earlier. 2) There were no authorizations for maintenance technicians. 3) The one remaining technician was currently in the base hospital after have been certified as crazy. 4) The audience thought so little of the station they were threatening to have it cited for fraud, waste and abuse.

He wrote up a report, presumably lengthy, of all the things he found wrong and forwarded it to home base at the 7122nd in Germany. The 7122nd sent an inspection

When Roger Maynard arrived to manage the Air Force station at Karamursel, Turkey, it was, to put the nicest face possible on it, a mess. It took two years, but Maynard managed to improve it to the point it won the award as "Best Station" in the 7122d Broadcasting Squadron. One of his innovations was to construct this studio on the beach belonging to the base. It was used, according to Maynard, to cover boat races and other important aquatic events. True not doubt . . . just as true as the suspicion that those binoculars were used by the announcers to watch passing sail boats rather than bikini-clad beauties. (Photo courtesy Roger Maynard)

AFN Bremerhaven, located in a port city near the North Sea, is probably the most isolated of the AFN stations. Still it gets its share of headquarters visitors who are most often more interested in the world's freshest fish dinners and in window shopping in nearby Hamburg's famed Reeperbahn than they are in inspecting the station. (Photo courtesy of Dave Stewart)

AFRTS announcers often ran into unexpected situations when they set out for interviews. Dave Stewart, manager of AFN Bremerhaven, thought he was going to be interviewing singer Jose Feliciano. When he arrived, it turned out to be a press conference, and, because Dave is fluent in German, he was conned into acting as interpreter for the press conference. That's Dave with the microphone, looking like he would prefer to be elsewhere. (Photo courtesy Dave Stewart)

team who turned out to be extremely helpful and slowly things began to turn around. Within two years, the station was named the best in the squadron; an honor it received twice. Engineers were finally assigned and the station began to do a job for the community.

A remote studio was built at the beach nearby and dubbed "The Bikini Watch Studio." The disk jockey shows frequently originated from there during the warm months and it also served well when covering water sports. Sports booths were built at the football and softball fields and coverage provided of the Mediterranean league games.

The most important change Maynard made was to provide such complete service to the audience that the audience turned around and became allies. Proof of this lies in the fact that for four years running, Karamursel — like all other stations — ran campaigns to collect funds for the Combined Federal Campaign. Each year the Karamursel audience led the list of per capita contributions of all locations in Europe.

IN GREECE the AFRTS station was also controlled by the Air Force and was located at the Hellenikon Air Force Base which was co-located with the Athens civil airport. The past tense is being used because at this writing the future of the station is very much in doubt due to the possibility of base closings in various areas. Currently one of the most popular programs running on the station is called "Top Ten Rumors of the Day."

Kevin Aandahl, the lone Navy staffer adrift in a sea of Air Force blue uniforms, was a young broadcaster in the late 70s and early 80s. He recalls that the station was a major source of information for the American ambassador during the crisis periods of those days. The Ambassador spent hours on the phone to the station during the abortive attempt to rescue the 52 hostages being held by the Iranians. Hellenikon became the focus of the world's attention when it became the first landing point of the Algerian Airlines plane bringing the 52 to freedom.

Like its sister 7122nd Broadcasting Squadron stations, Hellenikon covered the local community events including the Mediterranean Sports Conference games and it frequently fed play-by-play reports to the stations at the visiting team's bases around the Med in Italy, Spain and Turkey. It also held money raising events for the Combined Federal Campaign, the most unusual of which was the annual "AFRS Athens Bacchus Bash" in which chartered busses, each hosted by a staff member, would take listeners to a local wine fest in Dafni. The Air Force staff never did let their lone Navy member live down the fact he disgraced the Navy by depositing the results of a long night of wine sampling in his own lap during the long bus ride back to Athens.

In 1981 Athens was hit by a major earthquake and it must be reported that the station was slow to react. The quake hit in the middle of the night and awakened

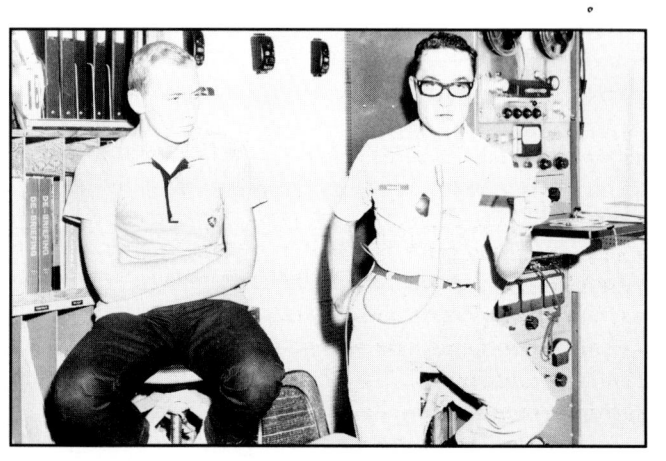

Until General Charles deGaulle pulled French forces out of the military arm of NATO, AFN maintained a network in France with headquarters in Orleans and studios in Verdun and Poitiers. Each studio fed about 20 small 50 watt FM transmitters. This is the Poitiers station which occupied the second floor and enjoyed several advantges. Bordeaux was just down the road, the lovely Loire valley was just up the road, Cognac country was about 30 minutes away and Belon oysters could be found for around 50 cents a dozen fresh out of the nearby Atlantic. Its prime attraction was that it was so far away from Frankfurt, it was seldom visited by the brass; nor was the cranky French telephone system of much use in receiving instructions you didn't want to hear. (Photo courtesy of John Liwski)

Before A2C Adrian Cronauer went to Vietnam where he became famous for his "Gooooooooood morning, Vietnam" he was a popular DJ at Iraklion AB, Crete. That's Adrian holding the mike while holding himself up on the lecturn as he co-hosts a dance party. Helping him is A1C Barry Gump.

In the other picture, he is doing his Saturday afternoon disk jockey show called "Beach Party" which actually orginated near the beach. The door was left open and anyone could walk in—as did the guy sitting on the left.

Still active in the Armed Forces Broadcasters Association, Cronauer is a successful Washington broadcast attorney and you probably wouldn't recognize him from these pictures. Today he sports a spiffy beard. (U.S. Air Force Photos)

Above left, Finally! The new station on top of Santa Rita Hill at Lajes Field, Terceira, Azores, is completed after a few minor mishaps. Like, some of the concrete slabs fell down during construction. A hole dug for footings got hungry one night and ate a cow. Glass for the control room windows kept arriving from the Portugese mainland in small pieces, requiring the General to send his own plane to pick it up. It turned out nicely, though. (U.S. Air Force Photo) Above right, Shortly after the new Azores studios opened, John Bradley, Carnie Babbino and weatherman Dick Kohls (L to R) prepare to do the evening news. It is uncertain why they did it standing up although it is possible that by doing so they could get a running start on irate viewers. (Photo courtesy of John Bradley) Left, View from the control room of one of the three daily newscasts produced at Lajes each day. News is a critical commodity on an island 500 miles from nowhere. (U.S. Air Force Photo)

staff members all over town. They immediately hit the road to the station. There they found the overnight disk jockey doing a Venus Flytrap type of Disco show. All the lights were off. Candles were burning. Music was pumping out disco music at maximum volume with such a strong beat that the DJ had not even noticed the quake. As the staff poured into the station, she was surprised when they told her the quake had lifted her tone arm and moved it three tracks across the record. At this point they began calming the listeners and played a significant role in keeping a shaken population informed.

IN THE AZORES John Bradley is trying to get his combined radio and television station built and is not having an easy time of it. Not the least of his problems is the fact they can't seem to get the antenna footings sunk because of the solid rock at their location. Finally a deep hole was completed and he arrived one morning to find an Azorean farmer and his whole family standing at the edge of the hole, holding a wake for the family cow which had committed suicide by jumping into the pit during the night. That was the American story. The Azorean story was that it was pure negligence to let the finest, most expensive, pure-bred cow in the archipelago fall in the pit. The legal people had to work that one out. Bradley had other problems, one of which was that 90-mile winds blew the uncompleted walls of the station down one night, leaving nothing but a pile of broken cement.

Later, as the interior work progressed, the problem became one of getting the large panes of glass for the control room windows delivered by boat. Boxes of shards kept arriving. After several attempts to obtain something besides broken glass, General Smith, the local commander who considered it HIS station, used his own aircraft to fly to Lisbon and get the glass. He personally loaded it aboard and supervised its installation.

In those early days of propeller driven aircraft, the Azores were a refuelling stop for Europe-bound aircraft and the field handled an immense number of planes. Many of them contained VIPs including Presidents, Generals, Admirals and Prime Ministers. Although Bradley knew this was HIS station, General Smith continued to labor under the misapprehension that it was his, so an accommodation was worked out. When VIPs arrived, there would generally be a three or four hour delay for refueling. Bradley's crew would film the arrival and when the film was developed, Bradley would send a secret coded message to the General that all was well. The General would then suggest a visit to HIS station.

Bradley would be waiting to show the group around and when they reached the control room would casually remark that his station would have the story on the news that night. By a remarkable coincidence, the film would be cued up on the projector and would the Ambassador/General/Admiral/Prime Minister care to see it? None ever said no.

The FAR EAST NETWORK has had its share of war stories and memorable characters. Dick Hiner tells many of them. Hiner served as station manager at Chitose, Japan, as Director of Operations and Station Manager in Tokyo, Chief of the Thailand Network and Chief of AFRTS to the Secretary of the Air Force Office of Information. All of this Air Force experience has led to his present position as Chief of Navy Broadcasting. Go figure! He also served as President of the Armed Forces Broadcasters Association.

He recalls the day he was called at headquarters in Korat, Thailand, by the security police who announced that they had just busted the entire staff of the station at another location on drug charges. He flew there immediately to find the equipment still operating but with only

119

one other living creature in the building — the station mascot, a confused-looking monkey which was sitting at the station manager's desk. His first thought was to have him hold up his paw and be enlisted to run the station, knowing he couldn't do much worse.

The State Department caused any number of problems for the Thailand Network of six stations. The Viet Nam war was going on next door and the State Department was insistent that no mention be made on the air of bombing runs originating in Thailand. This was a case of if-you-ignore-it-everyone-else- will-too, a theory frequently put to the test and inevitably failing. The Thais most certainly had an excellent idea of why there were 50,000 Americans in their lovely country and could hardly ignore waves of aircraft taking off and landing.

On the other hand, the State Department was so happy to have an AFRTS station that the embassy in Vientiane, Laos, surreptitiously ran a wire across the Mekong River from the station in Udorn so they could listen. Although this was illegal for any number of reasons, the AFRTS group in Thailand was happy that their fellow Americans in Laos could get the news. This happy state of affairs continued until some misguided striped-pants type at the Embassy in Vientiane became a critic and complained to the embassy in Bangkok that Paul Harvey was talking about B-52 raids out of Thailand. And while he was at it, he complained about several other things he didn't like about the station's programming.

From that moment on, the Vientiane embassy began experiencing extreme difficulties with their reception. Hiner is of the opinion that those damn, no-good Commie insurgents would sneak out every night and cut the line.

AFRTS was only permitted to broadcast television in Thailand if the signal did not go beyond the perimeter of the base. Buzz Rizer, famed for his innovative thinking, was in- country at that time, and invented a miraculous device called a "corner reflector antenna" which supposedly stopped the signal at the base fence. Any technician on earth will tell you that this is impossible. So will Rizer. What's important is that the Thais didn't know that and permitted the network of six stations and several unmanned relays to operate.

Hiner, by now back in Washington, was called in to the office of the Assistant Secretary of the Air Force for International Affairs. He had just returned from the Far East and Hiner expected nothing good from the meeting. He was right.

"Hiner," the Secretary said, looking at him as if every moment would be his last, "I just returned from Thailand and I got your TV signal on the third floor of a Korat whorehouse. What do you intend to do about it?" Hiner mumbled something about to the best of his knowledge, serving Korat whorehouses wasn't part of the mission. Ahhh, it might even be a violation of agreements with the Thai government. Oh, yeah, and John Broger, the chief of the Office of Information and Education was a very religious man and would no doubt order God to strike them all dead if he ever found out they were serving a den of sin and iniquity. He then promised to see what he could do.

As Hiner left the room, the Secretary said, "I'll be back in Thailand in a few weeks and I had better not be able to see it then." Hiner quickly advised the station to back down on their transmitter power. This must have worked because he didn't hear from the Secretary again. That's strange, too, because the man was over 70 and no doubt had plenty of time to watch television.

Before being stationed in Thailand, Hiner was assigned to Chitose, Japan. Chitose is on the northernmost island of Hokkaido, far removed from Far East Network headquarters in Tokyo. This distance allowed him the latitude to put his active imagination to work without some desk jockey telling him "you can't do that." One of his first acts was to find a record which must be the most obnoxious recording of all time. It will be recalled by fans of nauseating music as "My Love" by Mrs. Miller, an elderly lady who couldn't carry a tune in a bucket and whose recordings caused music lovers everywhere to try to end their own lives. Hiner put the song on a loop of tape and announced it would play continuously until the base contributed $3,000 to the Combined Federal Campaign. It worked like a charm.

The TV station at Chitose featured the ubiquitous DAGE equipment and used small sun guns as studio lights. Executives of the Hokkaido Broadcasting Corpo-

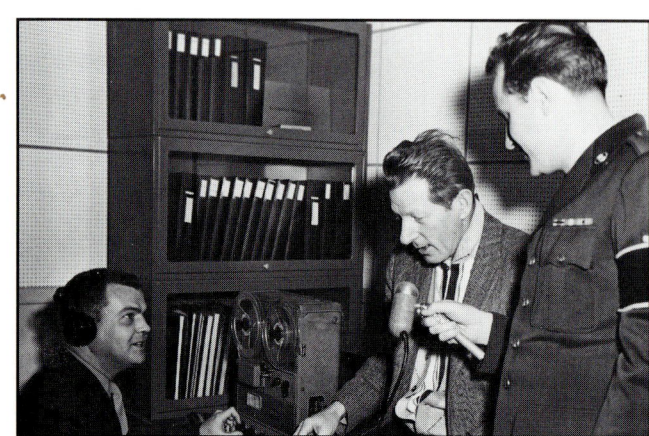

Today, Dick Cassin is a Department of Defense civilian and chief of radio instruction at the DoD Information School where his job is to learn new announcers to talk good, as in "Yestiddy I cudn't spell anouncer and tiday I are one." Back in 1962-3 he was an announcer with the Far East Network and is seen on the scene there interviewing the late Danny Kaye and Raymond Burr. The engineer helping on the Kaye interview is Marine SSgt Frank Almeda. (Photos courtesy of Dick Cassin)

ration saw it and were so struck by the Americans' poverty that they said to bring a truck by their studios in Sapporo and fill it with whatever was needed. This may be the first instance of foreign aid to America but the staff wasn't too proud to accept. Quickly Chitose was the best equipped station in the network and they were smart enough not to say a word to anyone. After enjoying their new-found wealth for a while, they were visited by Henry Yaskal, the feared chief engineer of the network, who had stood in place while the network was built around him. Yaskal bumped his head on the new Kleig lights and stumbled over the professionally designed and built news set and said to take it all back immediately. Hiner explained that this would be an insult to the Japanese but Yaskal said do it anyway. They took some back — but not enough to insult anybody.

A final note from Chitose is recalled by Marv Coyner who served in the Far East just before Hiner. Coyner tells of the rough, tough Marine Gunnery Sergeant named Augerino. Augie was a marine's marine who knew all about storming beaches but was something of a neophyte when it came to broadcasting. One day he was operating the camera when the director said over the headset, "Truck right." Nothing. "Truck right!!" Nothing. "Gunny! TRUCK RIGHT!" "Whadda ya mean," Augie whispered in his headset. "Move your camera to the right," the director said. "Okay," said Augie. So he picked it up, tripod and all, and carried it over to the right side of the studio.

Coyner also tells of another marine, this one a trainer at Far East Network headquarters in Tokyo. To teach new broadcasters breath control and articulation, he insisted that they be able to memorize and get through the following paragraph without a mistake.

Try it. You'll hate it.

Theophilis Thistle, the successful thistle sifter in sifting a sieve of unsifted thistles through the thick of his thumb thrust three thousand thistles through the thick of his thumb. Now if Theophilis Thistle, the successful thistle sifter in sifting a sieve of unsifted thistles through the thick of his thumb thrust three thousand thistles through the thick of his thumb, how many thistles did Theophilis Thistle, the successful thistle sifter have left?

Coyner is another of the AFRTS legends and is the second winner of the most prestigious of the special AFRTS honors — the Tom Lewis Award given out each year to the AFRTS member who has contributed most to the system. He is the man who started AFRTS' southernmost station. Bored as a public affairs NCO with a construction unit in McMurdo, Antarctica, he discovered that some of the technicians had hooked a tape recorder up to a small transmitter and would play music over it from time to time. He talked them out of their rig, installed it into a broom closet and began regular broadcasting. He convinced the command to apply to AFRTS for the regular radio service. It was approved and today the most isolated area on earth can not only hear radio instead of static, they now have television as well. Sometimes it is necessary to air drop the videotapes when the weather closes in but the men wintering over can follow the trials and tribulations of General Hospital, thanks to an innovative Marv Coyner.

All we know about this picture is that it was taken at FEN in Tokyo in 1954 and that this guy, whoever he is, should know better than to fool around with frigid women. Or, perhaps business was slow at the station that day and he is demonstrating that "there's snow business like slow business."

STATION BREAK NUMBER THREE

A Brief Pause for Station Tribulation

The stations in Alaska are the exception to the rule that AFRTS does not broadcast within the United States. Although the headquarters of the Alaska network are at Elmendorf Air Base, it only feeds a signal to truly isolated bases. These locations also feed the network, including a daily report of what is happening in their area. One station surprised the network by reporting: "Here's what's happening at King Salmon today..." The announcer then shut off the microphone and didn't say another word. The runner up report came from Shimya, which reported: "A lot of things are happening today. But not at Shimya."

★

Bob Cranston has probably worked in one way or another on every major project AFRTS has every been involved with up until the time of his final retirement as chief of the American Forces Information Service. One of his projects was to consolidate the various service broadcasters under that organization. It was not totally successful as the services fought to maintain autonomy over their own broadcasting outlets and each formed their own broadcast service which they maintain to this day. It really didn't matter except it put a layer of control between the top and the on-the-ground activities which could sometimes slow actions down. However, the fact AFIS controls the money flow and sets policy gives it de facto control. Cranston recalls that during the battle to obtain consolidation, the plan had to be carried around to the various sections in the Pentagon which were involved before the document went on to the Secretary of Defense for final approval. Cranston entrusted the job of hand carrying the document and getting the necessary approvals to a young colonel. He did a wonderful job, too. His name was, and still is, Colin Powell and he has since been given even more important tasks.

★

Every station, military or civilian, seems to be blessed (or cursed) by a listener who by her perhaps overheated hormones falls in love with one or all of the staff and makes life miserable by constant letters or phones calls. "Cheri," last name and address unknown, has kept up a constant stream of postcards for more than 20 years to AFN announcers, each card professing undying love.

One lady listener to AFN Nurnberg was convinced that Karl Haas, the Director of Fine Arts for WJR, Detroit, whose daily classical musical show was heard over AFRTS on transcriptions, was sending her secret messages. Letters to him at the Nurnberg station appeared two or three times a day, detailing what the poor misguided woman thought Haas (who actually was living in Detroit in total ignorance of his maniacal fan) was saying to her. Finally she became convinced that he had asked her to marry her and to meet him in the London airport. Off she flew and waited at Heathrow for a week without money. Finally the bobbies bottled up the boobie and returned her to Germany where, it was later discovered, they put her in a rubber room.

Kevin Aandahl, the only Navy man on the staff of AFRTS Athens recalls "Emmy" who has to be the most outrageous of all the poor misguided souls who fall madly in love with voices in the air. Emmy was a young teen-age Greek girl who fell in love with the entire staff, one at a time. She became the ultimate telephone pest. At one point she was logged as calling the station more than one hundred times a day. When told that the phone was needed for business, she would only answer that she had to speak to her love-of-the-moment "to plan our future." It got so bad at one point that during live broadcasts, the phones were all taken off the hook. She would still continue to dial and, when the phones were replaced, there was Emmy's call a nanosecond later.

The staff tried everything including sweet reason and, that failing, sending a stream of obscenities out over the Greek phone system to Emmy's tender ears. Nothing worked. Finally a Greek engineer at the station was able to locate her parents who admitted that, yes, Emmy was a bit of a problem but they had a solution. Why not, they asked, have one of the staff marry her? They even hinted a rather sizable dowry could be arranged. According to Aandahl, they also admitted she was terminally ugly even by Air Force standards.

The war continued. Dave Crispin of the Athens staff discovered Emmy liked cats. He produced a sound effect of a cat being flushed down a toilet and would play it every time she called. No luck. He then developed a 2,000 decibel blast and would blow out her eardrums when the phone rang. She would call back immediately after her ears stopped ringing.

The station finally called in the airbase authorities and Emmy and her parents were called in along with the base legal staff. Yup, Emmy was truly ugly. And the parents were warned that if they kept letting her near a phone, they would all go to jail. When last seen, the father was kicking Emmy down the street toward the main gate.

It solved the problem, however. There wasn't another phone call for almost an hour — just about the time it took for Emmy to get home. And they continued to come throughout Aandahl's three year tour.

★

Marv Coyner is the legendary Navy broadcaster who went on the air with his very own radio station in Antarctica. Little did he know at the time that he would one day receive AFRTS's most prestigious non-military honor, the Tom Lewis Award. He had a great acceptance speech prepared. At an AFRTS worldwide conference, all his contemporaries watching, he walked to the podium, speech in hand...and dropped it behind the stage. It turned out shorter and less prepared than planned.

On a somewhat warmer assignment later, at Guantanamo Bay, Cuba, Coyner was the first enlisted commander of the station there. He tells of a gag one of his people inflicted on the audience. Art Riccio produced announcements telling the listeners and viewers that AFRTS had chosen Guantanamo as a test station to determine whether the audience would like to see X-rated movies. The announcements ran for several weeks and the base residents, depending on their preferences, were either terribly upset or terribly impatient. The announcement told them all to tune in at Midnight, March 31 for the first showing of the X-rated movie test.

Came the big day and the base was wide awake at midnight. Lights burned everywhere as the audience heard, "The time has come. It is now 12:01, April first. Happy April Fool's Day."

★

Norm Medlin served in the Air Force throughout his military broadcast career and today is on the Public Affairs staff of the U.S. Army Europe headquarters in Heidelberg. His memories of his first assignment as a broadcaster are of Nouasseur Air Base in Morocco in 1962. The station left a lot to be desired in the way of staff and equipment. "Basic" is a word the comes to mind. Still, probably any music was preferable to the audience other than the five- tone caterwauling on the Moroccan airwaves.

Shortly after he arrived, the station began broadcasting the sounds of office noises and muffled conversations over the programming that was supposed to be going out. No one could figure out where the sounds were coming from. They went on for weeks and the staff was going berserk trying to find the source. Later, station manager Bill Miller was in the administrative office on the base and noticed a tangle of wires on a desk. A few questions later the mystery was solved. An administrative clerk had bought a build-it-yourself radio kit and after playing with it a bit, abandoned it. He also forgot to turn it off and it just kept sending out a signal which unfortunately was on the same frequency as the station.

The biggest problem was getting the latest news. Stars and Stripes took three days to get there. Shortwave reception was generally terrible. Phone feeds from the station in Madrid sounded like they were coming over a barbed wire fence. A regular newswire service couldn't be obtained because of a Moroccan regulation that civilian and military circuits could not be mixed or connected. Consequently, news broadcasts were chancy at best.

Comes now the Cuban missile crisis. Nouasseur is a Strategic Air Command base and many of the residents were convinced they were scheduled to be ground zero if a war started. Mothers were keeping the kids home from school so the family could all be vaporized together. The station became the only source of information and they didn't have much although they were able to partially follow the progress of the Soviet ships headed for the "this far and no farther" line imposed by President Kennedy.

Being well trained military personnel, the staff wasn't frightened. Not, that is, until the day the Command Post of the SAC base, charged with providing nuclear retaliation in case of war, called the station to ask them whatinhell was going on with that mess in Cuba.

★

Bill Swindle was Army and he, too, has had his share of assignments during a long career. Retired now, he recalls the worst as being with the station in Teheran, Iran, where at one time Americans were the "Great Station" and not the "Great Satan." He recalls mayonnaise being $10.00 a jar and having to soak all vegetables and fruit in Clorox. Water was trucked in for the Americans to avoid dysentery and they were not permitted to drive their own cars. They also learned, if they had daughters, to keep a baseball bat handy. Not that the girls wanted to play baseball. It was just that there was a tradition in Iran that if a man took a fancy to a young girl, he could kidnap her or rape her and local law then took over. The law said she would have to marry the man. Bill's daughter was 12 at the time and he and his wife, Mary, were offered a very nice Persian carpet for her. Needless to say, they turned down the carpet and bought a baseball bat to protect her instead.

★

Doug Frey was a Major and commander of the AFN station in Berlin when the station decided to participate in a community-day parade. The station announced that the AFN Berlin Marching Band would be in the parade. Because Berlin was still officially an occupied city, the rule was that no troops could be added to those already there without the permission of the four occupying powers. In the translation of the announcement, it came out that this was a "military band" and therefore a new unit. The Russians immediately protested.

Came parade day and the Russian observers were on hand with cameras to document this proof of American perfidy. And there came the band. All community members — young, old, military, civilian — and all carrying radios tuned to AFN Berlin which broadcast march music to which the group kept in step. A wireless circuit back to the station gave the marching orders and the station put them on the air so the marchers could hear them.

It is still not certain that the Russians understood what "right shoulder radios" means.

★

Bob Harlan without doubt is the best known of the AFN broadcasters, having been there for 35 plus years. Even a number of former network commanders give Bob credit for teaching them the intricacies of their job and, in his soft-spoken Southern way, diplomatically showing them the ropes. In researching this volume, almost without exception material from former AFN staff members mention that they learned more from Harlan than from anyone else in their careers. But as professional as he is, strange things happened to him as they do to everyone with a low voice and a microphone.

For example, he and Sergeant Mel Riddle were assigned to produce a documentary on the concept of Air Mobility of the Infantry. They drew a deuce and a half from the motor pool and on a miserable March day, took off for Giebelstadt Air Field, 190 miles from Frankfurt and so well hidden that the American Army didn't find it until the war was over. They spent the day interviewing troops, landing, taking off, putting together a word picture of this new airmobile technique.

When it was over, Harlan complained to their C-82 pilot about the long drive facing them getting home to Frankfurt. "Hey, no problem," sez he. "Just back your truck up the ramp." Thirty minutes later the dashing duo and their truck were landing in Frankfurt.

The problem was that the motor pool sergeant was adamant that he wasn't about to accept no damn cockamamie trip ticket that showed 190 miles outbound and 10 miles back. Whaddid they think, he was stupid or sumpthin'?

Very early in his career, he married his sweetheart P.B., a true belle of the South, totally charming and with a delightful Southern accent which could add syllables to single syllable words. Bob was assigned to record the official English language tapes for use in teaching English to anyone who wished to learn. It required a male and female voice for the conversations so he naturally asked his new bride to participate.

To this day it is not unusual to hear an older German who still believes the pronoun is "you all" and who sounds a little like Johnny Cash.

★

Bob Matthes was the resident genius of the spot announcement at AFN and turned out spots by the thousands during his twenty some years there. One of his favorite tricks was to grab visiting celebrities and have them cut announcements that fit in with their stage persona. It just happened that he was assigned to do a series on having the military send your paycheck to the bank rather than getting it delivered through the sometimes shaky postal system. Then into town came singer Johnny Paycheck. A natural. Paycheck was delighted to help out recorded a series of clever spots which he loved because they had him saying, "a Paycheck is much better than Cash." He couldn't wait to get home to tell Johnny Cash what he had done. Matthes had scored again. Problem was, he had to take the spot off the air.

Shortly after it began, Paycheck was sent to jail. For writing bad checks.

★

ANNOUNCER: ...but seriously folks, it's time now to get back to our story. Enough ribaldry. We move again through time and space. The time: 1961. The place: Vietnam.

FRANK BRESEE is the host of *Golden Days of Radio* which is currently the longest running program on AFRTS. Frank began his radio career as an extremely young actor and if you remember the Red Ryder radio serial, you heard Frank as "Little Beaver." A lifelong fan of radio programs, his collection of transcriptions from the golden days of radio is one of the largest, if not THE largest, in the world.

His daily program began as a specially prepared feature for AFN but quickly was added to the entire AFRTS lineup of programming sent out worldwide. Each weekday it presents the kind of programs no longer heard anywhere else in contemporary radio — a chance to hear, from those golden years, the likes of Bob Hope, Jack Benny, Red Skelton, Baby Snooks, Eddie Cantor, Groucho Marx, The Great Gildersleeve, Edgar Bergen and Charlie McCarthy, Abbott and Costello as well as drama from series such as "Sam Spade," "The Whistler," "Lux Radio Theater" and dozens of others.

A former president of The Pacific Pioneers Broadcasters, Frank is perhaps the prime mover in keeping interest in nostalgic radio alive and well.

(Editor's Note: Readers who were overly enthusiastic about studying a picture of Mr. Bresee should be aware that he is the person to the rear in the accompanying picture. Those who failed to notice there is a second person pictured should know that she is his lovely wife, Bobbie.) (Harry Langdon Photography)

AFRTS managers would, like television managers everywhere, often shrug off criticism of their children's programming by remarking that parents only wanted TV to act as "an electronic baby sitter.,"

This picture, taken in Ramstein AFB base housing proves that this simply isn't true. Children are obviously interested in the station identification slides.

Chances are also excellent that these children — who are now in their thirties — spend considerable time writing letters to their local television stations complaining about children's programming. (Photo Courtesy of AFRTS)

Right, two views. After a series of totally inadequate homes, the Saigon headquarters station—which had been a hotel and a BOQ—moved into this slightly more adequate facility. It later moved into an even better one, but at this time this seemed like the height of luxury. Below, Maybe it wasn't exactly the "Tonight Show", but this is an example of the kind of programming the staff of the Saigon headqaurters station of AFVN produced for their troop audience. (Photos courtesy of Roger Maynard)

CHAPTER EIGHTEEN

In which the country ignores Joseph Heller's comment that "I'd like to see the government get out of war altogether and leave the whole field to private industry." AFRTS goes to Vietnam and gives a whole new meaning to the term "On the Air."

WHAT LIGHT...WHAT TUNNEL?

Hundreds of thousands of young men and women had no idea John F. Kennedy was talking to them when he said in his inaugural speech in January 1961: "Ask not what your country can do for you; ask what you can do for your country." It turned out that what they could do was go fight in a country many of them had never heard of. At that time the U. S. military mission to Vietnam was limited by law to 685 people. A year later it was 2,646 and six months after that, 5,576.

As their older brothers or fathers had before them, they began building their own radio stations back in the boondocks and in Saigon. The Saigon station operated from the bachelor enlisted quarters and the men who built it required the nerves of a burglar. In truth, it required them to be burglars as the equipment was informally requisitioned. Translated, that means the supply depot wrote it off as "evaporated" and the new owners explained it to inspectors as "found on post."

The North Vietnamese government also recognized the need for radio among the Americans and in July, 1962, began broadcasting American music heavily laced with propaganda. Forty days later, American Forces Radio, Saigon, went on the air.

General Paul D. Harkins, Commander of the Military Assistance Command, Vietnam, (MACV), had recognized the need and had received permission from the South Vietnamese government to use 820 KC in the Saigon area and received four other frequencies for use elsewhere in the country. AFRS in the Philippines donated a World War II tactical transmitter and it was installed at the Vietnamese Radio Communications site at Phu Tho.

Harkins signed the station on and in his opening remarks said, "Today many American servicemen are again far away from their homes and families in many locations throughout the world. The need for Armed Forces Radio, therefore, continues and is perhaps even more important in light of the complexities of today's world."

One of those complexities was getting equipment. Most of it came from military sources. Operating personnel were also hard to come by. Initially the staff was five official full-time members and several part-time volunteers. Somehow they managed to keep the station operating 18 hours a day, seven days a week. News was received — atmospheric conditions permitting — by shortwave from AFRTS leased transmitters in California. The quality varied from reasonably awful to awfully awful. It improved somewhat after the AFRTS station in the Philippines arranged to rebroadcast several hours of news each day using a Voice of America shortwave transmitter in the Philippines.

Bob Andresen was a member of that early-day staff. The studios were then in the Rex Hotel in Saigon and he clearly remembers a gentleman in an Hawaiian shirt walking into the newsroom one day, flashing some impressive credentials, and leaving a piece of copy declaring that the city was under martial law by order of President Ngo Dien Diem. The station stayed on the air all night, constantly repeating the message. The next day rebel troops blasted their way into the Presidential Palace two blocks away and within moments Diem and his brother were dead. The station, to avoid panicking the public, was forbidden to broadcast news of the assassination or the battle raging in the streets outside their windows although the sound of gunfire could be heard clearly every time they opened their microphones.

Part of the Saigon station stayed in the Rex Hotel but in August of 1964 the studios and administrative sections were moved to the Brink BOQ which gave it more space. At the same time the quality, the strength and the morale of the station simultaneously improved as the antique World War II transmitter was replaced by a brand new Bauer transmitter. It was from here that Andresen had to call the MACV Information Officer in the middle of the night, November 22, to advise that President Kennedy had been shot in Dallas. Both men were so shaken that it didn't occur to them that this transcended normal broadcast procedures. As a result, PFC Lee Hansen, the morning DJ, continued to rock 'n' roll as normal while Andresen kept breaking in with last minute bulletins on the events in Dallas. It was perhaps not their finest performance but certainly typifies the confusion everywhere in those first terrible hours.

American troops continued to pour into Vietnam and to provide service it became necessary to increase coverage in outlying areas. The State Department provided twenty "Provincial Radio Station Kits" which had been intended for the South Vietnamese government. In exchange they gave the Vietnamese a powerful transmitter and the small 50 watt stations were installed throughout the South to cover most troop concentrations. They were better than nothing, but not much. A team from Washington made an inspection tour and in their report noted that "effectiveness of AFRT radio broadcasts to personnel stationed outside the Saigon coverage area is greatly reduced by technical deficiencies of the equipment available."

Before much could be done to improve the situation, the Saigon station suffered a technical deficiency in the form of a Viet Cong attack. On Christmas Eve, in the middle of a holiday program, the VC set off 250 pounds of plastic explosive at the Brink BOQ. Two died, scores were wounded, the hotel was badly damaged and the station was in ruins. An auxiliary unit was able to restore service quickly to the troops which, by now, totaled 23,000.

Studies continued to be made, including one by the Military Assistance Command, and another by a team from the States headed by the ubiquitous Colonel Robert Cranston who, along with engineer John Scales, of AFRTS Los Angeles, prepared the definitive study. This led to the eventual establishment of high powered stations covering larger areas from Da Nang, Qui Nhon, Pleiku and Cam Ranh Bay. Later additional stations were added. Programming expanded from 18 to 24-hours a day and FM service went on line in Saigon during the afternoon and evening hours.

With radio now a permanent part of life in Vietnam, it was time to move on and bring in television — both on the air and in the air. While plans were being negotiated between the U. S. and Vietnamese governments for permits and frequencies for ground facilities, the Joint Chiefs of Staff gave the assignment to the Navy Oceanographic Air Survey Unit at Patuxent River, Maryland, to develop flying television stations. This had been tried before and had failed. The so-called "Indiana Project" had equipped a battered old C-54 with a couple of videotape playback units and a transmitter and sent it up over a three state area with the idea of feeding educational material to the schools below. It can only be described as moderately successful technically. No one ever did figure out how to keep the antennas from burning up and the project was eventually scrapped.

The Navy used C-121 aircraft, the military designation for the civilian Super Constellation which was a propeller-driven aircraft with its twin tails attached to the fuselage by an extremely narrow portion of the plane's body. The flying stations were dubbed "Blue Eagles" and Blue Eagle One was equipped for radio transmissions only. It arrived in time to broadcast live coverage of the 1965 World Series. The Navy in the meantime continued construction and testing of Blue Eagles Two and Three. They contained two television transmitters of 200 watts each, a 10,000 watt AM radio transmitter, a 1,000 watt FM transmitter, single sideband shortwave with a four channel teletype hook up as well as two video tape recorders, six audio tape recorders, two 16mm film projectors and a small, two person, live studio with a remote controlled camera.

The antenna problem was solved and as Blue Eagles two and three were delivered, work began on four and five which were to follow shortly. The purpose of the two television transmitters was that AFRTS programming, including locally produced programs, would be transmitted to the troops on Channel 11 and Vietnamese programs, on Channel 9, would be sent down to the Vietnamese citizens below as a contribution to U. S. objectives including rural pacification, urban stability, a show of free-world support and a strong show of the U. S. presence in Vietnam. The latter programs were totally separate from the AFRTS output which, since its formation, made every effort to maintain absolute freedom from propaganda programming in order to preserve audience credibility.

Broadcasting from aircraft was an entirely new concept in AFRTS operations. Being on the air was one thing. Being in it was quite another. One of those who spent many hours there is former Air Force Sergeant Roger Maynard. Maynard, who keeps popping up in unlikely places throughout this narrative and who ended his AFRTS career as a civilian deputy to the Chief of Navy Broadcasting, tells what it was like running a television station with 10,000 feet of open air beneath your chair:

I arrived at Tan Son Nhut airport in Saigon several weeks after the start of the Blue Eagle operation and was met by Sergeant Major Pat McCusker, Air Force Master Sergeant Shelly Blunt and Staff Sergeant Gene Leroy. After I reported in, they showed me around and pointed out Blue Eagle Two which was sitting there with its tail section missing the result of a recent Viet Cong mortar attack.

I was scheduled to fly every other night, but I frequently had back-to-back missions. AFRTS people did not get flying pay although, much later, we did receive hazardous duty pay. We could not even get blood chits, the multi-language messages on waterproof fabric that said a ransom would be paid if we were returned to our units.

A normal duty day went like this: we load up the boxes of 16mm film for the night and then visit the Saigon Hollywood district. Using local facilities, we produce on 2-inch video tape spot announcements, video news intros, news film clips and all the other video we need for the schedule. We also load up the USIA film for the night. Then we stop by the UPI and AP offices and pick up any new news slides or late news releases when available. Then we drive our stolen Weapons Carrier out to Tan Son Nhut airport.

Even then the C-121 was considered an old airplane. Although well maintained and flown by excellent Navy pilots, we had our heart-stopping moments. The procedure was that two Blue Eagles would be rolled out. If anything caused the first aircraft to abort the mission, if it could still taxi it would move close to the second. The AFRTS person on board would slide down a 20-foot rope to the ground, then the crew would lower the programming boxes of film and tape and it would be transferred to the second Blue Eagle for immediate take-off. This happened more than a dozen times in my case.

Take-off was the most hazardous and most vulnerable moment. Because of our weight our climb was extremely slow and placed us directly over unsecured territory. Landings could be bad as well. One night a 50-calibre round came up through the seat of the chair next to me which happened to be the chair I normally sat in.

At takeoff, the planes were already 5,000 pounds

overweight; we were warned that the first hour was critical. If we lost an engine, we would go down with no chance to dump our fuel or get rid of the heavy equipment which included a very heavy generator back in the thin, weak, tail section.

Once airborne, we cranked the antenna down, played the sign-on tape and began the live news from the studio. Unlike Dan Rather, we had to wear seat belts because the frequent downdrafts caused our scripts to lift off the desk and do strange things to our stomachs. This rapid change in altitude caused the air pressure to change. Those who remember the old 2-inch Ampex recorders will recall that the record/playback head was a spinning disk. The tape passing it was held by vacuum in a semi-circular shape so the edge of the spinning circular heads could match the full width of the tape. When the pressure changed, the vacuum would fail to work and frequently the spinning head would slice the tape into two neat pieces.

Other moments recalled vividly was the eerie feeling of flying over the Delta region and seeing fire-fights going on below; flares lighting up large areas while our Dragon Ships strafed whatever moved. And it really got our attention when flights of F-4Ds dived past us on their way to rocket attacks below. Most unnerving of all was being called from Paris Control, the Tan Son Nhut tower, to tell us to please move to another area because the Army was presently lobbing 105mm artillery through our position. Twice we lost an engine, neither time could we maintain altitude and once there was also a lot of smoke coming into the fuselage from the wing section. Both times we followed the Saigon River back to base in case we had to ditch but both times we were able to make it with harder than usual landings.

While still continuing to fly occasional Blue Eagle missions, Maynard began working with the planners and designers for the ground-based station being built in Saigon. The antenna tower was to become the tallest man-made structure in Vietnam. The guarded compound was to have its own power supplies and in many ways be self sufficient. The project was a pet of the Commander in Chief, General William Westmoreland, who would frequently come by to check on progress.

A more frequent visitor was a drunken one-star general who kept demanding entrance toward the final part of the construction so he could come in and watch tapes of AFRTS shows. On sign-on night the word was passed that there were to be no visitors.

Just before sign-on came the knock on the door. Whoops! It was General Westmoreland. Come in, sir. Obviously we didn't mean YOU! He promised to stay out of everybody's way and did until three minutes before sign-on. Another knock on the door. Seeing everyone was extremely busy, Westmoreland answered it and there was the drunken one-star General. The staff could overhear the conversation very well and learned what a genuine four-star butt-chewing sounded like. It isn't anything a person would like to receive.

Early members of the staff included the commander, Army LTC DeForest Ballou III, executive officer Navy Lieutenant Commander Cleveland and the senior enlisted man, Master Chief Ed Halley. Others, including

One of five "Blue Eagles" sits at Ton Son Nhut airport, waiting for nightfall to start bombing the peaceful Vietnam countryside with news sports and "Gilligan's Island." Built and flown by the Navy, operated by broadcasters from AFRTS(AFVN) from all services, these Super Constellations (C-121s) flew every night transmitting both American and Vietnamese programs. Some nights the crew could watch firefights going on beneath them while those beneath them were watching programs coming down from the Blue Eagle overhead. They took direct hits on a number of occasions and mortar fire on the ground put one out of action for some time. (Photo courtesy of Roger Maynard)

a number sent down from the Far East Network, included Gunnery Sergeant Eddie Stein, SSgt Ron Graebert and SP4 John T. Mikesch, an outstanding TV director who had formerly been president of the Alaskan Television Network. Present for a short time, during which his father visited him, was John Steinbeck IV.

While their counterparts in the States waited for the next rating book find out what their audience thought of them, the Saigon staff had quicker and more accurate indications of their popularity. One morning the news director found on his desk a bullet which had come through the roof. Windows shattered by bullet holes were constantly being repaired and, later, the station was severely damaged when a car bomb went off.

Using the more primitive vans of World War II and Korea as models, trailer studios were constructed under the supervision of T-ASA, the Television-Audio Support Agency, at the Sacramento Army Depot. This is the organization charged with procuring, testing and shipping of all the non-military standard broadcast equipment required by AFRTS stations world-wide. It also supplies engineering assistance of all types to AFRTS including design and, in some cases, installation. The vans T-ASA came up with included a complete studio and associated transmitting equipment, a 5KW transmitter, two film chains, a slide projector and multiplexer, a complete audio and video console and both studio and mobile cameras. Its tower could be raised 120 feet and the unit contained two trailer mounted 45-kilowatt generators. TV programming continued to be furnished by AFRTS and by the headquarters station in Saigon. Locally produced programming was fairly simple but enthusiastic. Radio programming improved tremendously as the network was connected, in the spring of 1966, to the underwater cable which linked it to AFRTS Los Angeles. Now it was possible to hear major events live without the bothersome fading and squawking of the shortwave signal.

The first television vans arrived and, under the command of Captain Willis Haas, began to be put in place. The first went in on Vung Chua Mountain in Qui Nhon province. General Westmoreland cut the videotape ribbon. Haas noted that the local Post Exchange had already sold 1,000 television sets in anticipation of the event. Don't even try to figure out why one of the most popular programs was *Combat*.

The second van was installed on Monkey Mountain in Da Nang. As American ground stations continued to proliferate in the north, the need for U.S. television from the Blue Eagles became less. A mortar attack at their base at Tan Son Nhut had put several out of action for an extended period but the remaining birds continued to transmit Vietnamese programming to the southern part of the country.

The Saigon station gang got a Christmas greeting from the Viet Cong as bullets whistled through their Christmas party but work continued on placing the vans in strategic locations. The third van went on Dragon Mountain in Pleiku in February. Van four was put on Non Tre Island near Nha Trang and served both Nha Trang and Cam Ranh Bay. Van five was installed in Hue and number six went to Tuy Hoa. All were operational by May. The final van was kept in Saigon as a spare and used to train new replacements who were to be assigned to the field. There were now a total of 22 transmitters throughout the country but there were still areas which did not receive an adequate signal. To solve this problem, five additional transmitters, three of them radiating 50,000 watts, were installed. The 50KW stations were at Cat Lo, serving Saigon and the Delta area, Pleiku and Cam Ranh Bay. As they became operational on June 1, 1967, AFRTS Vietnam became an actual network and adopted the name "American Forces Vietnam Network"—AFVN. At this point, almost 100 percent of the troops could hear radio and, by October when television became fully operational, at least 85 percent of the troops could watch.

Of course, there were times when they were too busy. The AFKN staff had its busy moments as well. Da Nang's transmitter site received mortar fragments and considerable damage which was quickly repaired. The Nha Trang television site came under attack and the mess hall and adjacent lounge were destroyed.

Seven broadcast vans arrived in Vietnam and were installed under the direction of Captain Willis Haas. They were much more sophisticated than the World War II and Korea vans. Each contained a complete studio and transmitting equipment, film chains, slide projector, multiplexer and a 5KW transmitter powered by two 45 kilowatt generators. The antenna was telescopic and could be raised 125 feet. Shown here are two views of van number three which Haas installed on top of Dragon Mountain at Pleiku. As the vans and other more permanent facilities throughout the country began to be linked together, radio and television operations became a network in fact and became AFVN—American Forces Vietnam Network. (Photos courtesy of Roger Maynard)

Then, in 1968, came the Tet Offensive. Saigon was well prepared and had created two complete staffs so it could operate even if the station suffered heavy casualties. On January 31, the station atop Vung Chua Mountain came under heavy enemy fire but no casualties were sustained. That night, North Vietnamese regular army troops, who were in control of portions of Hue, attacked the AFVN station there with small arms and light mortar fire.

It started at 2:30 in the morning. Army Sergeant John Anderson and his staff were awakened by explosions. Moments later, bullets were smashing through the barracks windows and they discovered the area was completely surrounded. Until this moment, Anderson had not considered he and his men were in any danger. The worst that had happened were "a couple of mortar attacks." They felt their area was secure. Besides, Anderson said later — much later — he only had a month to go on his tour and six months to retirement. What could happen?

Plenty, as it turned out. Although there were no injuries in the initial attack, the attack continued for the next five days. A marine sergeant was killed. Every man had been wounded at least once. When their ammunition was used except for about 100 rounds, with water gone, their ability to resist was gone. The remaining AFRTS staff decided their best hope was to try to reach a friendly unit stationed about a mile away. They made a dash for it — and ran directly into a group of North Vietnamese Army regulars and Viet Cong.

Their next five years were spent in a North Vietnam prison camp suffering all the horrors that entails. Anderson's health was never the same after release. He went through open heart surgery while, now a civilian, he served as the AFRTS station manager in Stuttgart, Germany. Television viewers with long memories will recall the dramatic moment when the released prisoners were coming down the aircraft ramp as their loved ones cheered. One dropped to his knees and kissed the ground.

That was John Anderson. He, at last, had found the light at the end of his tunnel.

When the Tet offensive was finally contained, AFVN repaired its damage and, as the years passed, became very much like its sister networks. Equipment continued to be improved. Staffing grew and the programming became ever more sophisticated. Special events teams roamed the country covering every kind of event from Bob Hope troop shows to a midnight mass by Archbishop Terrance Cooke. An agreement was reached with CBS to receive the Walter Cronkeit evening news each day and the audience's favorite news story was when President Nixon announced he would reduce troop strength in Vietnam by 25,000.

At its peak, the network, headquartered in Saigon, had seven detatchments capable of live broadcasting scattered across the country. Five broadcast both radio and television. Two were purely TV.

ELSEWHERE IN SOUTHEAST ASIA

AFKN had a sister network known as AFTN — the American Forces Thailand Network which was essentially an Air Force Operation operating from air bases throughout Thailand. They were manned, and womaned, outlets in such today forgotten locations as Nakhon Phanom (generally referred to as NKP), Takali, Ubon, Udorn and U-Tapao. The headquarters of the network was in Korat. There were also repeater stations elsewhere in Thailand, formerly known as Siam. The pronunciation of many was perfectly explained by Yul Brynner in his portrayal of the King of Siam in *The King and I* as "It's a puzzlement."

Thailand is one of the loveliest lands on earth. It is difficult to find a former staff member of AFRTS who does not still harbor happy memories of this soft and perfumed country. The memories they recall — except for one — are happy. Although only minutes as the jet flies from the war in Vietnam, the men and women in Thailand were able to enjoy their stay in this fabled place. Only the planes taking off at regular intervals for the war front intruded on their total enjoyment of life here in Lotus Land.

Air Force Sergeant Denny McKell was there. He continued his career in the Air Force, served with the 7122nd Broadcast Squadron in Germany, civilianized, worked with the Squadron to set up the production facility at Ramstein Air Base there and today is the program chief at the Southern European Network in Italy. He is also, as his name would imply, a wild Irishman. In Thailand, he served as the sports anchor on the evening news. His partner at the news desk was Dave Nuttall who, as we shall see, was later punished by the air force for a slight transgression by being promoted to Captain. Nuttall recalls turning the desk over to McKell one night to do the sports. As he stepped out of the picture, he released the station mascot — a friendly monkey. Because the studio was air conditioned for the sake of the equipment — never of the staff — the chilled monkey suddenly appeared on camera and disappeared down the front of McKell's shirt. McKell never missed a box score but the audience must have been fascinated as his shirt front, like Dolly Parton doing aerobics, took on a life of its own. So interesting was it that it became a nightly ritual.

MSGT Skip Manke stands in front of the entrance to Armed Forces Thailand Network at Udorn, Thailand, prior to the tragedy which was soon to overtake it. Udorn was one of the primary stations of the network which was headquartered at Korat under the command of LTC Dave La Follette. (Photo courtesy of John Bradley)

The Deputy Chief, and later Chief, of AFTN was Dick Hiner, last seen here working in Chitose, Japan and next to be heard of as Chief of AFRTS for the Secretary of the Air Force Office of Information and, later still, to became Director of Navy Broadcasting.

Shocking as it may be, Hiner points out that when men are sent to a far-away land such as Thailand, they do not necessarily leave their sex drive at home. He explains his memory of the way this truism worked out in real life:

When you arrive in Thailand, you receive a lecture from a doctor about VD, in which he emphasizes that masturbation will not make you crazy, blind or grow hair on your palms. Then you receive a "Health Coupon Book," in case you don't believe him. If you are in a bar downtown and are particularly enamored with a member of the opposite sex from the host country, you may escort the young lady out for an an agreed upon sum of baht. And, oh yes, you must leave a numbered coupon from your health book with the bar's proprietor. If after the required weekly health examination conducted upon these young ladies, one is found to have a case of cupid's exema, all the coupons deposited in her box are sent to the base medical authorities. These numbers were then broadcast over AFTN and published in the Daily Bulletin.

One afternoon, our DJ in Korat was reading the numbers over the air when he read his own. He was astonished. After practically choking to death, there was a short silence. Then he nervously announced that we were going to hear "three songs in a row." He dashed across the street to the clinic and was back before the third song finished.

However, he did stand up for the remainder of his shift.

On April 10, 1970, the station manager at Udorn, Master Sergeant Jack Lynch, was having another pain in his shoulder and thought a former case of lumbago might be coming back. He decided to run over to the dispensary to have it checked out. There, he heard an unmistakable noise. Looking back toward the station, he could see a column of ominous black smoke. He ran back. He recalls now: "I went around the corner and there was just nothing there. The station was enveloped in smoke and flame and it looked like the whole thing was destroyed."

It was. A crippled Phantom F-4 jet returning from a mission over Vietnam had plowed through a nearby building and crashed directly into the station. Truly into it. The nose was inside the back door and the tail was

On April 10, 1970, an F-4 Phantom Jet took enemy fire over Vietnam and tried to make it back to its base at Udorn, Thailand. Unable to make the runway, both crew members ejected safely and the crippled aircraft plunged into the AFRTS (AFTN) station on the base.

Nine members of the staff died in the flaming inferno.

The F-4 plowed directly through the studio building seen here with smoke from the resulting fire rising into the air. The control equipment for the station was located in a mobile van, the rear of which is to the left. A small portion of the aircraft is seen in the foreground.)

The control center of the station, in the van, is seen here with flames still rising,

Flames have now engulfed both the van, foreground, and the roof of the studio/office building behind it. Firefighting teams were on the scene in minutes but damage was too extensive to control.

Several days later, this is all that remains of the once proud Udorn station. (Photos courtesy of U.S. Air Force)

inside the front door. The body of the aircraft was directly across Lynch's desk.

Death's throw of the dice came up lucky for three of the other staff members as well. Sergeant Gary Sumrall normally worked the daytime shift from eight a.m. until five p.m. On this day, his boss had asked whether he would take the swing shift to cover for another staff member who was leaving on temporary duty. It made no difference to Sumrall. At 2 p.m. he was lying on his bunk listening to the radio when he heard the F-4 pass overhead. Suddenly the radio went off the air just as the Barbara Randolf show started at 2:05. Sumrall looked out and saw the rising cloud of black smoke and on his way to the station, found one of the two pilots who had ejected seconds before the impact. Both survived.

Ray Hacket survived as well. He left the station at 2 p.m. to pick up the mail. He saw the Phantom jet in his rear view mirror as it plowed into the building he had just left.

Program director Larry Salwell normally arrived at 2 p.m. to pick up his mail. On this day he overslept a few moments and was walking toward the station several minutes later than usual when the tragedy happened.

Of the thirteen man staff, only four were left. At times like these, it is difficult to know exactly what to do. Lynch notified his headquarters in Korat and found that Dick Hiner was the ranking officer. The Commander, Lieutenant Colonel Dave LaFollette, was back in the States on leave. CMSGT John Ushold was the ranking NCO. He and Hiner, began the awful task of notifying the chain of command, planning a memorial service and helping notify next of kin. They also began planning to restore Udorn to service as quickly as possible, knowing that a rapid restoration of service was essential to the mental health of the survivors who grieved over the tragedy which wiped out the lives of nine of their fellow workers and close friends.

IN MEMORIAM
TSgt Frank Ryan
TSgt Roy Walker
TSgt Jack Hawley
SSgt Jim Howard
SSgt Al Potter
SSgt Ed Strain
Sgt John Rose
A/1C Andress McCartney
A/1C Tom Waterman

Within twenty-four hours, Hiner had brought the survivors down to Korat, given them their own radio studio, arranged for land lines to Udorn and put them on the air serving the Udorn base from Korat. Two mobile homes were borrowed from the Chase Manhattan Bank which had planned to use them to set up a branch. They now became the new Udorn station. Equipment, most of which was totally destroyed, was gathered from throughout the AFRTS system. TSgt T. J. Davis, a programmer at Korat, pulled off a minor miracle when he produced a C-47 aircraft, complete with crew, ready to ferry replacement equipment to Udorn. When Hiner asked him where he got it, he merely replied, "Don't ask."

Because it had no markings of any kind, Hiner suspects that Air America was without an aircraft for awhile.

The Udorn station was back on the air with radio being fed from Korat within 30 hours. It took about 30 days to begin live broadcasts using the new equipment furnished from other stations. And soon after that, television returned to the air, although with limited capability.

To this day, the tragedy of Udorn remains an example of the AFRTS spirit which, that April day, became quite real.

BACK IN VIETNAM...

By 1971, the war was groaning to a halt. United States forces had begun massive withdrawals and as the troops left, stations began closing. Tuy Hoa was first, followed by Can Tho. Service continued from Saigon to the remaining stations but the equipment from those closed was sent to other AFRTS outlets or returned to T-ASA in Sacramento. As the drawdown of troops intensified, broadcast detatchments were relocated or reorganized. Nature seemed to be on the side of the enemy as well. In October Typhoon Hester knocked down both the FM and TV towers at Da Nang and it was days before service could be restored.

Staff cuts continued into 1972, as did station closings. Hue left the air in February. So did Qui Nhon. The government of South Vietnam was the recipient of their equipment. In April Cam Ranh Bay went off the air for good and the staff was moved to Nha Trang. The entire AFRTS group in Vietnam continued to try to provide the kind of service of which they had become so proud. They and their predecessors had built a real network...a network of which any area could be proud. The words of Secretary of Defense Melvin Laird on the occasion of the network's tenth anniversary were appreciated but provided scant comfort:

I congratulate all personnel, past and present, of the American Forces Vietnam Network as the Network marks its tenth anniversary. The dedicated efforts of these hundreds of skilled men and women to provide current, comprehensive and accurate information are well known and appreciated by all who have served in the Republic of Vietnam. These efforts have been recognized at all command levels as truly professional. Now, as we continue to decrease American involvement in this area of the world, the work of the American Forces Vietnam Network is no less important. For those thousands who felt a little closer to home because you were there, I thank you for a job well done.

Nice, but the end was in sight. There, see it? The light at the end of the tunnel? That's home for the draftees, a new assignment for regulars. The equipment was being packed and shipped off or given to the Vietnamese. Enough was left to leave an automated FM station on the air, managed by a few Department of Defense civilians, but AFVN was no more.

Colonel Felix Casipit, the last commander, sent a final message to headquarters in Washington on March 23, 1973. The final line read:

AFVN ceased to be as of 2400 hours 22 March 73.

John Broger, long time head of the agency controlling AFRTS and the central figure in a long line of morale shattering flaps with those he supervised and the media, is prepared to be interviewed on the American Forces Korea Network in Seoul in 1973. Helping him with the microphone is Sergeant Major Clay Lacy. Standing by are SSgt Dave Stewart and network commander, LTC Howard Myrick. (Photo courtesy of Dave Stewart)

CHAPTER NINETEEN

In which we learn that some commanders agree with Elvis Presley who said,"I don't know anything about music. In my line, you don't have to." Many, we find, are critics like the one described by Kenneth Tynan as "a person who knows the way, but doesn't know how to drive."

EXTRA CENSORY PERCEPTIONS

AFRTS managers, wherever they operate, face constant pressure from the local commanders from whose bases or areas they broadcast. Small stations take direct hits from a base commander. Larger stations or networks, serving wide geographic areas, get hit by bigger artillery including multi-starred generals. Almost invariably, the flak is fired by the local public affairs officer who more often than not considers the AFRTS outlet as his or her personal public address system and feels, as the duly appointed spokesman for the command, that he speaks for his boss, the commander.

The situation becomes a tricky one for the AFRTS management which in most cases has considerably more experience in the broadcast arena than the Public Affairs Officer — the PAO — or his boss. Unless the situation between the two deteriorates beyond all redemption, the AFRTS person in almost every case attempts to settle any dispute without escalating it to a higher headquarters. After all, the station is almost always dependent on the local command for such amenities as water, electricity, a building, toilet paper and light bulbs.

The disputes run from honest disagreements over good taste to serious attempts at censorship to differing interpretations of what constitutes a possible security violation.

In most cases, the AFRTS manager is on the side of the angels in that Department of Defence policies on censorship are very clear: unless it could endanger life and limb or unless it is completely offensive to the host country laws or customs, it can be run. The basic rule of thumb is that a service person overseas has the same rights of citizenship as those at home and that applies to what he or she can hear on radio or television. It is a proud heritage and one that causes both amazement and envy from the foreign broadcasters, almost all of whom work for government-controlled stations. It passes all credence when they see and hear an American station — and one owned by the U.S. Government — broadcasting commentary of all shades of opinion and often blasting its own leaders. That AFRTS would play anti-war songs during the Viet Nam era ranked as incredible. And when it carried the entire endless Watergate hearings, it ranked right up there on the believability scale with Bigfoot and Flying Saucers.

This freedom has given AFRTS credibility that many foreign broadcasters would kill for and because it has gained it, often painfully, station and network managers tend to put careers at risk to fight off those who — often with the best of intentions — would dilute it. Although by its agreement with host countries, AFRTS does not address the local population, it would be absurd to suggest that the locals who speak English or merely like American music don't look and listen. Foreign newspapers frequently quote AFRTS newscasts and during crisis situations such as the Kennedy assassination, the Cuban missile showdown or similar tense moments, AFRTS is the first place many local people turn, knowing that they will be getting uncensored news and getting it fast.

Public Affairs officers know this, and the overzealous ones frequently try to use AFRTS' hard-won credibility for their own purposes.

Sometimes they win. Almost always they outrank the station management by several grades. Usually the commander on whose staff they serve backs them up, right or wrong. In some cases, they write the efficiency reports for the AFRTS person and can in this way influence the person's future career in the service. Sometimes they win because they are right.

Sometimes they lose. Often the AFRTS manager can reason with them. Sometimes the local commander agrees with the stand taken by the station or network. Other times the flap escalates to a higher command which rules against the local PAO or commander. And once in awhile the AFRTS manager ignores orders and does what he or she feels must be done. Very often the manager wins.

Sometimes they lose and become a twenty-years-in-grade Second Lieutenant in charge of issuing fishing licenses in the Gobi Desert.

SOME WINNERS AND LOSERS

Al Edick at Southern European Broadcasters in

Italy was both a winner and a loser in 1961. December 7 was approaching and he wanted to do a program remembering the 20th anniversary of Pearl Harbor. The staff worked on the show for days when the local Public Affairs Officer dropped in. He said he had heard about the project and wanted it stopped. Now! He announced there would be no recognition of the date. He said the Japanese were our allies and forget the whole thing.

Edick won half a point. The PAO agreed that the station could do a program — but only if it did not use the word "Japanese" or the phrase "Day of Infamy." The program went on and it referred only to "the enemy." It turned out to be an excellent program and Edick was a partial winner. So was the PAO, Major Alfred Yamazaki.

Each year, American embassies in countries with AFRTS outlets are tasked to provide a list of "Host Country Sensitivities" which are provided in a Confidential document to stations, networks and the AFRTS broadcast center. Depending on the country and America's relationship with it at the time, the list can run from dozens of items to a simple, "Avoid insulting the host government." It is universally thought by AFRTS managers that the lowest and most stupid underling at the Embassy is charged with providing this information. It is also widely held that the reason for the document being classified is that no one wants it to see public scrutiny due to the incredibly ridiculous restrictions listed therein. Like laws, these restrictions, once inserted into the document, are seldom changed or removed even though they are no longer valid due to changed circumstances. In honesty, the topmost management of AFRTS has tried to get the State Department to relax and recently some of the older and more trivial restrictions, if not dropped, have been ignored. For example, for years AFRTS stations in Iceland were told not to broadcast programs which would encourage dog ownership among the population. Editors of the television package going out to stations had to notify Iceland every time they sent out a *Lassie* Movie. This nonsense went on for years until a program executive at AFRTS Los Angeles noticed a picture of the Prime Minister of Iceland greeting some dignitaries with his faithful dog beside him. The restriction was quietly dropped and so far the guy in the Reykjavik embassy hasn't noticed. Other restrictions continue to be received and stations notified when they pop up in the programming. As often as not, however, the station will opt to ignore the whole thing on the simple theory that it's easier to apologize than call the embassy and get permission to run it. This technique does not apply to those items which are of genuine concern to the host nation and these are most often of a religious or cultural nature. Christmas carols are not popular in Moslem countries. Lutheran hymns don't go well in Catholic Spain. Try not to say anything nice about Turks when broadcasting in Greece or about Greeks when broadcasting in Turkey.

Not all restrictions were to be taken lightly. In the early days of the station in the Azores, governed by and a part of Portugal, the government actually was one man — Dr. Antonio Olivera Salazar. For years he ruled Portugal with an iron hand and his secret police were everywhere. When the State Department said, "Don't criticize the government or anything Portuguese," they meant it. More than one American who did so, found him or herself on the next plane or boat out. The station exercised extreme caution when talking about the host country.

The most effective managers have found the best way to handle these problems is to learn as much about the country as possible, and then act like a well-mannered guest. Roger Maynard who managed stations in some of the really tough places such as Turkey and the Far East did just that when he arrived to take over the station in Iwakuni, Japan, which was about to celebrate one of its anniversaries. Part of the promotion was a new slogan: "FEN Iwakuni is going Radio Active." That's an okay slogan in Oshkosh, perhaps, but not in Iwakuni. The nearest off-base audience is in Hiroshima.

He then turned right around and got into trouble for playing Frankie Laine's version of "On the Road to Mandalay," the lyrics of which were written by Rudyard Kipling and contain the line "by the great God Buddha." According to the State Department No-No list, references to Buddha are simply not done, dontjaknow.

AFRTS through the years has considered censoring music before it is released to the overseas outlets as serious as censoring news at the source. Quite rightly, it feels that it is the prerogative of the people on the scene to make the determination as to whether the material is appropriate. The general rule is that if the service person could hear it on the radio or television at home, he or she should have the same privilege while serving overseas. This concept has never met with wild enthusiasm from local overseas commanders, many of whom are both exceptionally conservative and totally out of step with the musical tastes of their troops. Almost every song that reaches the music charts in the States ends up overseas, either as a part of the weekly shipment of music or included in one of the long-form DJ shows which make up the bulk of the programming and from which stations can extract individual songs not included in the music package. Songs which are patently offensive or totally inappropriate are often included in the shipments but the radio section of AFRTS in Los Angeles which is responsible for selecting and shipping the material alerts the overseas outlets, who then make their own determination about playing it.

Two incidents in different parts of Spain in 1976 illustrate the seriousness with which commanders try to dictate the musical tastes of the station in their area and the troops serving under them.

The first took place at Torrejon Air Base near Madrid. Captain Bob Lenox, the commander, and SMSgt John Antoniewicz, the station manager, were both attending a conference of the 7122nd Broadcast Squadron in Germany and Dave Nuttall, an experience broadcaster and a Sergeant, was in charge. Nuttall was well known to his compatriots as the man who had been put in jail in Turkey for killing a cow with his car. He was also remembered for a radio first — an interview from the back of an elephant in Thailand.

One of the station DJs discovered a novelty record played on *The Dr. Demento Show*, a program featuring weird songs and provided by AFRTS. The song was called "Irving — the 149th Fastest Gun in the West." It got a couple of air plays before the local commander, Colonel B. R. (Ray) Battle heard it and decided it was an insult to Jewish traditions and stereotypical. He ordered, in writing, that it be taken off the air. Word of his order reached Air Force four-star General Dutch Huyser who told Battle that even if he was right, he was stupid to put such an order in writing. The whole thing caused such a flap that Nuttall, who had been scheduled to be sent to a routine assignment in Germany was diverted and shipped to faraway, exotic Omaha. That turned out to be real punishment. He was able to finish his college degree and ended up an Air Force Captain.

On the southern coast of Spain at Rota, the Navy was operating a station at that time, managed by Lieutenant Paul Hanson. His relationship with two consecutive Navy Base commanders had been excellent and he initiated a radio program on which the commander would appear and answer direct phone questions from listeners. It was very popular and demonstrated to everyone that the boss was interested in their problems.

When a new commander, Captain Charles Roe, came aboard, things changed. Roe would no longer field questions on the spot but insisted they be submitted in writing for answer the following week. With the spontaneity gone, the program quickly lost favor with the audience. Captain Roe then took to calling Hanson at home to report that he was receiving numerous complaints about the programming although he could not or would not identify who was doing the complaining. Hanson strongly suspected that the complaints mostly originated with Roe. Hanson investigated each complaint and could find nothing objectionable or anything that did not meet contemporary American broadcast standards.

Then two programs released by AFRTS — *Wolfman Jack* and *Charlie Tuna* — began featuring two novelty songs. One was *Don't Touch Me There* by Tubes. The other was *The Shaving Cream Song*" a re-write of a song originally titled *Sweet Violets* and containing a number of grunts normally heard only in toilets and sung with glee by generations of boy scouts. Captain Roe was outraged and told Hanson so but had the good sense not to order them off the air. Hanson, for his part, did not wish to irritate Roe by explaining that AFRTS regulations forbid removing portions of a program and he couldn't take the songs out of the shows if he wanted to. He did know that novelty songs have a short life and they would soon disappear — which they did.

Then the Paul Anka song, *You're Having My Baby* appeared in AFRTS programs and in the States became a top ten hit. Naturally it would, containing as it did allusions to such musical material as abortion, pre-marital sex and pregnancy. Roe called Hanson into his office and forbid him to play the song again under any circumstances. Hanson had no choice but to explain that while he could keep his own local disk jockeys from playing it, he had absolutely no control over whether the song appeared in an AFRTS pre-recorded program. Roe's answer was succinct and to the point: "I don't care what they say. I don't want to hear that God damned song on the air again."

With bravery over and above the call of duty, the Lieutenant told the Captain the only way he could get it off the air was to ask AFRTS to declare it as sensitive to the host nation. But, he said, he didn't think that was too great an idea seeing as how the Spanish stations were all playing it.

Roe knew how to handle the situation. Several days later Hanson found himself reassigned to Air Traffic Control and the base protocol officer found himself in charge of a radio station in spite of the fact he knew diddly-squat about broadcasting. What he did know was when Captain Roe said take a song off the air, he would take it off the air. This he did, along with a list of other songs which the Captain, for whatever reason, found objectionable.

At this time, the various services were in the process of forming their own broadcast headquarters. The Navy, the Army and the Air Force were setting up headquarters which would manage their own stations through an office of the Assistant Secretary of Defense for Public Affairs. The Navy was well in the lead in preparing for the change and Navy Broadcasting Service was now operational under Jordan (Buzz) Rizer, a man with long experience in military broadcasting. Rizer hit the roof when he heard about Hanson. So did Rizer's boss, Admiral David Cooney, the Navy Chief of Information. They saw Hanson's transfer as a jurisdictional matter as well as intolerable interference with their prerogative to assign their own personnel. In addition, the situation raised questions of censorship and the possibility of infringing on long standing agreements with the music industry. While that was happening in Washington, the personnel of the base were busily penning letters to congressmen raising hell about the censorship of their music.

Several admirals, visiting Rota, took Roe aside and made it clear they were tired of the controversy and to settle it immediately. Hanson returned from his annual leave and discovered he was once again assigned as station manager. He tried to turn the assignment down on the grounds he had only thirty days left in-country but the Captain gave him no choice.

The whole flapdoodle over whether or not to play a song may strike some as trivial but in fact it provided the basis for avoiding numerous future problems. More important than that, it insured the jurisdiction of the various broadcast services over their own personnel and gave the broadcasters the ability to resist command pressure without injury to their future careers. Thanks to Rizer and Cooney at the Navy Broadcasting Service and Bob Cranston at the American Forces Information Service, they knew they could obtain needed support from on high and would not be left dangling in the wind as Hanson almost was.

Future AFRTS managers overseas were to routinely run programs the military hierarchy found objectionable including the CBS documentary, *The Selling of the Pentagon* and *Vietnam — the Uncounted Enemy* over which General William Westmoreland sued that network for millions. He lost, as did more and more commanders and others in positions of authority who considered their personal prejudices and tastes to be the final arbiter of what should and should not be seen or heard. The deciding factor on whether or not to run something more and more became the choice of the AFRTS managers who became less and less concerned that perhaps they were laying their careers on the line.

Lieutenant Colonel Pete Barrett, jeopardized his when, as Commander of the American Forces Network Korea (AFKN), he was given a direct order by the Public

Affairs Officer for Major General Robert Kingston, the command's chief of staff, not to run a video clip of the Village People singing *In the Navy*. In spite of that, it got on the air and the fit hit the shan. Barrett then discovered that the General had ordered it off the air as a favor to Rear Admiral Warren Hamm, Commander, Naval Forces Korea who felt the song did not reflect positively on the U.S. Navy. Barrett pointed out all the reasons why the order to ban it was not legal and also pointed out it was then number three on the music charts. As a sidelight, he mentioned in a letter to the Public Affairs Officer of Eighth Army that apparently the rest of the Navy wasn't all that concerned because they loaned an aircraft carrier to Bob Hope to tape an NBC special and, right there on the flight deck were the Village People singing *In the Navy*. He further said that AFKN would no doubt be receiving the Hope special and he intended running it.

It must have worked. Barrett became a Colonel and deputy director of AFRTS Los Angeles.

EVERYBODY WANTS TA GET IN DA ACT!!
— Jimmy Durante

To a person in a position of power, whether military or governmental, the chance to control a network of broadcasting stations whose combined voices reach around the world is often too heady to resist. Over the years, those trying to get in the act never ceased trying. In some cases their interference was beneficial. More often, it caused staff disruption, degraded service, low morale, endless hours of busy-work and loss of credibility with audiences and the press. In almost every case the cause was a power-grab, an ego-trip or a total lack of knowledge about AFRTS and its mission or limited comprehension of the basics of broadcasting.

Even the United States Congress fell into the ignorance trap in 1977. It was no doubt a dull day on Capitol Hill when a congressman, possibly just returned from an all-expense paid trip to exotic places where he had heard AFRTS, recalled in his wisdom that radio and television stations charged large amounts for commercials and made money in the process. Why, he reasoned, should the Congress budget for AFRTS when all they had to do was start selling time and soon would be not only self supporting but pouring excess funds into the treasury?

For those not familiar with where money comes from to pay for salaries, paper clips, request forms for toilet paper and other necessities of government operation, a quick explanation. There are "appropriated funds" which are monies from the federal budget, approved by congress and which most government agencies and departments use to operate and pay their employees. Then there are "non-appropriated funds" which are dollars generated by the agency concerned. Example: **Stars and Stripes**, which through book and newspaper sales generates its own funds. Military clubs, through the sale of food, booze and slot machines are also non-appropriated. AFRTS networks and stations, although having no revenue producing activities, were in some cases — particularly in Europe — a strange mixture of appropriated and non-appropriated funding. The American Forces Network and the Southern European Broadcasting network both received non-appropriated funding from revenue producing military sources in Europe. Generally this was used for civilian payrolls. With the plan, as outlined by congress, the salons on the Hill planned to make all of AFRTS non-appropriated fund agencies; in other words, self-supporting.

Just to make sure AFRTS got the congressional message, Congress cut $1.1 million from the AFRTS budget feeling that this would move the broadcasters to get out there and start selling spot announcements. This got their attention all right.

When Congress speaks, people listen. Particularly people whose source of revenue is that august body. Soon stations and networks around the world were being asked by AFRTS to prepare statements of what impact such a system would have on their local operation. Reply by yesterday's close of business, please.

AFRTS found the replies to be remarkably similar. After adding the impact on its own operation of obtaining and duplicating program material, the salient points were:

★ AFRTS pays approximately $8-million dollars a year for programs. If it were commercial, it would have to pay commercial prices at an estimated cost of $120-million a year based on the size of its audience.

★ Much of the current programming being distributed is right off the network and is not for sale. It is unlikely that the current sponsors would pay to have their announcements shown in Adak, Antarctica or Diego Garcia or in countries where they do not distribute their product.

★ Many of the larger countries where AFRTS shares the airwaves with the local government stations such as Germany, Italy, Japan, Korea, Panama and others, have commercial networks of their own. Getting permission to go into competition with them is only slightly less probable than seeing a "Support Your Local Police" bumper sticker on Zsa Zsa Gabor's Rolls Royce.

★ Selling $120-million worth of commercial time to pay for the programming would require a sales staff roughly as large as the Third Armored Division. And even if they could sell that much advertising, which they couldn't, the government would find itself illegally endorsing products ranging from deodorant to hemorrhoid cures.

At this time Bob Cranston was director of AFRTS in the office of the Assistant Secretary of Defense for Public Affairs, and used his contacts which he had built over the years to help convince the congress that their zeal, while commendable, was also ill-advised, unworkable and poorly thought out. And also stupid. One of his most telling arguments was collecting letters from the heads of the various entertainment unions and guilds such as AFTRA advising that in the event the plan was implemented, the unions would insist on payment of residual fees for use of the entertainment programs. He also polled a number of producers and suppliers of programming. They were unanamous that should AFRTS go commercial, costs would rise something on the order of ten-fold for program material. Even foreign governments got in the act, advising that they would shut down AFRTS in their countries if the organization went into competition for sales with the local broadcasters.

Cranston and Captain John Worthington prepared a report outlining the arguments against the congressional mandate. The report was fairly lengthy, but boiled to its essentials it said: *Dear Congress: This is a dumb idea. Sincerely yours."* A copy was sent to Bill Baldwin, a long time popular announcer, former president of the Pacific Pioneer Broadcasters and an official with the American Federation of Television and Radio Artists (AFTRA.) He was in Washington on AFTRA business, read the report and immediately saw the implications. If this were somehow to get by Congress, it would be bye-bye AFRTS. He called on Senator Barry Goldwater who also happened to be a General in the Air Force Reserve and he, too, saw the impossibility of the scheme.

Goldwater appeared before the congressional committee investigating the plan, and later on the floor of the Senate where he introduced an amendment killing the plan, and said that no action should be taken before a complete study could be done. The committee agreed and a study panel was formed, including Bob Light of Southern California Broadcasters (and a former AFN commander) and Jack McQueen of Pacific Pioneer Broadcasters. They quickly declared the congressional plan to be without merit.

The storm was over. A grateful Defense Department presented Bill Baldwin a nice plaque identifying him as DOD's "Sixty Million Dollar Man."

...A HOUSE DIVIDED

Not all of the problems caused by higher commands were external. Many of them were home grown. The 1960's and the early 1970's were the "Broger era," remembered with a shudder by AFRTS staff members as a time of internal turmoil over news management, job insecurity, political power plays and general discontent and uncertainty.. Credit must be given to the stations and networks around the world for continuing to provide professional service to the audiences as the in-fighting at a higher level raged around them. Few in the audience, unless they happened to be regular readers of Jack Anderson, realized their local station staffs were performing their duties in a forward manner while protecting their backsides.

The storm of controversy was created by and centered around John Broger. Broger headed that Pentagon office in charge of internal information for the Armed Forces which, during this period, was under the Assistant Secretary of Defense of Manpower and Reserve Affairs. It was the predecessor of today's American Forces Information Service and supervised military publications, internal information policies and military broadcasting. Its power to help by providing a supply of credible news and information through dozens of media sources was awesome. Its power to withhold information by various means and control what service personnel overseas were permitted to know was frightening. Broger's attempts at thought control through control of news sources will be covered more thoroughly in the next chapter.

Broger was brought into government service by Admiral Radford, then Chairman of the Joint Chiefs of Staff, who admired a paper Broger had written called "Militant Liberty." It was an espousal of extreme far right ideas which made the John Birch Society look like a bunch of fiery-eyed radicals. This was the era of McCarthyism in which the militant right blamed everything from unwanted pregnancies to grasshopper plagues on the hordes of Godless Commies hiding under every bed, ready to take over the world. Everyone who disagreed with the far right was immediately branded a closet Communist or, at best, a Commie dupe.

There was a brief period in which a group within the military hierarchy attempted to indoctrinate troops with the John Birch far right philosophy. The plan was named "Pro Blue" and John Broger, according to wire service reports at the time, helped author it. It was implemented by Major General Edwin A. Walker, commanding general of the 24th Infantry Division then headquartered in Augsburg, Germany. Troops were force fed the philosophy at mandatory "training" sessions which continued until **The Overseas Weekly**, a troop-oriented independent newspaper began running accounts about Walker's attempts to brainwash his soldiers. It was too much for the Pentagon and a congressional committee chaired by Congressman Moss of California. Walker was relieved of command.

Bob Harlan, program director for AFN Germany stationed in Munich, was one who took his career in his hands. Walker's public affairs officer, Major Arch Roberts, kept insisting that Harlan's station broadcast similar material to that being force fed to the troops and Harlan kept refusing. Even the Moss report to the Congress commented that he was correct in doing so.

Broger, who had remained in his Pentagon foxhole, remained unscathed and even had his revenge several years later. **The Overseas Weekly**, which enjoyed a large circulation overseas and was appreciated by the troops for its consistent stand for their rights and its obvious joy in tweaking the noses of the brass, applied for permission to begin publishing and distributing in Vietnam and throughout the Far East. Broger said no way. Far Right, okay. Far East, forget it. He explained there was no more room on military newsstands but failed to explain how the Saigon post exchange magazine racks found room for *True Romances, Lady's Circle, True Love, Better Homes and Gardens* and *Patterns, Fashions and Hair-dos*. The paper went so far as to offer to build extensions on the magazine racks. Eventually, after threatening to sue on the basis that the action was denying American troops the freedom of thought which they were fighting and dying for to bring to others, **The Overseas Weekly** forced Broger to back down.

As his assistant, Broger selected a young man named Henry Valentino whose exposure to mass media had been as a lieutenant running an AFRTS station in Saudi Arabia. He joined the civil service ranks as a GS-9 and, according the the AFRTS chapter of the American Federation of Government Employees, received such intense support from his new boss, he was soon a GS-15. The deputy was Colonel Earl Browning and the three of them set up shop in the Pentagon. The AFRTS group and other members of the staff were quartered across the

Potomac in Rosslyn in the Pomponio Building. Heading up the AFRTS group as Broger's deputy for radio and television was Lieutenant Colonel Bob Cranston, last seen here as commander of AFN in Europe.

Broger and Valentino seldom came to Rosslyn. Instead, they would hold their meetings in the Pentagon which meant the majority of those attending had to cross the river and be away from their office for long hours. Meetings usually consisted of Broger pontificating and at times started with a prayer session. It soon became obvious to the Rosslyn group that they were powerless. No decisions were permitted to be made there. The Pentagon group made the decisions and announced them to the remainder of the staff at the meetings. The staff was permitted to comment and even disagree although Cranston can't recall a single decision ever being changed once it was made.

One day Valentino presented what he termed his "concept of operations and cost reduction." It had never before been discussed and was presented as a full-blown, completed plan. The orders were to implement it. The basis of the plan was to automate all the radio stations worldwide. The staff was told to figure out how many people could be reassigned to other jobs by using automation and then translate this into a reduced dollar figure.

Station automation was still in its earliest stages in commercial broadcasting and there was no experience base to use in developing cost saving formulas. Also there were so many variables that it was almost impossible to come up with figures anywhere near accurate. Should you assume the equipment would work 100 percent of the time or do you need enough people to operate if it goes down? How much engineering support will the equipment require? How do you figure in leaves and sicknesses of the staff? Can you assume there will be immediate replacements for departing staff? Will arriving staff be trained in automated equipment?

Cranston insisted it was too early to get into the automation business. Some stations had sufficient personnel but many did not. Some had enough funding. Others didn't. Some commanders in the field liked the idea but the majority did not. No matter, said Broger and Valentino. Press on.

A group was sent to Bellingham, Washington, which was the home of International Good Music, a company headed by a gentleman named Rogan Jones who happened to be a good friend of Henry Valentino. It also happened, perhaps by coincidence, that the company's newest vice president was Dominick Puccio who had recently retired as an Army Lieutenant Colonel and an aide to Broger. IGM made station automation equipment and proceeded to demonstrate it for the group. One of the stipulations everyone agreed on was that there had to be random selection of the music. Simply stated, each musical selection would have an electronic address and could be called up by the operator in any order. The music was contained on huge oversized reels of tape and when one reel was playing, another would be automatically searching for the next selection. The group stood watching the machine do its thing. Music played nicely from reel one. Reel two spun frantically and as it neared the selected music, the brakes started to slow it down.

Flames and smoke belched forth as the friction on the brake lining generated tremendous heat. IGM people said, yes, we do have this small problem but not to worry, it will soon be solved.

Returning to Washington, the group found that teams were being formed to roam the world, inspecting stations and trying to determine how much could be realized in savings.

At about this point, Broger embarked on a concentrated effort to enlarge his empire. He took over the Audio-Visual activities of the three services. He took on support for the National Guard and the Reserve. He had an expensive color studio installed in his Pentagon office with the intention of personally doing an interview program with Washington officials explaining policy to the troops overseas. He next closed the New York Armed Forces News Bureau and moved it to Washington. It was then equipped with semi-automated equipment which would record incoming radio newscasts and play them back to the overseas outlets.

With the latter, a bell went off and for very good reason the staff and overseas commanders began to suspect that this might well be the beginning of control of the news. His announced plan was to feed the automated stations overseas with hourly newscasts direct from his headquarters. Overseas stations would only be able to present news of local military events while all national and international news would be pre-digested and spoon fed to the world.

Time now to test the IGM equipment which had stopped smoking. Frankfurt and Stuttgart, Germany, were the sites selected for the test and the automated monster was set up in Frankfurt to feed Stuttgart and the local Frankfurt station with an actual on-air test. Colonel Bob Cranston and Lieutenant Colonel Harry Bangs were there to observe as was Jack Brown, the program chief from AFRTS in Hollywood. There, too, were Mr. Broger and Mr. Valentino and Air Force representatives.

In a blinding display of tunnel vision, Broger, after arriving in Frankfurt, explained the concept to program director Bob Harlan. "Surely," he said with a straight face, "your audience would be better satisfied knowing the news came from the seat of government."

Interest was high on the part of the AFN staff whose job it was to make the strictly experimental prototype machine work. Interest was also high on the part of major commanders and their public affairs officers who, after a series of stories in **The Overseas Weekly**, became concerned about the management of news out of Washington.

Perhaps most concerned of all were the totally confused members of the radio audience in Frankfurt and Stuttgart. The machine malfunctioned frequently and real live people would have to jump in and continue the broadcasts. The brake linings continued to smoke and burn. The staff hated the damn thing and, although they were never caught sabotaging the monster, they did develop a couple of tricks. One was to always pick a song that was near the end of the upcoming reel so the heavy tape reel would develop tremendous inertia and cause it to do interesting things when the brakes tried to stop it. In another test, Giesela Breitkopf, the AFN music librarian who had been with the network from its very earliest days and knew personally every one of the six million songs in the library, was pitted in a speed test against the machine. Both were given a song title. Giesela walked down the record stacks and calmly picked it out in a matter of seconds. The machine was the clear loser in the contest. Besides, Giesela didn't smoke.

It was clear to everyone that the best description of the test was "disaster." The group from Washington finally returned from whence they came and the Rosslyn group was told they would not need to write much of an after-action report. That would be done in the Pentagon. The Pentagon report called the disaster an unqualified success and recommended that the plan be adopted world wide. The people in Europe, hearing about this, met with the top commanders there and convinced them to let the Secretary of Defense know what really happened. They did and when Broger's paper arrived on the SecDef's desk, he said "NO."

That was the end of the automation plan for the moment. It had been doomed from the start but an untold amount of money had been poured into trying to whip a dead horse back to life.

The instigators of the project soon found a bigger and deader horse and began beating it for all they were worth. They called it "keeping the troops informed." The rest of the world called it "news management."

MUSIC: OMINOUS CHORD.
ANNOUNCER: STAY TUNED. THAT STORY NEXT!

This is a mystery photo. It is no mystery that it was taken at AFRTS' 30th Anniversary. It's no mystery that pictured L to R are humorist and Gene Autry sidekick Pat Buttram, Los Angeles Mayor Sam Yorty, Colonel Bob Cranston and singer Bing Crosby.

The mystery is: Why are Pat Buttram and Mayor Yorty holding hands?

Bing Crosby and Dorothy Lamour are honored guests at the AFRTS 30th Anniversary Party. Both contributed many hours of their talents during the war years to provide entertainment to the men and women serving overseas. Standing behind them are Colonel Bob Cranston and the man who started it all, Colonel Tom Lewis. (Photos courtesy of Robert Cranston)

Certainly no news story had more impact on the decade of the 60s than the assassination of president Kennedy. All networks and stations covered it completely as did every commercial broadcaster. The difference at AFRTS was seeing the shattering impact the event had on local populations

AFN in Frankfurt, as did most AFRTS stations, set up a method where locals could come by to express their grief, perhaps sign a book or leave a bouquet of flowers. The exterior shot shows the people of Hoechst visiting the guard room at the castle gate which had been turned into a visiting room temporarily. The young captain at the left of the picture is Buzz Rizer. LTC Bob Cranston greets a visitor at the door, (center.)

Inside the temporary memorial, Cranston watches a citizen of Hoechst add his name to the book. (Photo courtesy of Robert Cranston)

CHAPTER TWENTY

If in other lands the press is censored, we must redouble our efforts here to keep it free.
— President Franklin Delano Roosevelt

The fact is that censorship always defeats its own purpose, for it creates, in the end, the kind of society that is incapable of exercising real discretion.
— Henry Steele Commager

You have too much news. I want to be entertained. I don't want to see or hear about rape and incest and drugs. I can get all that at home.
— Viewer complaint letter to AFRTS

ELSEWHERE IN THE NEWS

Nothing can make a broadcaster's hackles rise faster than mentioning the words, "News Management." The very thought that a station or a network could be forced to broadcast anything but unbiased, uncensored truth; that its newscasts could be made less than accurate; that truth might be shaded by omissions of fact, cause professional news people to throw tantrums, books, typewriters, fits and -up.

All these things happened at AFRTS outlets around the world on a bleak day in 1963. Since the United States Information Agency (USIA), of which the Voice of America is a part, was formed in 1948, the relationship between the two groups has been cordial but wary. By law, USIA and Voice of America directed their efforts toward local populations around the world in an attempt to win friends for America and explain American policies. Not to put too fine a point on it, VOA was and is a propaganda outlet for the U. S. Government. Nor is there anything wrong with that. Almost all major governments have similar services and many of them, like VOA, provide exceptional broadcasts of interest to the foreign listeners.

The broadcasts of the AFRTS outlets are quite different. They are designed to appeal to the American listener or viewer, even though he or she may be overseas. While many non-Americans have become accustomed to government controlled broadcast news, American audiences have never been exposed to it and would be unlikely to recognize it if they heard it. American news people shudder at any such manipulation as do, in all honesty, the many professional practitioners, both military and civilian, in the public affairs business.

In 1963, the Voice of America was broadcasting to the world by shortwave a twice daily "summary of the news" which it called *Today's Analysis of Events from Washington*. One of the most disgusting episodes in AFRTS's history began one day when word was received at AFRTS stations world-wide that they were ordered by John Broger's office in the Pentagon to air these reports twice daily. Broger was sympathetic and explained that he had received his orders from the chief of USIA, broadcast-deity Edward R. Morrow. According to Broger's office, Morrow had read a report prepared by the Moss Committee which pointed out that, particularly in Europe, AFRTS was listened to more frequently than the national stations. This being the case, said Broger, USIA asked to make full use of this bonus audience and gave Broger no choice but to order it aired on all AFRTS stations.

Inside AFRTS the outrage was violent but there was little anyone could do but comply. Bob Cranston was commander at AFN when the order was received and he was infuriated. At the time, AFN had regional networks in Germany and France interconnected but frequently splitting apart for separate programming. The French government had assigned a French broadcaster to act as liaison with AFN headquarters. Cranston played a copy of *Today's Analysis of Events from Washington* for the Frenchman and said something subtle to the effect that, "You certainly wouldn't want this played in France, would you?" "*Non! Non! Non!*" said the Gallic liaison officer and Cranston quickly notified Broger that troops stationed in France would unfortunately be denied the benefits of hearing exactly how Washington was analyzing the news on that particular day.

Next to take up the cudgel in defense of the AFRTS position was station WTOP in Washington. This came as no surprise, in that the General Manager was John Hayes and John Hayes had been AFN's World War II commander as well as a good friend of Cranston's. The on-air editorials came thick and fast over WTOP, which took USIA to task for exceeding its charter by trying to force-feed Americans a diet of pre-digested news pablum. From April through November the station kept hammering away at this news management. Others took up the issue. **Broadcast Magazine** reported it. Jack Anderson made an issue of it. Reporter Sara McClendon at a news conference asked President Kennedy why he permitted it and it was obvious that Kennedy had never been briefed. He said he would look into it.

Until this time, it appeared that Broger was not in favor and was working on the side of the angels. To the press and to AFRTS management, it seemed that Ed Morrow was the villain of the piece. John Hayes finally met with Morrow and asked about it. In a letter to Cranston, Hayes said he had learned that it had been Broger who pointed out to Morrow the extra audience the military could provide and offered the AFRTS facilities to USIA. Morrow, of course, had no objection but by now the Congress did. They applied the heat and, as quickly as it had begun, the program disappeared.

From its earliest days under Tom Lewis, it has been an article of faith at AFRTS that it would treat its listeners and viewers as intelligent adults and present the news quickly, accurately and without bias. In addition, every newsroom had posted in it memoranda from various high Pentagon officials with pious statements that there was to be absolutely NO, repeat, NO manipulation of the news, no censorship, no holding of release while the news turned to history. "Maximum disclosure with minimum delay" was one phrase frequently quoted. Secretary of Defense Robert McNamara was particularly adamant that headquarters higher than AFRTS would not diddle around with the release of military news. If the news was embarrassing, tough. In most cases public affairs officers who had the releasing authority for military stories originating in their area agreed wholeheartedly with these regulations in theory. And in some cases they even followed the guidelines. In other cases, well, they had careers to consider and if the general wanted a story killed, was it worth risking a promotion for the pleasure of telling a general he couldn't have his way? For the most part, stories of an embarrassing nature eventually got released, although seldom fast enough to satisfy station news staffs. The really good Public Affairs Officers, and there were many, got stories out in a hurry before the rumor mill had a chance to turn them into public relations disasters. The not so good ones often sat on stories until long after the damage was done and the stories had grown totally out of proportion to their importance.

At other times, strictures were placed on the reporting of stories that perhaps made sense at the time but which in retrospect seem somewhat silly. An example is when a B-52 bomber with a load of nuclear bombs crashed into the Mediterranean while refueling over the tiny town of Palomares, Spain. The word quickly went out to AFRTS stations. Hey, gang, those things aren't bombs. They're "devices" and that, bygawd, is what you'll call them. For months newscasts reported on the search for the devices and continued to do so until all the bombs were found and retrieved. If words are weapons, this type of weaponry was forged by the same kind of mind which could call a retreat a "strategic withdrawal."

Today, in this age of information overload, most officers are more aware of the media and its part in the military scheme of things. This, however, is a fairly recent development. Bob Cranston was born into and continued to work in broadcasting and public affairs throughout his entire career. It was something of a shock to the AFN staff after his departure to find that his successor was either totally ignorant of broadcasting and media or gave a fine imitation of a man who was. At that time the European **Stars and Stripes** newspaper put out a very early edition carrying the next day's dateline for distribution in distant areas much as Sunday papers appear on newsstands Saturday afternoon. A copy was delivered to AFN so a spot announcement could be prepared plugging the center feature page common to all editions. When Colonel Phinney noticed one Monday that he was in possession of Tuesday's newspaper, excitement ran high. "Do we get this every day?" he asked. "Yes," he was told. "Fantastic," he cried. "Now we don't need the wire services. We've already got tomorrow's news." It is still uncertain whether the explanation as to why his plan wouldn't work ever truly sunk in.

Receiving news copy by any reliable means was impossible in the earlier days of the overseas operations. Shortwave from the Armed Forces News Bureau in New York, and later in Washington, was chancy at best. Sun spots would wipe out reception for hours or days on end. American wire services such as AP and UPI did not furnish a radio wire to many areas and copy was received in newspaper style. This meant a complete rewrite of every story. Smaller stations frequently could not afford to pay for the wire services, depending on that year's budget.

John Bradley, out in the Azores, had a fairly heavy news schedule including two half-hour television newscasts, four 15-minute radio casts, a half-hour simulcast and five minute radio newscasts on the hour. What he didn't have was a reliable source of news although it was received by the Communication Squadron on base which had fine equipment. But when atmospherics were bad, that was too bad. There's bad news tonight, folks...no news!

On the perfectly sensible theory that the good Lord helps those who help themselves, Bradley helped himself to a news source. With possibly a little help from the Blue Mafia at the air field, a teletype receiver, complete with the equipment needed to pick signals out of the air and convert them to words, appeared one day on the station doorstep like an orphan child. Immediately falling in love with the little tyke, the staff gave it a room of its own—behind a hidden door marked "Record Storage. Do Not Enter." This was to prevent the normal stream of visitors from disturbing it. Then they discovered that their baby loved to chew out words from a Reuters news circuit which, by an amazing coincidence, passed directly over the Azores on its way to South America. Bradley, of course, never knew for sure that any of his staff might possibly have used those words as an emergency news source. But if those clear-eyed, true-blue, dedicated men did, it was in the best tradition of the American heritage of providing comfort and help to fellow citizens.

Besides, the statute of limitations has long since expired.

If you were to swim from Bradley's place through the Straits of Gibraltar, continue past Sicily and the sole of Italy, and turn left for a couple of hundred miles, you would come across Al Edick who is still trying to get on the air "any day now." Because the Southern European Network only fed the housing area and post by closed circuit at first, Edick wanted to make sure everyone was able to get late breaking news. His gem of an Italian engineer, Frank Paris, built a Rube Goldberg antenna using wire, a bicycle wheel with spokes and the drive shaft from a fatally injured helicopter. It worked fine and was generally able to drag in the news. Getting the news out to the audience who couldn't hear the station presented a more difficult problem. Again, Frank Paris to the rescue.

He built a playback/record unit with one minute of tape loop constructed to stop when the loop was complete. Announcers would constantly update the news tape and listeners could phone in and hear a minute of late news. All the announcer had to do was punch the record button and read. It turned out that Edick's idea and Paris' machine were almost too popular. The darn thing never stopped. The moment the announcer would get ready to update the news, it would start rolling as someone else called in. Screams of "Gawddddammmittt!" were the most common sound originating from the vicinity of the machine. They got even louder when a second one was added for sports news.

Edick's most vivid memory of his attempt to serve his audience with news is of the morning the announcer, with a few minutes to go before doing his morning closed circuit broadcast news, set down the news script on his desk and tried to record the phone loop. It kept rolling as more people called in. Finally, with seconds to go, he rushed directly into the studio in time to present the following:

ANNOUNCER: Central European Time is 10 hours. Now, here is _____ with the news.

NEWSMAN: (Long Pause) Oh, hell! I forgot it.

PEEKING THROUGH THE IRON CURTAIN

If AFN gave AFRTS fits because of its insistence that it should chart its own course, the AFN station in Berlin gave AFN headquarters a constant dose of its own medicine. "We're unique" was its cry from its earliest days. There could be no doubting that. The Berlin station is located 110 miles inside what was then East Germany. The city itself was controlled by the four occupying powers and was the last such occupied territory remaining — until 1990 — after World War II. The station benefitted tremendously from the fact that the German government, through payment of occupation funds, paid many of the bills. Consequently, the radio station and, later, the television outlet, were somewhat on the lavish side compared to sister outlets in the remainder of Germany. The radio station originally was in the lovely home of a former German banker, just down the street from boxer Max Schmelling's home. More important than the quarters was the fact that the station could afford an adequate staff of news personnel. This was important because there was an overriding necessity to keep the audience informed about what was going on around them. Politically the area was volatile. For much of its broadcast life, the station was on the very front line of the cold war. East Germany and the Russians were a stone's throw away. Worse, the audience felt cut off from the rest of the world. Access and travel to and from Berlin was either by air, an uncomfortable overnight ride on an Army train or a long drive, filled with bureaucratic booby-traps, to the Western zones. It was like living on an island and, although Berlin is an exciting big city, it tended to close in on the Americans after a time.

The AFN Berlin news team included military personnel, civilians and a German national named Vladimir Benz. They produced a steady stream of stories for use locally and by the network. For many years it was a political rite of passage for U. S. politicians to visit Berlin, have their picture taken, perhaps make a speech, get a briefing from the Berlin commander and, always, be

This peaceful scene is the border crossing between Germany and Czechoslovakia at Waidhaus-Rozvadov the morning of August 23, 1968. During the night, Soviet troops and tanks have invaded Czechoslovakia and from where the author is standing, tanks can be heard just over the hill in the background.

Shirley Temple Black had been attending a conference in Prague when the invasion occurred and he was waiting for her return, hoping to get a first person story of what happened the night before. Instead he got a story he didn't expect.

The Czech border guard to the left of the barrier made a dash for the West seconds after this picture was taken. His Russian counterpart, on the right, shot him dead just yards from the Western border.

Ambassador Black graciously provided an interview later that night back in Nuernburg. (Photo John Pilger, **Stars & Stripes***)*

interviewed by AFN. The news teams over the years covered visits by Lyndon Johnson, Richard Nixon, Gerald Ford, Jimmy Carter, Ronald Reagan and they were there when John Kennedy told the world, "Ich bin ein Berliner."

During the terrible period during the Berlin Airlift which lasted from June 24, 1948 until September 30, 1949, the eyes of the world were on Berlin as Allied planes kept the city alive by flying in food, fuel and medicines. Berlin newsmen covered every aspect of the story day after day and the weary pilots used the station as a homing beacon from which they could get both good music and a navigational signal as well. The transmitter was just off the runway at Templehof Airfield.

As the cold war continued, tensions between the two ideologies remained. Bob Cranston recalls how AFN inadvertently caused embarrassment to the on-going political infighting between the Russians and the U.S., British and French. As AFN's commander, he was briefed on a sudden increase in tension and the possibility that the Russians would again try to close down land access to Berlin. Stories about Berlin would be cleared by the public affairs officer in Berlin, Colonel Lou Breault. This sent the news staff, and news director Ed DeFontaine, into a rage. It reeked so of news management that DeFontaine offered his resignation. Cranston was sympathetic and agreed that stories off the AP and UPI wires could be used. One such story, from the wires, said that the U. S. was putting out patrols on the Berlin-Helmstadt autobahn.

Breault was furious and Cranston received an invitation from General Watson, the commanding general in Berlin and the U. S. representative on the four-power committee which governed the city. General Watson was a reasonable man and explained in detail how such a relatively minor story could cause big headaches. The Americans, he explained, had been announcing locally that they had "military assistance vehicles" on the autobahn. When the Russians complained to Watson that the Americans were patrolling, Watson innocently explained they were not patrols but military assistance vehicles. The Russians pulled no punches and called him a liar. "We heard on your own radio station that they are patrols," they said.

Both Cranston and DeFontaine couldn't help but agree that from then on all stories concerning the divided city would be cleared by the Public Affairs Office. The flap soon cooled down but it provided Cranston and his staff with a wonderful example to use with PAOs when explaining why AFRTS stations should be put into the picture early so that this sort of thing wouldn't happen.

East-West tensions continued to escalate. Thousands of Germans from the Russian controlled sector were moving to the Western zones of the city. This brain-drain became intolerable and on August 13, 1961, the unthinkable happened. AFN newsman Dick Rosse was the first American journalist on the scene at Potsdammer Platz as the Russians and the East Germans began rolling out barbed wire to close off the exit from their zone. This was the beginning of the Berlin Wall which for nearly thirty years would symbolize the division between the two systems and would divide families and friends living in the same city. For months Rosse covered the events as the grim wall went up and access to and from the East Zone was halted.

He, too, had to submit his copy to the Public Affairs Office and, at times, it became totally frustrating. One such occasion was when he was told not to use a story that the Russians had closed the Friedrichstrasse crossing, later to be dubbed "Checkpoint Charlie," by placing tanks across the access. Vladimir Benz reported it directly from the scene but a major in the Public Affairs office refused to believe it although Benz, a highly professional journalist, was there and the major wasn't. Eventually Rosse was moved into the Berlin Command Headquarters and his copy had to be cleared by both the military and the State Department staffs. He recalls he felt like a school boy, handing in book reports.

But now, to quote Paul Harvey, "for the resssst of the story!" Rosse finally got to write "endit" at the conclusion of the story he had worked on so hard and so long. In November, 1990, he was assigned by Mutual Broadcasting to return to Berlin. From the same spot on which he had stood 28 years earlier, he and Vladimir Benz reported to the world the scene as the East Germans began to tear down the hated wall. And, once again, he was the first broadcast reporter on the scene.

Now to Munich...beer, bratwurst and beatings

In 1962, Don Marsh was the civilian news correspondent for AFN, stationed in Munich. At that time, AFN was suffering a surfeit of money and was able to post correspondents throughout Europe. Steve McQueen, Jim Garner and others were making the feature film *The Great Escape* in Munich and Marsh had become acquainted with them. At the same time the students at the University were being naughty and each night would gather along Leopoldstrasse, disrupting traffic and the German sense of order. Police moving down the street on horseback, pushed the crowds of students out of the way or merely walked over them if they refused to move. Each night the scene got a little uglier.

Marsh was there every night, covering the story for AFN. One night he found himself being rather savagely beaten by the police as he watched the action from a side street.

Soon after, he ran into James Garner on the street and Garner was furious. Marsh recorded Garner's comments describing how Garner and a lady friend had been out for a drink when the police accosted them. He said the police had demanded his passport which was in his wallet containing about $1,000 and which soon disappeared. He later got his money back but at the time was more concerned over the treatment of the students by the police. On the tape he unfavorably compared the police to some police brutality he had seen in Japan during student rioting there and said what he had seen in Munich reminded him of what it must have been like in Nazi Germany during the 30's.

Marsh put together the story, using Garner's quotes, and it ran on AFN's *Weekend World*, the most listened to show on the network. The Germans went ballistic. They didn't mind being compared to Japanese but they sure as hell didn't want to be called Nazis. Complicating the matter was the fact Garner had been a member of the 24th Division which had fought its way through Germany during World War II and he was about to be honored at ceremonies at the Division headquarters in Augsburg. Worse, the Bonn government was suggesting he would be happier outside of Germany, which he might well have been, but that would make it difficult to finish the film which was now only partially completed. Although he initially resisted, the pressure became too great and he apologized for his remarks. The film did get completed, as anyone who owns a television set can tell you, and it was a blockbuster. One could wonder whether Garner was acting as he dug his way out of a mock German prison camp or whether he really wanted out.

YOU'LL LOVE KOREA...IT'S A RIOT

Possibly the most difficult type of coverage an AFRTS station handles is local rioting. It's a problem that pops up frequently because stations are located in some of the most volatile places on earth. Where there is political instability there are sure to be riots. The stations are torn between their challenge to present all the news and their responsibility to protect the safety of their American audience, while at the same time knowing their stories can inflame the host nation on which the station's right to exist depends.

Often the fighting between the Public Affairs Officers who are responsible to the military and State Department and the AFRTS news staff reaches approximately the same ferocity as the fighting going on in the streets around them. Most often the underlying disagreement is actually about which of the two sides is more qualified to make fast, accurate news judgements during a crisis situation.

Lieutenant Colonel Peter Barrett, commander of the American Forces Korea Network, went through a number of news flaps during his tenure as network boss. Highly experienced, he had been executive officer at AFN and was to move on as deputy commander of AFRTS in Los Angeles. Because of the explosive nature of the political situation in Korea, the station was probably more closely "guided" by the Public Affairs office than any other. Their guidance, which in fact were orders, was the cause of frustration to the AFKN staff. Barrett and his deputy, Captain John Cowan, kept up a running attempt to recapture the decision making process. Sometimes they won an argument; often they lost.

Some of the guidance bordered on the absurd. They were forbidden to run widely syndicated film clips of a new subway in North Korea because, declared the PAO, this made the North Koreans look good and they were the enemy. **Stars and Stripes** had just run a picture spread on it but that made no difference. Kill it!

When the world table tennis championships were held in Pyongyang, North Korea, the U.S. participated. This was the first time a U.S. group had visited the country since the Korean War and AFKN considered it a major story. The PAO and the acting Commander in Chief of American Forces, Korea, Lieutenant General Rosencrans, did not. Permission was finally obtained to run the scores but Rosencrans refused to allow video coverage of "that North Korean propaganda bullshit."

This type of news-influencing decisions continued throughout Barrett's stay. Near the end of his tour of duty, all hell broke loose in Korea. In October, 1979, the chief of the Korean CIA, Kim Jae Kyu, making a strong political statement, gunned down and killed the President of Korea, Park Chung Hee. The country began to seethe with unrest and street demonstrations erupted with increasing frequency. The marching orders given the Network were to play down the riots as much as possible and, in no event, broadcast news about them until the Korean media had done so. Barrett and staff were kept under tight rein for months as the local political situation deteriorated and at one point he considered asking to be relieved.

Things became extremely dicey when there was a military coup in December following the assassination. In May there was a major military uprising in Kwang-Ju and martial law was declared. At its height, American schools were closed and personnel were told to stay in their homes. It became essential for AFKN to be able to report on events quickly and without clearing every story

John C. Broger, ran the armed forces information programs under a variety of institutional names for many years. Many of his subordinates considered him arbitrary and dictatorial. A man of firm opinions not easily changed by facts or logic, he frequently found himself the target of columnists such as Jack Anderson who delighted in printing scathing stories about "the military's media czar."

Dave Stewart, now a public affairs officer for the United States Army Europe, recalls one of the management techniques recommended by Broger and his assistant, Henry Valentino. They wanted local stations to use more of the programming provided by AFRTS and the Broger shop. The affiliates wanted more local time for their own use. The Broger/Valentino solution: Cut the staff down to a bare minimum, then they will be forced to use less locally produced programs.

When Broger finally left government to devote his time to religious broadcasting, Bob Cranston took over the office and because of his wide knowledge of the needs of overseas broadcasters, the situation eased tremendously. (Photo courtesy U.S. Air Force)

through several layers of bureaucracy in order to help protect the lives and property of the audience it was there to serve. Barrett decided to lay his career on the line. He asked for, and received, an appointment to meet with the Chief of Staff, Major General Kenneth Dohleman. Dohleman had been the instigator of the "wait until the Korean media use the story" order.

Barrett pointed out that the order had the effect of giving the Koreans direct censorship of AFKN news and was in conflict with Department of Defense directives ordering a free flow of information. It set an "intolerable precedent," he said. He then said that the Korean media do not receive AP Radio, UPI Radio News or U.S. network news feeds and thus would never use stories available to AFKN. He pointed out that the credibility of AFKN was being destroyed when it failed to report on events that the audience could see going on all around them. He also noted as a sidelight that, good as they were, his news staff did not speak Korean and had no way of knowing what the local Korean stations were doing.

Barrett carried the day and restrictions were finally eased. In a letter to Brigadier General Robert Sullivan, the Army Chief of Public Affairs, he said that he hoped the conflicts had been satisfactorily resolved but anticipated that he would "have an early return from Korea."

He didn't and instead was awarded AFRTS's highest honor the following year — the coveted Colonel Tom Lewis Award, given to the AFRTS member who has contributed the most to the system during the previous year.

Not all AFRTS newsmen were totally happy with the type of journalism they were, by force of circumstances, allowed to practice. By training and inclination, they wanted to break big stories and play Bernstein and Woodward. By the nature of the organization this just wasn't the way the world worked. New news people invariably come in raring to "rip the lid off this town." Eventually they come to realize that their function is to see that the viewer and listener is given the same access to the same news as their contemporaries at home. Essentially, AFRTS news people are news disseminators, not news gatherers. They naturally in the course of their duties gather news with a military slant but, as government employees, are quite rightly not normally permitted to be in competition with the commercial news services.

It hurts sometimes, but that's the way it is. Almost always when a public affairs officer puts an embargo on a military story, there is an excellent reason for it although the reason may not be apparent, or acceptable, to the news person. When PAOs try to kill stories from other sources, such as wire services or U.S. radio and television newscasts, their efforts seldom are successful when the story is going on all around the people they are trying to keep it from. Someone caught up in a riot is reasonably certain to notice it. There is simply no way to hide a major story today in this age of instant communications. Satellite signals, news magazines, live network newscasts around the clock and English language newspapers published everywhere are available to everyone, no matter where on the globe they are.

It was not always thus. In 1967, AFN News Director David Mynatt hired a new managing editor, William Slatter. Slatter had been news director of aggressive, news-oriented station WDSU-TV in New Orleans. Slatter, a thoroughly professional newsman, had some difficulty in adapting to the military way of doing things. But he did manage to "rip the lid off." At a luncheon of the Frankfurt Press Club, he was the guest speaker and proceeded to lay out all his complaints about the military system in some detail. His major complaint was that stories concerning the military had to be cleared by the U.S. Army Public Affairs Office in Heidelberg. The boss there was the same Colonel Lou Breault who had formerly served so well in the Berlin Command.

Among the examples given by Slatter was that a story about allegations of graft in the military exchange system had been held up for 26 hours until the exchange system could furnish a statement. Although reminded by a superior that no harm was really done by waiting a day, Slatter contended that the issue was whether AFN could "let the accused party control the news." He also complained that a story concerning a stabbing of a white officer outside a Frankfurt bar which catered to black troops had been held up for a day and a half while the Army investigated the incident.

Because the speech was delivered at a press club luncheon, the wire services picked up the story and it soon hit Washington. The wire service reporters quoted Mynatt as saying that while Slatter was an "experienced and capable newsman and a fine managing editor, he has not been in Europe long enough to understand the uniquely sensitive position of a military operated news medium in a foreign land." Colonel Breault did not deny that stories were sometimes delayed, and explained that "by regulation AFN is not intended to be the mouthpiece for any command, individual or group. Well-informed and temperate expressions of opinion about improvement of practices in the military establishment have a place in AFN broadcasts."

When the story, and the rebuttals, reached Washington on the wire services, there was a delay in retransmitting it back to Europe, and the rest of the world, on the Armed Forces News Bureau wires. Because of its nature, an underling held it for some hours before John Broger was contacted and who ordered it released immediately. He had excellent reasons for doing this. The Secretary of Defense, Robert McNamara had just months before issued orders that there would be no news manipulation and stories embarrassing to the military would not be delayed for that reason alone. McNamara read the wire story account of Slatter's remarks and, because it seemed to run contrary to his memorandum, sent two colonels to Frankfurt to investigate. Bob Cranston, who always seemed to get this kind of weird job, headed the team.

The team could find no clear case of censorship in any of the newscasts since the McNamara memorandum had been published. They found Slatter's accusations not to be an attempt at brainwashing troops but rather faults in the system. These faults triggered recommendations to the Secretary that the AFRTS networks be free to report all the news that was fit to broadcast by generally understood broadcast standards.

The Secretary agreed and since 1969, when columnist Jack Anderson printed a number of charges that John Broger and his staff were censoring certain stories, the problems have become manageable. Broger and his group are long gone. Bob Cranston took over that office after retirement from the service and his understanding of the essential dichotomy between the requirements of the military and the need-to-know of the audience made life easier for everyone.

Each Secretary of Defense generally publishes an update of the original McNamara memorandum. Secretary Dick Cheney's, in effect currently, makes the policy quite clear. It contains five parts and reads:

✯ *Information will be made fully and readily available, consistent with statutory requirements, unless its release is precluded by current and valid security classification. The provisions of the Freedom of Information Act will be supported in both letter and spirit.*

✯ *A free flow of general and military information will be made available, without censorship or propaganda, to the men and women of the Armed Forces and their dependents.*

✯ *Information will not be classified or otherwise withheld to protect the government from criticism or embarrassment.*

✯ *Information will be withheld only when disclosure would adversely affect national security or threaten the safety or privacy of the men and women of the Armed Forces.*

✯ *The Department's obligation to provide the public with information on its major programs may require detailed public affairs planning and coordination within the Department and with other government agencies. The sole purpose of such activity is to expedite the flow of information to the public: propaganda has no place in Department of Defense public affairs programs.*

❖ ❖ ❖ ❖ ❖

ANNOUNCER: DID ANYONE EVER TELL YOU THAT YOU SHOULD BE IN HOLLYWOOD? KEEP READING AND YOU WILL BE.

❖ ❖ ❖ ❖ ❖

During the 60s at a National Association of Broadcasters convention Los Angeles, AFRTS put its best foot forward and broadcast its shortwave service to the world directly from the convention hall. Seen here: T/Sgt. George Hudak at the microphone, Sgt. Clark Knowlden at the control board. (Taking no chances, AFRTS labeled the equipment just in case someone at the NAB didn't recognize it.) (Photo courtesy of AFRTS)

In 1967 Johnny Grant's show on AFRTS was called "Small World." Jane Russell hardly seems like an appropriate guest for a show with a name like that—but she was.

Ed Ames (R) was still a member of the Ames Brothers when this was taken while appearing on the Bill Stewart Show. Later, as a single, he became an even bigger star on television and on Broadway. Always gracious to AFRTS, he was one of the featured performers years later at the AFN 45th Anniversary Gala in Germany and stopped the show with his great voice.

Movie and TV personality Ruta Lee had her own show called "Music and Me" on AFRTS in the late 1960s. We know the gentleman who is her guest this particular day is famed song writer/composer Jimmy McHugh but no one can remember the name of the dog in her lap.

Andy Mansfield's program was named "America's Popular Music" and the name really fit the day he had one of the great band singers of all time on his show. Visiting this day is "liltin'" Martha Tilton who had helped make the Benny Goodman Orchestra the musical phenomenon it was.

Staff members at the Hollywood studios never knew who they were going to bump into in the halls when they came to work. Shown here are a few of the hundreds of guests who appeared on the various disk jockey programs produced at the AFRTS Hollywood facilities over the years.

Right, Bud Widom (R) was a mainstay, in and out of the military, of AFRTS for many years. Now, in the 60s, he is back in Hollywood with his own show, Bud's Bandwagon, and with his guest this day, Eddie Fisher. (Photos Courtesy of AFRTS)

150

CHAPTER TWENTY-ONE

In which we learn that AFRTS could have been what Gen. David Sarnoff was talking about when he said, "We're in the same position as a plumber laying a pipe. We're not responsible for what goes through the pipe."

Hollywood is where they place you under contract instead of under observation.
-- Walter Winchell (1897-1972)

HOORAY FOR HOLLYWOOD
(and points East)

Until this point in the AFRTS story, the thrust has been on the so-called "field stations" -- the AFRTS outlets around the globe. These are the suppliers of the memories the listeners and viewers who have served overseas have of the AFRTS system. They supplied the voice of home, the local news, the place to sell your car or told you what was on at the movies. Probably few in the audience stopped to think that the station got orders from above; that someone had to supply their equipment; that someone else had to choose and contract for the Stateside programming; that someone had to deal with the host government to secure permission to broadcast and obtain frequencies; that someone had to set basic policy to avoid airwave anarchy.

Describing a governmental organizational setup makes for the same kind of exciting reading as *The Minutes of the Meetings of the Ladies' Quilting Society of the Warrensburg, Missouri, Reformed Church of Our Lady of Perpetual Snarls -- 1906-1910*. Still, there may be those who care where AFRTS fits in the government scheme of things. So, taking a deep breath and starting at the top of the pyramid:

1. The **President of the United States,** who has practically no connection at all except to write a nice letter about what a great job AFRTS is doing on each fifth anniversary. He also affects the AFRTS operation when he directs an economy move to either cut budgets or cut personnel. During the Holidays, he usually uses the system to send greetings to the troops stationed around the world.

2. The **Secretary of Defense,** under whom -- very indirectly -- the whole ball of wax rolls.

3. The **Assistant Secretary of Defense for Public Affairs** is the most direct top gun of the whole shootin' match. He is the spokesman for the DoD, and sets the general policy guidelines for the people one level lower who make the specific rules and recommend policy. They are:

4. The **American Forces Information Service.** This group, located in Alexandria, Virginia, but within a stone's throw of the Pentagon, are the direct descendants of the groups which have had immediate operational control of AFRTS since its founding. Apparently chosen by a secret government bureau known as the Department of Confusion and Obfuscation, an incomplete list of some of the former names of predecessor organizations include:

- *Information and Education Division of Special Services.*
- *Army-Air Force Troop Information and Education Division.*
- *Armed Forces Information and Education.*
- *Office of Armed Forces and Education.*
- *Directorate for Armed Forces Information and Education.*
- *IAF—Information for the Armed Forces*

It has had a rocky history and has taken its share of lumps from the press and the congress in the past, in particular during the autocratic management days of John Broger and Henry Valentino. Since their departure, and under the former management of Bob Cranston and, today, under Jordan "Buzz" Rizer, AFIS functions in a considerably less explosive atmosphere. Sub-units reporting to Rizer and his group are:

- Armed Forces Radio and Television Service (AFRTS). AFRTS directs the broadcast policy of the Department of Defense and supervises the operation of AFRTS-BC, the Broadcast Center in Sun Valley, a Los Angeles suburb. The director is Mel Russell.

- American Forces Press and Publications Service which includes policy control of *Stars and Stripes* and most military publications.
- Radio and Television Production Office (RTPO) which acquires and approves spot announcements for use on AFRTS -- non-commercial, of course.
- Armed Forces Digest is the electronic news branch operating out of Washington. It supplies both television and radio material on military subjects to broadcast outlets.
- Defense Audio Visual Policy.
- Print Media Policy.

In addition, AFIS offers considerable input to the operations of the Defense Information School (DINFOS) located currently at Fort Benjamin Harrison, Indiana. This school trains members of every service, and some civilians, in public affairs, public information, broadcasting and both print and electronic journalism. With few exceptions, every entry level broadcaster will pass through DINFOS prior to his or her first assignment to an AFRTS station or network. There are also a number of advanced and refresher courses taught for senior broadcasters and broadcast management, both officer and enlisted. With the congressional mandate of 1991 closing numerous bases, including Fort Benjamin Harrison, the future home of DINFOS is still an unknown as of this writing but it is anticipated it will be somewhere on or near the East Coast. Although not officially a part of the AFRTS family, DINFOS graduates from the broadcasting school most often end up with an AFRTS assignment and the school's contribution to military broadcasting cannot be ignored.

A bit about the military's Little Red Schoolhouse

During World War II, the military had the benefit of all the communications experts it could use. When the war ended the need for their expertise remained, although they didn't. With the cold war, the draft and a population tired of war, the need to inform both the public and the military was critical. In 1945, Jack Lockhart of the Scripps-Howard newspaper chain submitted a report to the Chief of Army Information calling for the establishment of a public relations program taught by trained personnel and with the backing of the military establishment. Following his recommendation, the Army Information School was established at Carlisle Barracks, Carlisle, Pennsylvania on January 23, 1946.

At first it was only for officers and by October three classes had been graduated. The school fought off attempts to close it and by 1947 the Navy, Air Force and Marine Corps were sending small groups to attend. Concurrently, the Navy had a Journalist school at the Great Lakes Training Center near Chicago and the Air Force duplicated efforts with a school at Craig Air Force Base, Alabama. In 1948, this redundancy came to the attention of Secretary of Defense James Forrestal and he directed

Tom Lewis returned often to visit the scenes of his triumphs long after he had retired. Shown here, center, he is flanked (L) by AFRTS program chief Jack Brown and Dorothy McAdam who had been with Lewis from the earliest days of World War II and became a key member of the Industry Liaison office of AFRTS, which is responsible for obtaining broadcast rights from program owners and suppliers.

On the right is Ruby Schad, who for many years was the staff assistant to the commander and who was the person to ask no matter what you wanted to know about AFRTS. Next to Ruby, the man most agreed really ran the place — Bob de la Torre. As commanders came and went, as rules and techniques changed, Bob provided the continuity of knowledge that prevented the wheel from being reinvented with each change. (Photo Courtesy Dorothy McAdam)

the formation of the Armed Forces Information School, closing the Army Information School.

The new institution was moved to Fort Slocum, New York in April 1951 with larger quarters and an increased staff. This also meant it cost more to operate and the eagle-eyed budget people in Congress wondered if such a school was necessary. By June 1953 they decided it wasn't, and they closed it. The Department of Defense then reopened the Army Information School. Navy and Marine students were sent to Great Lakes and Air Force and Army personnel to the Army School. Some Navy and Air Force Officers were sent to special courses at Boston University and Army Officers to the University of Wisconsin. At this time, Robert McNamara, a fast man with a pencil, was Secretary of Defense and it didn't take him long to recognize that all this was hardly saving money. He ordered the reopening of a Joint Service School, to be called the Defense Information School (DINFOS.) It already had 18 years of experience and had graduated 23,000 students.

The location was to be Fort Benjamin Harrison in Indianapolis, then the home of the Adjutant General's School and the Finance School. It was a tight squeeze fitting the new bunch in but somehow it was done. The location also obviated getting top-notch speakers from the communications world in New York on a moment's notice. The number of courses was increased and today include the Information Officer Basic Course, Broadcast Officer Course and the Information Officer Refresher Course. Enlisted students are assigned to the Basic Military Journalist Course, the Advanced Military Journalist Course or the Broadcast Specialist Course. Officers and enlisted personnel heading overseas as station or network managers also are sent to a specialized Broadcast Management Course. The curricula of these courses is under constant review and changed to meet the changing needs of the services and the available technology. For example, in recent years much greater attention has been paid to electronic journalism. Instructors, almost without exception, have spent time in the various overseas outlets and can quickly teach new students the unique requirements of broadcasting in foreign countries.

They can be tough, too. DINFOS today has the second highest fail rate of any military school. It is not unusual to see only a third of a class finally graduate. Instructors point out neither Paul Harvey nor Walter Winchell would last a week. AFRTS managers are grateful for this kind of tough training. After all, once these students are assigned to them, they stay the full length of their military tour. Unlike the civilian world, a military manager can't fire a staff member if a better qualified candidate appears at the door.

The Gadget Getters

The equipment used by AFRTS stations is, almost without exception, normal commercial broadcast gear and, as such, is not a standard part of the military

Jack Brown really got around. Here is Jack with movie star Clark Gable. Also seen: Clark Gable with radio star Jack Brown. For those wondering, that's Jack on the left. or maybe that's him on the right. It's one or the other. (Photo Courtesy of Jack Brown. Or Clark Gable.)

equipment inventory. It is purchased, received, tested and shipped to the user by the Television-Audio Support Activity (T-ASA) located in Sacramento, California. Although officially an activity of the U.S. Army Information Systems Engineer Command, T-ASA and AFIS work very closely together in the procurement of equipment and T-ASA provides invaluable engineering assistance in the installation and design of overseas stations. When specialized equipment, such as a mobile unit, is required, T-ASA engineers generally design and construct it to specifications laid down by the user service. A small number of AFRTS members are attached in order to provide quick interaction between the two organizations. T-ASA is the Radio Shack of the AFRTS system. When a transistor, a recorder, an integrated system or a complete station is needed, T-ASA is the one-stop shopping center. When a station's Electronic Holographic Ohm-eating Framus circuitry burns up in order to protect the fuse, T-ASA gets a midnight panic call and can often have a replacement on the way the next morning.

In addition to these functions, AFIS provides over-all policy guidance to the Army Broadcasting Service, the Air Force Broadcasting Service and the Navy Broadcasting Service. Although each of the service broadcasting organizations reports to their individual service Secretary, AFIS, through the Assistant Secretary of Defense for Public Affairs, exercises considerable control over them. AFIS and its AFRTS component, approves or disapproves the formation of new outlets. It determines the configuration of networks. It concurs or non-concurs in the selection of network and station commanders, works to standardize equipment and certifies equipment for use within the system. It handles major negotiations with host governments. Staffing levels are developed by AFIS as are training standards and objectives. The big stick is control of the Department of Defense budget allocated to service broadcasting which is administered by AFIS.

Each service broadcasting group is responsible for a specified area. The Army handles AFN in Northern Europe and SEB in Southern Europe. Also under Army Broadcasting Service is SCN in Panama, AFKN in Korea and Kwajalein in the Marshall Islands. Its headquarters is in Alexandria, Virginia. Under the wings of the Air Force Broadcasting Service are broadcast activities in Turkey, Spain, Greece, the Middle East, the Azores, Alaska, Greenland and areas of the Pacific. Headquarters is in San Antonio, Texas. The Navy's responsibilities are for Iceland, Rota (Spain), Diego Garcia in the Indian Ocean and those parts of the Mediterranean basin with shore-based naval facilities plus the hundreds of ships now equipped for both television and radio. Headquarters is in Washington, D.C.

"Buzz" Rizer was the man who organized the Navy Broadcasting System beginning in 1971 and developed the on-board broadcasting systems. His deputy was Gerry Fry, today the director of programming for AFRTS in Los Angeles. So successful was it that Congress decreed that the other services should follow suit and both the Army and Air Force systems began operation in 1980. Today each works under the same set of Department of Defense directives although each has certain service-unique situations which are taken into account by AFIS.

(Pssst. When are we going to Hollywood?)
Leaving the big ivory tower in the sky, we bid a reluctant farewell to the seat of Government and set course for gay, glamorous Hollywood. The year is -- take your pick-- sometime in the late 40's or early 50's. The place is the corner of McCadden Place and Romaine, one block east of Highland Avenue, and one block south of Santa Monica Boulevard. There on the Northeast corner is the home of AFRTS Hollywood. To be honest, this is not exactly the heart of Hollywood. Maybe the gall bladder or the spleen, but not the heart. It's a neighborhood of small shops, many with "For Rent" signs in the windows. There's a school nearby with the usual quota of graffiti on every available wall and populated by tiny terrorists intent on adding more to the walls of the AFRTS building. General Service Studios are just down the street. I Love Lucy shoots there. So does I Married Joan.

The AFRTS building was formerly a drop forge factory which was rented by the military when it stopped dropping forges or forging drops or doing whatever a drop forge factory does. The interior has been remodeled by the famed Low Bid Construction company which has spared all expense to make it reach the lower limits of adequacy. Private office space was at a premium originally but now a second floor has been added as well as a mezzanine containing fairly spacious offices for the commander, the deputy commander, the director of programming and one or two others of the top echelon. The automatic answer from a staff member asked by a visitor, "How do you get to the mezzanine?" is, "Kiss ass."

The ground floor contains all the broadcast equipment because the second floor is too puny to hold it. Here are television master control, the recording rooms, radio studios, film and record vaults and the radio news room. Upstairs are offices for such varied activities as procurement, contracting and programming. Space is tight and space fillers such as thousands of reels of television programming are stored under contract at another location. The roof is a constant problem and during the famed Los Angeles rain storms, the roof always seems to leak over the main transformer which furnishes great gobs of electricity to the building and when wet produces spectacular displays of lightning and flying sparks. Splashing through the puddles to turn off the main circuit breaker is a job usually given to the newest, and most easily replaced, person on the staff.

This was the place the stations and networks overseas looked to for help, advice, sympathy and -- most important -- the bulk of their programming. They knew about Washington but that atmosphere was so rarified that little thought was given to it by the field. Hollywood was where they dealt when they wanted something or when they had a complaint. To some of the Hollywood staff who had never been farther overseas than Catalina Island, getting the programming out to stations in places they had never heard of was a ho-hum kind of a job. To others, it was a never ending crusade for excellence.

Noteworthy among these was Bob de la Torre who joined the staff during wartime and stayed until his death many years later. Although a civilian, he spent more time as acting commander than some commanders did; he stayed at his desk and made the decisions while the nominal bosses checked out the local golf courses knowing their tour of duty would take them elsewhere soon. He and the commander's secretary, Ruby Schad, knew where every body was buried, where every problem lurked and how to get anything done that was worth doing.

For many of the early and mid years, the gentleman last seen working as a Navy broadcaster on the Philippine Island of Samar, Jack Brown, was a key staff member in Hollywood. He was, and is, a consummate broadcaster and became Director of Programming for AFRTS. Under his guidance, AFRTS followed the national trends and was able to provide its overseas affiliates with a near perfect approximation of what was happening at home. His job was frequently made difficult by the entrenched bureaucracy, both military and civilian, which by training and inclination believe "change" and "evil" to be synonymous. To these people, Brown had a disturbing tendency to be right and because of his calm, reasoned and well constructed plans, AFRTS was able to continue to provide the kind of programming the overseas audience deserved and expected. Now officially retired, he is hard at work as an officer of the Pacific Pioneer Broadcasters, the roster of which reads like a Who's Who of broadcasting.

Also retired, though not forgotten, is Bob Vinson. Largely because of his wide contacts in the industry, he was able to lead AFRTS into the television business. His successor in the position of Director of Industry Relations, Vince Harris, has been equally effective in the critical job of obtaining broadcast rights to the top rated shows at a price the sometimes inadequate budget can afford. Harris, and his assistant, Dorothy McAdam, are now the employees with the longest terms of service and their corporate memory is priceless in preventing the reinvention of the wheel as new commanders come aboard every few years. Thanks to the esteem in which the broadcast industry has held these people for almost the entire lifetime of AFRTS, they have been able to maintain the cooperation of the suppliers of programs and the craft guilds and unions which began during World War II. Proof of their success is the fact AFRTS is able to provide more than 90 percent of the top 75 television programs each year to overseas American audiences. In addition the system provides around the clock live satellite coverage of real-time sports events, news and specials of all types. Toss in a good sized dash of specialized programming -- childrens', religious, ethnic, daytime drama, feature films -- and you end up with a programming selection and mix that any U. S. program director would kill for. What's more, it costs the American taxpayer approximately eight cents on the dollar if figured on the basis of what commercial stations of comparable size would have to pay for the same programming.

A SHORT STROLL THROUGH THE HOLLYWOOD YEARS

THE 1940'S. No television yet, but great radio. Programs went out on giant 16-inch transcription disks, initially

In the 1960s, the whole approach to radio changed and as the lavish programs of earlier days disappeared to television, radio became a disk jockey medium for the most part. AFRTS produced this type of show using the best in the business from the Los Angeles area, all of who would come to the AFRTS studios and record a week's worth of programs in one day.

Getting them all together for a picture wasn't easy, but this is the majority of the DJs whose recorded programs went out to the troops and their families each week back in the 60s.

(L to R) Bill Stewart, Andy Mansfield, Joe Allison, Jim Pewter, Herman Griffith (who introduced soul music to AFRTS), Roger Carroll (who opened his show by saying "Hi, I'm Roger Carroll and I play records" except at Christmas when he said, "Hi, I'm Roger Record and I play Carols) and at the tail end of the line perhaps the hardest working announcer in Hollywood at the time, Gary Owens, who was the voice on countless TV programs as well as radio. (Photo Courtesy of AFRTS)

with the commercials included. Music library shipments began of individual songs and Major Meredith Willson was the Music Man who selected them. The AFRTS-produced shows such as *Command Performance, Mail Call, Jubilee, G. I. Journal, Melody Roundup, G. I. Jive* and *Jill's Jukebox* received so much mail that special employees had to be hired to handle it.

During the War days, AFRS produced 22 hours weekly of this special programming. A few continued for a few years after the war but production was scaled down tremendously after V-J Day. Commercial television began during this decade and long-time radio stars began moving to the new medium. Some predicted the end of radio. AFRS continued to send out approximately 60-hours a week of radio programs to overseas affiliates but, for the most part, it was much more modest in content than the star-studded extravaganzas of wartime.

THE 1950'S. Radio didn't die, but it certainly changed. Tod Storz, a Texas radio man, noticed while sitting in a bus terminal that people played the same songs over and over on the juke box. From this observation, he began programming the same songs over and over on his radio station -- and Top 40 radio was born. This format, the transistor radio and Rock and Roll dictated a change in AFRS programming as well and the programs became more and more disk jockey oriented. Nor was it possible to ignore television which began in a small way and, like a 900 pound gorilla, soon was a commanding presence in the operation of the system. (Now the "T" was added and AFRS became AFRTS. At first this stood for <u>Armed</u> Forces Radio and Television Service. In 1969, someone thought this was a little militant and the name was changed to the <u>American</u> Forces Radio and Television Service. People continued to refer to it as "Armed" and after several years of being continually called by the incorrect name, "American" disappeared and "Armed" reappeared, causing great glee and profits to printers of business cards and letterheads.)

By June of 1958 there were 19 AFRTS television stations on the air broken down into five groups, or circuits, each of which received 40 to 50 hours of kinescope programming each week. Los Angeles now had the mission of removing the commercials and inserting information spot announcements. It also was charged with supplying programs in sequence. Until now programming had been given free by the owners on a haphazard basis and it was seldom that the same program would show up in the same period every week.

THE 1960'S. The radio decade started out for AFRTS just as it did for commercial stations -- as a hodge-podge of rock and roll disk jockeys mixed with more traditional types of programming. *Arthur Godfrey, Don McNeill's Breakfast Club* and *Gunsmoke* shared the airwaves with hours of fast-talking DJs playing Chubby Checker and Bill Haley records.

The 16-inch transcription disappeared during this decade and was replaced by the 12-inch disk. And, to its horror, AFRTS, which was working its collective tail off to provide 60 hours of programming a week on the new disks, discovered that the stations were only using less than 45 percent of them. The rest of their schedule was being filled by locally produced programming in each of hundreds of locations. They felt this was a tremendous duplication of effort, and the Assistant Secretary of Defense in charge of such things felt the same way. He laid down the law in a 1964 memorandum. From now on, it was decreed, every station would broadcast a minimum of 18 hours, expandable to 24 if wanted. There was to be five minutes of news on the hour every hour, with expanded newscasts at key times such as 7 a.m., noon and 6 p.m. Local DJ shows were to be scheduled at peak listening hours to provide time for local information and would total approximately 6 and a half to seven hours a day. This left about 80-hours a week of air time and AFRTS established this as the length of the program unit shipped to each station each week. Henry Valentino, working with John Broger out of the Washington headquarters, had a simple way of enforcing the limited-local-time edict. He merely advised stations that if they had that much manpower to waste, he would cut their staff. It never came to that. The stations continued to grab every second of local time they could...and worked longer hours.

These figures remain valid today although through the years, a number of overseas stations have figured out that by going to a 24-hours a day schedule, they have an extra six hours a day to play with. Some have hidden their smirks behind a straight face and declared midnight to 6 a.m. as "prime time" in which they play a portion of the AFRTS radio unit, thus leaving time during daytime hours for their own local programs. When local programming seems to AFIS to be approaching unrealistic boundaries, an informal query to the station as to whether it has surplus personnel usually gets the station back on the track. The present management of AFRTS is also much more sympathetic to the utilization of local time. It also usually agrees to the normal response from the station that it is their duty to properly train personnel and that can't be done unless they are given time on the air.

During the 1960's, America's music continued to change. Soul music joined AFRTS with *The Herman Griffith* show. DJs became more outrageous and the stuffier of the overseas commanders became more outraged as AFRTS continued to mirror what was being heard in the States. Overseas stations were now required to fill in a quarterly report giving audience reaction to all AFRTS shows. Programs rating consistently low were dropped and others added. Most were of the DJ variety. And most DJs were from the Los Angeles area for simple logistical reasons. It became impossible with the growing trend toward talking over records, playing jingles and sequeing from one song to another to edit out the local or untimely references. Most of the off-air DJ programs were filled with chatter directed toward their live local audience, none of it of any interest or use to the AFRTS listener overseas. Consequently, the DJs were now brought in to the AFRTS studios and produced their programs there. It was a difficult task for them because they had no way of knowing when the individual overseas stations would be playing them; whether it was hot or cold where heard, night or day, summer or winter, rainy or bright.

As the decade ended, there were 48 television outlets. And, FM radio came to AFRTS. Stations were added as quickly as host country approval could be obtained. Now many stations had two outlets. At first, in a reversal of the norm, FM music consisted of Musak-like tapes which caused considerable dissatisfaction among the young troops who wanted to groove to stereo contemporary music on their boom boxes. The only place they could hear that was on AM in mono. It took some time, and some griping by the older folks who liked "that damned elevator music," but the FM service slowly took on a today-sound.

THE 1970'S. The radio priority unit began, a group of programs featuring the very latest music and shipped immediately to the stations so that listeners there could hear the newest songs at the same time as the folks at home. This seems to be of critical importance to young listeners, although none can explain exactly why. The first such show was *Wolfman Jack* whose real name is Bob Smith and who became a cult figure during this period. When introduced to Lieutenant Colonel Bob Cranston by Jack Brown, he called Cranston "Lieutenant." After Brown explained a bit of military protocol to him, he thereafter referred to Cranston as "Colonel, baby." His show was so different that a few of the overseas outlets refused to run it at first. Soon he became the most popular program on the system. When a German entertainment magazine ran a contest to find the most listened to radio program in Europe, it turned out to be Wolfman Jack, much to the embarrassment of the foreign broadcasters. Over the next fifteen years his popularity slowly declined. Other interests consumed his time and the quality of the show finally reached the point where it was necessary to cancel it. Although he wrote a "Dear Ronnie" letter to President Reagan complaining that AFRTS was treating him unfairly, the program went off and the letters of complaint were too few to count.

The most popular weekly program without doubt was Casey Kasem's *American Top 40*. The producer, Watermark Productions, would get an advance three days before the publication of the *Billboard Magazine* HOT 100. An AFRTS radio producer would be in the studio as the program was produced and add Department of Defense information announcements in place of the commercials. The program was then sent out on tape each Friday, going directly from the production house to the airport. AFRTS stations, except for a few in extremely isolated locations, were able to play the program the same week it was heard in the States.

This decade was also one of rapid growth in the television area. Mini-TV, a system whereby a playback unit and a monitor was supplied to isolated locations far removed from the nearest station was begun. Weekly shipments of videotapes were sent to them. Alaska and Europe were the first to begin this service. The Navy outfitted more ships with one-inch videotape playback units which fed monitors throughout the vessel. They also had the capability to play back 16mm film and this, too, was supplied. Circuits were developed so a group of ships could exchange programming on a systematic basis. Ship One would play a week's programming. That would go to Ship Two while Ship One started the following week's programming. Then, like musical chairs, Two would ship to Three and One would ship to Two. Eventually six or eight ships on the circuit would all be in synchronization and getting programming, in sequence, each week. Sounds simple enough, except that the logistics became a nightmare as ships were sent to and fro at a moment's notice while the Fleet Post Office tried to catch up with them to deliver next week's *All in the Family*. Somehow they performed miracles and by 1977, there were 34 land stations and 159 ships receiving the regular weekly program shipment. That same year the Mini-TV sites began receiving 28 hours of programming plus a special priority sporting event through nine circuits and 72 outlets. Two years later television had grown to 40 total circuits serving 283 outlets.

The 1980's. The radio music unit now reflected the increased number of minorities in the service. There was more Rhythm and Blues, more Soul, more Latino. Pleasing all the minorities was of high priority at the Hollywood program center but of fairly low success. As most outlets had only a single channel, pleasing such a diverse audience throughout the day became an impossibility although each group had specialized shows designed for their musical tastes at some time during the day.

In 1981, the Armed Forces News Bureau which had been moved from New York to Washington and renamed AFRTS-W, was moved to AFRTS Los Angeles. With it as chief came George Balamaci, and his operation continues to this day to send out a steady stream of news and sports to the world. All the major network and news service newscasts are recorded and re-transmitted on a rotating basis. A six second delay circuit is often used in order to give the operator a few seconds to cut out the commercials on the fly and insert public service announcements.

Perhaps the most important event of the decade was the entrance into the satellite era. It actually began in

At last...a new home for the Los Angeles crew. This is the new and efficient home of AFRTS, located in Sun Valley, a suburb of Los Angeles. (Photo Courtesy of AFRTS)

1979 when the studio in Elmendorf Air Base, Alaska, linked its transmission by satellite to isolated Galena, King Salmon and Shemya. So widespread was the signal that portions of their daily schedule were picked up and retransmitted from such far flung locations as Adak, Panama and Guantanamo Bay, Cuba.

The Los Angeles program center began feeding television by satellite in July, 1980. It started modestly, feeding an hour of ABC and NBC television network news to Alaska, Panama and Guantanamo Bay. In August, 45 minutes of CBS news was added and Roosevelt Roads, Puerto Rico, joined the satellite network. By year's end, feeds from CNN and *Good Morning America* were being scheduled as well.

The 1980's were without doubt the ten years which saw the greatest number of changes in AFRTS's history. Broadcast technology changed rapidly during these years and AFRTS moved with it. Black & white television disappeared from the system and color became the standard. Film disappeared to be replaced first by 2-inch videotape, then by 1-inch and 3/4 inch and finally Betacam SP for land based stations and Betamax two-hour cassettes for ships. A number of stereo programs were added to the FM service. The watershed change in the whole AFRTS system, and the one which has made the most difference in the quality of service, however, is SATNET.

SATNET, standing pretty obviously for Satellite Network, now covers the world from Los Angeles. Twenty-four hours a day, mostly in real time, satellites over the U. S., the Atlantic and the Indian Ocean are fed a constant stream of news, sports and time-sensitive programs. The three birds cover most of the earth. But it's one thing to put the signal up there, it's quite another to get permission to pick it up in the various receiving sites around the world. The fact that AFRTS stations are permitted to do so is essentially due to the far sighted preliminary work of Bob Cranston in selling the concept and arranging for funding to be made available as needed and to the efforts of two men: Mel Russell and Air Force Lieutenant Colonel Larry Pollack. Russell became chief of AFRTS at AFIS in Washington and Pollack has retired. Their monument is SATNET and the hours the latter two spent shuttling around the world negotiating for receiving rights from various countries make Secretary of State Jim Baker look like a homebody. Thanks to these guys, people around the world now all sit down to dinner at the same time. Half time. Because of them, when Joe Montana throws a touchdown bomb in the Super Bowl, cheers go up simultaneously from Adak to Zaire.

The only problem they were never able to solve is the inexorable march of the sun around the planet. Time zone differences cause *Good Morning, America* to appear at four in the afternoon in Europe and baseball games to begin at 3 a.m. somewhere on the globe.

All this paled to insignificance to the staff as they packed up and moved in mid-decade to new quarters in Sun Valley, a suburb of Los Angeles next door to Pacoima, the abandoned car capitol of the world. This meant a name change, of course, and the California group now became AFRTS-BC. BC stands for "Broadcast Center" and in no way indicates the era of the work produced there, which is purely late AD. The new building boasts a great deal of new equipment including the very latest video editing and off-line recording equipment. Outside it bristles with satellite receiving and transmitter dishes. Everything works, the roof doesn't leak, the floors don't creak and everyone has a parking space. Never once has anyone wallowed publicly in nostalgia and wish they were back in the Drop Forge Factory.

THE 1990's. As AFRTS ends its first half-century, this stroll through the years ends with a few gee-whiz facts and figures about the organization which, according to its director, "Buzz" Rizer, is the Department of Defense's "most important internal information medium."

✤Although numbers frequently shift with the political winds at home and abroad, AFRTS currently maintains manned radio, television or both in twenty countries. It supplies programs to unmanned, Mini-TV sites in 110 other countries.

✤The AFRTS Program Center in Sun Valley, California, contracts for, receives, edits, inserts public service announcements, checks for quality and content, arranges duplication, ships, tracks and receives back 90-hours of programming weekly to 23 different circuits supplying 160 land based stations and approximately 120 ships. (The number of ships varies constantly as only those

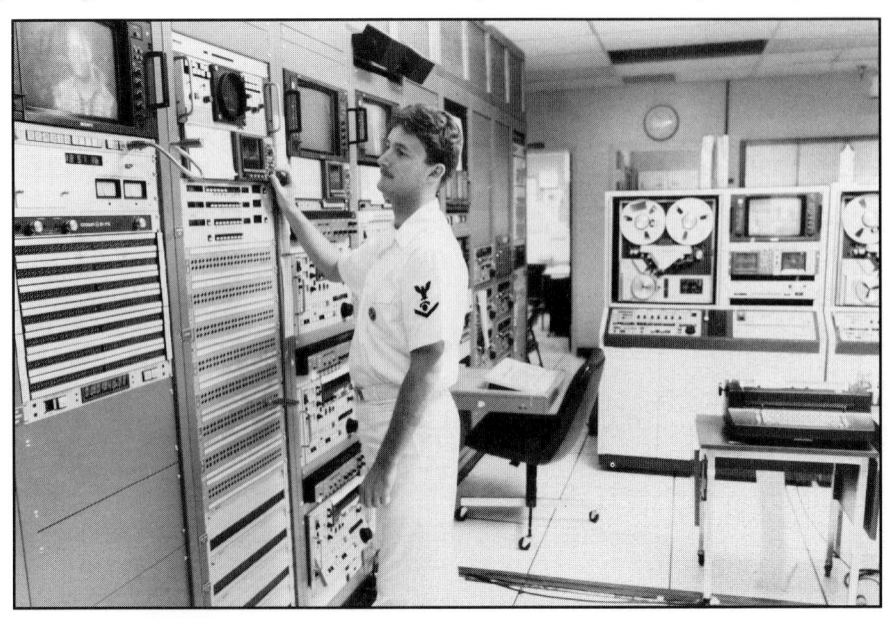

Program series are often supplied by the program distributor to AFRTS on videotape in batches of shows or as a complete season. Many programs — soap operas and timely programs are examples — are recorded in this room and given to the AFRTS producers to decommercialize and send to tape duplicating facilities for overseas shipment. Only a portion of this section can be seen here. There are banks of video recorders in various formats and through the patch panels and switching apparatus, programs can be recorded from domestic satellites or off direct lines from the major networks. (Photo Courtesy of AFRTS)

deployed are sent programming. Ships in U. S. ports can tune to shore stations.) At any given time, there are more than 80,000 videocassettes in circulation -- each of them accounted for and guarded to prevent unauthorized duplication.

❖Mini-TV systems are sent about 40 hours of programming each week. The number varies as small units activate, deactivate, or combine but the number hovers around 180 outlets.

❖The SATNET center at Sun Valley selects from all networks and cable systems a 24-hour a day diet of time-sensitive programming which it feeds to AFRTS television stations and networks through uplinks at Santa Paula, California and Southbury, Connecticut and downlinks from satellites over the Atlantic, the Pacific and the Indian Oceans. Programming includes all major network newscasts, sports events, CNN and any breaking news story. This material is encrypted to prevent unauthorized usage and is decrypted at the receiving end where it is integrated into local program schedules.

❖In order to get SATNET time-sensitive programming to the fleet, or other outlets unable to receive the signal, duplication facilities (DUPFACs) have been set up to receive, duplicate and deliver programming from Christchurch, New Zealand; Sigonella, Sicily; Bahrain; Rota, Spain; Diego Garcia in the Indian Ocean; Subic Bay, Philippines; Roosevelt Roads, Puerto Rico and Jedda, Saudi Arabia. This material is most often flown out directly to ships at sea where the crews can see, for example, Super Bowl with less than a 24-hour delay.

❖Radio remains highly important in that overseas listeners can get it during their duty day wherever they are, thanks to fatigue-jacket sized portable radios. AFRTS supplies stations 80-hours a week of programming; mostly disk jockey style programs produced in the Sun Valley headquarters featuring top DJ talent. It also supplies six hours weekly of individual music cuts selected from the popularity charts used by most radio stations. These are used by the overseas DJs to prepare their own local programs and are retained indefinitely in the station libraries.

❖Radio stations also have available to them several other music sources. SATNET uses its audio sub-carrier capability to deliver a voice channel, two multi-service channels and a two-channel stereo feed to stations with receiving capability. AFRTS also has space on INMARSAT (International Maritime Satellites) and feeds all overseas affiliates and some ships the Voice Channel including news and sports through this system which requires only a small receiving dish, unlike the giant dishes required for the SATNET television reception. The quality is excellent and long gone are the squeals, squawks and garble suffered for so long during the shortwave era.

❖Radio programming, aside from the live feeds, are supplied using various technologies including vinyl disks, tape and compact disks. Digital Audio Technology, DAT, is just over the horizon and will be used as quickly as the government, the music industry and the manufacturers settle their squabble over the advanced process. All recorded music is specially mastered and pressed for exclusive AFRTS use. Distinctive labels make it immediately recognizable and prevent it showing up for sale at the corner music store. The same holds true for videocassettes which are specially manufactured with a blue case, recognizable a block away as property of the government.

❖With the help of computers, AFRTS-BC keeps track of more than 3,500 television and 3.100 radio spot announcements and schedules them appropriately in the hundreds of availabilities required to be filled each day. The majority of these are on subjects requested by various military departments and are produced by two contractors, Northwest TeleProductions located in Minneapolis and Video Ventures in Miami. The coordination and supervision is provided by the Radio and Television Production Office (RTPO), a unit of AFIS in Washington. Some public service announcements, which are germane to the overseas audience, are obtained from industry sources such as the Ad Council.

ANNOUNCER: WELLLLLL...NOW YOU POSSIBLY KNOW MORE THAN YOU CARE TO KNOW ABOUT THE INNER WORKINGS OF AFRTS. SO NOW LET'S MOVE OUT AGAIN BOTH HITHER AND YON AROUND THE GLOBE AND SEE HOW THESE INNER WORKINGS WORK WHEN THEY ARE PUT INNER USE. A LITTLE TRAVELING MUSIC PLEASE...

Today in the new studios in Sun Valley, the AFRTS Broadcast Center features the latest state-of-the-art radio studios. No more 16-inch transcriptions. Now it's CDs, cassettes and stereo disks. (Photo courtesy AFRTS)

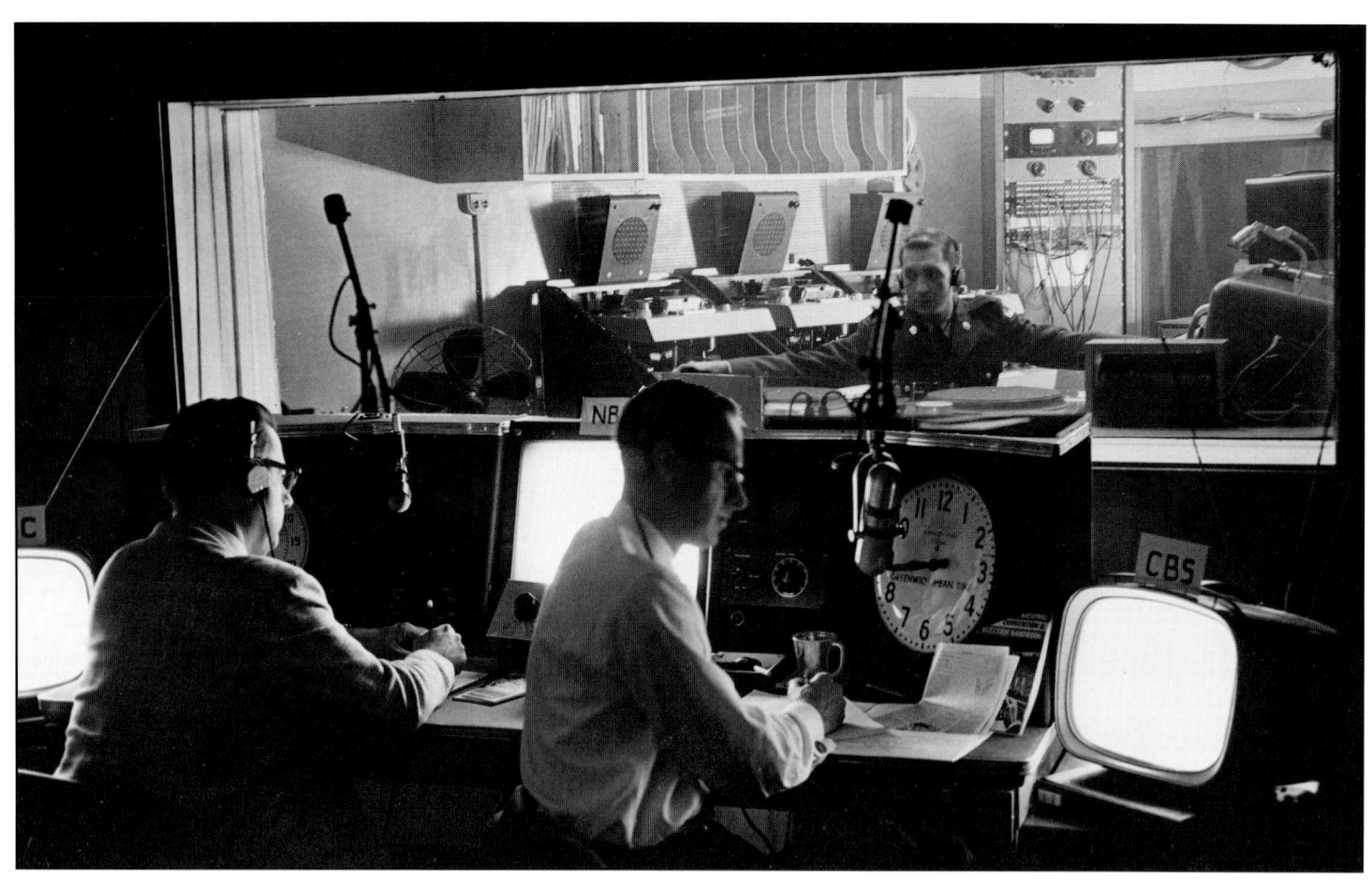

As time went on, the former drop forge factory on McCadden Place in Hollywood came to look more and more like a radio station and less like it had had a forge dropped on it. Equipment improved slowly but surely but there was never quite enough room to turn out the quantity of material required. Shortwave news, sports and special events were broadcast around the clock and entertainment disk jockeys were continually jockeying their disks waiting for studio space in which to do their thing. Staff members often found it easier to come to work on the Pasadena bus rather than try to find a parking space. And many of those didn't even live in Pasadena. (U. S. Army Photos)

CHAPTER TWENTY-TWO

Being a compilation of strange but true tales of broadcast activities around the world...many of them stranger than fiction.

"Why shouldn't truth be stranger than fiction? After all, fiction has to make sense."
—Mark Twain

AROUND THE WORLD IN MANY WAYS

In Italy...

Enough shillying and shallying. The time has come to discover the answer to the question, "Did SEN in Italy ever get on the air or did it continue to broadcast over power lines into people's toasters?"

Yes.

When this subject was last examined, the audience, the SEN staff and the local and theater commanders were all demanding over-the-air broadcasting. The Italian government-of-the-month was being balky. To understand why, it is necessary to understand that Italy, in the 60s, had basically two strong political parties and a bucketful of minor ones. The Christian Democrats were in power and were closely followed in numbers and control by the Communists who provided strong opposition. Shortly after World War II, the Christian Democrats had passed a law which said that all broadcasting in Italy would be controlled by the Government. Translated, that meant that they could legally, if slightly unethically, prevent the Communists from reaching the population on radio and television. This suited the Christian Democrats just fine although the law was neither Christian nor democratic.

Then along came the U.S. Forces and the Southern European Broadcasters with a request to be allowed to broadcast. The Italian government, while sympathetic, feared that if they allowed the Americans on their airwaves, the opposition — the Communists — would shake their fists, demand their rights and point out that the Government was allowing foreigners to broadcast while preventing good, loyal, Red Italians from doing so.

Although a number of meetings were held over the years between the American Embassy and the Italian broadcasting authorities, most came to nought. Finally, in 1967, the Embassy announced that an accommodation had been reached. The Italians had indicated that they would not give the U. S. permission to broadcast, the Embassy told SEB, but on the other hand had also indicated that if the U. S. went ahead and did it anyway, chances were no one would say anything. This is roughly analogous to the government of, say, Tanzania setting up a network in the U. S. and expecting the FCC to ignore it. It may be the loosest and most tenuous authority under which AFRTS ever operated anywhere in the world.

The U. S. authorities put more restrictions on SEB than did the Italians. They said broadcasts had to be on the FM band as far on top of the dial as possible so Italian receivers would be unlikely to find them. Power was limited to 50 watts. And there was to be no, repeat no, publicity of any kind including announcements in base or post newspapers. On receiving the news, joy reigned at SEB until manager Al Edick pointed out that SEB didn't have any transmitters, 50 watts or otherwise. To the rescue, came engineer Frank Paris who proceeded to build the first one in just a few days. Peeling the acetate off the aluminum disk of a 16-inch acetate recording and using that as the metal base, Paris found enough spare parts to construct a transmitter. It fit nicely on top of a filing cabinet in the military newspaper office on the second floor and featured a wire antenna run out the window. Its frequency was 106.

Because there was no advance warning that the station was on the air, Edick began calling people by phone and asking them to tune to 106. Most of them said they couldn't hear a thing. Two stories high just wasn't high enough for an antenna. Those that could hear it were delighted and began calling their friends and a pleased and surprised audience soon developed. The audience was not the only group who were surprised. So were the Criminal Investigation people, the CID.

Al Edick explains what happened this way:

"One of our civilian staffers, let's call him Herb Glover for want of a better name, lived in an Italian apartment building in town. Shortly after we had turned on our home-made transmitter, Herb was tuning in his radio and picked up some strange noises. He heard other voices and street sounds mixed with the SEB signal. He soon deduced that there was a second transmitter somewhere and it must be very close to his apartment. Using a small radio, he wandered around and finally found a "bug" hidden in the transom over a doorway in his building. When he disabled it, he was able to hear SEB again. Apparently, for reasons unknown, the CID was using it to conduct surveillance on someone in his building. He took the bug and

161

placed it in his young daughter's bedroom, becoming the first person I knew who could monitor his child's activities while listening to his radio. I always wondered what the CID thought when they heard childish noises coming from the hidden bug and what they did when they discovered it missing from the transom. For sure they must have done some scampering to change the frequencies of their remaining bugs.

Soon commercial transmitters were obtained, as were proper antennas and SEB (still known then as SEN) was on the air at Livorno, Aviano (near the Yugoslavian Border north of Venice,) and in Naples. The Navy put one on the air in Sigonella, Sicily, and SEB added outlets in Verona, Rimini, Gaeta, San Vito and on Sardinia. The connections were by land line telephone circuits and, while better than nothing, were of generally poor quality. Today the network uses satellites and the affiliates are linked together, both radio and television, with broadcast-quality circuits. Still, low quality is better than no quality which is what the network had previously and which it got again on January 18th, 1969. On that bleak day, they were notified by the Italian government to kindly vacate the Italian airwaves.

The order came down through Embassy channels from Rome. Although they didn't know at the time, it later turned out that someone had complained about the foreign broadcasts and the Italian government felt it had no choice but to shut them down. Instructions were: shut off all transmitters at noon, Monday January 20. The commanding general of the Southern European Task Force, Major General John Hughes, called a meeting in his headquarters in Vicenza about the situation on Sunday, the 19th. He phoned the embassy in Rome several times during the meeting and arranged for a head to head meeting with those damn State Department tea sipping, pussy-footing, cookie pushers at the embassy for Tuesday. He told Edick to attend and prepare to fight. He also pleaded with the embassy people by phone for an extension of at least one day. He pointed out that Mr. Nixon was going to be inaugurated as President at 5 p.m. local time on the 20th and it would look like the Italians, as a protest, had ordered SEB off the air just so people couldn't hear that event. No dice, said the embassy. Get off the air as ordered.

Edick headed for Rome and the network headed for extinction. As ordered, they signed off at noon on the 20th. First they played, *You Gotta Do What You're Told* which was followed by *The Sound of Silence*, both musically and actually.

Edick and the SETAF public affairs staff met at the embassy and got nowhere although they stressed that they had to get back on the air quickly. General Hughes called several times during the frustrating meeting and the embassy people were aware he was fighting mad and very interested. Back they went to Vicenza and two days went by with no word from Rome. General Hughes decided he had enough and he informed the embassy he was having the station sign back on the air Sunday, the 25th, approval or not. He told the staff he wanted "the best Goddamned version of the National Anthem played at sign on." Herb Glover, he who found the hidden bug, edited together a blockbuster using different recordings including the Boston Pops version and one by the Salt Lake Tabernacle Choir, both mixed with assorted sounds of fireworks, cannons and jet aircraft. The general loved it so much they made him his own personal copy.

The insidious thing about the entire episode is that it should never have happened. They never heard another word from anybody, U.S. or Italian, and remain on the air undisturbed to this day. Nor did they ever find out who filed a complaint in the first place. So joy returned to SEB. At least it did for a month. Then they had to broadcast the tragic news that their champion, General John Hughes, was dead in a fiery airplane crash at the airport in Milan.

IN THE AZORES...

More than one AFRTS person has found that, while they have a nice job, it does not always involve sitting in a comfortable chair playing music and saying brilliant things to an enthralled audience. Once in awhile, the assignments can get downright nasty. John Bradley again reappears in a few examples:

1957 was an International Geophysical Year and mother nature, sweet thing that she is, decided to be helpful and create a volcano for her earthling children to study. She picked a spot a few hundred feet off the coast of the Azorean island of Fayal which was about 90 miles away from Bradley and the AFRTS station on the island of Terciera.

For several weeks, Bradley filmed the spectacular event as the cone rose from the sea and spewed smoke and ash tens of thousands of

Volcanos are nature's way of making mortals humble. This one sprung up in 1957 in John Bradley's back yard off the island of Fayal in the Azores. He filmed it daily and the spectacular shots got wide coverage around the world. He also learned that the curriculum at Navy flight schools don't have lessons about not flying through clouds of volcanic ash. His pilot tried it and barely made the journey home before the engines ground themselves to bits. (Photo Courtesy John Bradley)

This is a typical AFRTS editing bay. Here programs as received from other sources are edited to remove commercials or other non-appropriate material and insert public service announcements more germane to the service person who will see the final product overseas. Dozens of duplicates of the videotape master produced here go to "circuits" consisting of two to six or more stations who pass then on from one to the other. Each week about 90 hours of programming is sent to some 23 circuits. (Photo Courtesy of AFRTS)

feet in the air. Filming was done from a Navy twin-engine amphibian designated a UF-1 and which was ideal for the job as it had a door from which the top half could be removed for easy filming. The films were returned to the Lajes base and, because this was a stop on the busy trans-Atlantic route, were soon out of the developer and on their way to networks in New York for nation-wide release. Copies were shown on AFRTS in the Azores within hours.

Just about the time the flights were becoming routine, the young pilot decided it would be an experience to fly through the cloud produced by the eruption. It turned out to indeed be an experience. The major solid ingredient of volcano effluvium is pumice which is a very efficient abrasive. The aircraft engines sucked up an unknown quantity of pumice as it sailed through the cloud and immediately started to make strange noises of the type anyone inside an aircraft would rather not hear. As the moving parts continued to chew themselves up, the engines ran rougher and rougher and so did the language of the passengers as they prepared for a possible water landing. The pilot managed to reach the first few feet of the runway. The plane required two new engines. The pilot disappeared from the island, never to be seen again.

When a SAC KC-97 went down north of the island, Bradley was aboard the 57th Air Rescue Squadron C-54 which was dispatched immediately. His trusty Bolex recorded a textbook rescue operation. The wreckage of the downed plane was found and filmed as it bobbed in the waves below. Then the radio operator picked up a signal from the crew as they floated in their life raft and were soon spotted and filmed. After dropping emergency equipment, the Air Rescue plane took off looking for water-level help and, sure enough, there was a Liberian freighter just over the horizon. They vectored it toward the raft and circled overhead for five hours until the ship arrived. Then Bradley was able to get pictures of the actual rescue and have them both on his evening newscast and on their way to the States the same night. It was a picture-perfect operation.

These are examples of why broadcasters generally love their work. While there can be long periods of boredom, just as in any other kind of job, there are those wonderful days which start out like any other and end up, sometimes days or weeks later, finding yourself in another city, country or improbable situation. Bradley went to bed one night and was aroused to go to Ghana to cover the evacuation of refugees fleeing out of the Congo. The trip took him from the Azores to Chateauroux, France, to Casablanca, Morocco; to Dakkar, Senegal; to Accra, Ghana. Then the return via Nouasseur, Morocco, back to Chateauroux and home to the Azores. All routine kinds of work except for a few things like being deathly ill much of the outbound flight from shots for diseases that hadn't even been discovered yet; for shooting pictures of Aeroflot planes bringing in people to that troubled area under the guise of relief workers and which were shown as proof to the United Nations of Russian intervention; of spending nights trying to sleep on the floor of parked aircraft; of flying while strapped to the floor well forward in the plane so it could accommodate a few extra evacuees and helping one elderly gentleman who suffered a severe heart attack shortly before landing in safety.

SOMEWHERE OVER THE PACIFIC

Other people riding in airplanes had nicer experiences. Dwight King, now editor of a magazine devoted to those who served in the China-Burma-India theater, was the bombardier on a B-29 flying out of Tinian with the 468th Bomb Group on bombing missions over Japan in 1945. He recalls a neighboring special bombing squadron had dropped "a big one" on August 6th and their commander had tried to explain just how big it really was. Somehow, it was too big to register.

On the 14th of August, he and the crew of the "Lady Be Good" took off for still another raid over Japan. They dropped their bombs on the Hikari Naval Arsenal and were partway back to Tinian. Someone tuned in the AFRTS station on Iwo Jima and put it on the intercom system so they could all hear some music from home.

Some GI's had it rough. Back in 1951 in Europe, Sergeant Mel Riddle was forced to interview Frank Sinatra and Ava Gardner while Sergeant Bob Harlan was ordered to do the same with Janet Leigh and Tony Curtis. It hardly seems fair. Other guys got to go to the rifle range and on field maneuvers.(U. S. Army Photographs)

Instead what they heard was the news they had been waiting for. The Japanese had surrendered because of the "big one" dropped from a plane from their Tinian home base.

Their mission that day turned out to be the last air raid of the war in the Pacific. And that broadcast is one no one who heard it will ever forget.

IN THE PHILIPPINES

The two major bases — both with radio and television AFRTS outlets — in the Philippines are Subic Bay and Clark Air Base. Early on, neither was considered a particularly great assignment. Dave Johnson, who had a long career at a number of Air Force stations, claims that those assigned there were either being tested for future greatness or punished for unknown past sins. In the mid- to late 60s, this was doubtless true.

Later both bases stood out from their surroundings like a diamond in a cow-plop. Beautifully maintained, surrounded with manicured, park-like grounds and carefully swept streets, the bases immediately changed character when leaving the main gate. Olongapo, outside of Subic, and Angeles City, outside of Clark, are the sort of places for which the word "slum" was invented. The stations are gone now, courtesy of a Philippine senate and Mount Pinatubo, a nearby volcano which many who served there feel was acting on orders of a vengeful God by raining fire and brimstone on the local inhabitants. No one knows what the AFRTS stations did to offend Him, if anything, but in any event they were caught along with the more than 600 Filipinos who died in the eruptions. Today the station at Clark is a ruin, it's roof caved in under the weight of the falling ash and the equipment inside destroyed. Subic, although damaged, managed to stay on the air for awhile but was ultimately doomed by the failure of the negotiations to maintain bases in the Philippines.

Johnson is now a civilian working for Northwest TeleProductions, an independent production company which has a contract to produce spot announcements for AFRTS. In the 1960s, he was an Air Force Sergeant at Clark. He recalls a meeting called by the commander for the whole staff one early morning. When the commander and his straphangers left, the group discovered a nearby snack bar which was well supplied with San Miguel beer. A party began which went on for hours. When the time came to begin preparing the evening newscast, the crew was in rather fragile condition and dragging themselves through their various duties. In attendance at the newscast was a VIP guest, the commander of the Weather unit which supplied weather news to the station and who was visiting from Hawaii. He was somewhat overcome to see on the rear projection screen behind the weather man a silhouette rise from the floor, stretch its arms, yawn, step out into the middle of the newscast and ask no one in particular whether the Commander's meeting was over yet.

Johnson also recalls a station manager who was blessed with such unusually gifted leadership that he took an intense interest in the most minor details of running the station. With awesome dedication, night after night he would call the control room from the

Officer's Club Happy Hour during the live newscast and try to direct it over the phone. Nor did this kind of dedication stop after he returned to the States. One night about three months after his departure, he phoned at the same time via the Pacific cable and continued to offer his help.

The staff, of course, appreciated this kind of commitment, just as they had that of a previous station manager who was blessed, or cursed, with the world's finest hearing. He claimed he could hear the scanning lines on his neighbor's television set. This being the case, they showed him every military courtesy — by blowing a high-frequency silent dog whistle whenever he appeared.

They also enjoyed showing their appreciation to the Program Director who didn't have a hair on his head. Each enjoyed spraying his dazzling dome with dulling spray every time he went on the air. They were even happier to do this after they found out the meanings of the words "caustic" and "trenchant" which he was fond of using in their regular evaluation reports.

One young airman assigned to the late shift at the station was in the habit of working in a tank top, shorts and flip-flop sandals. He knew no one would bother him about this unmilitary attire in the middle of the night. He was wrong. The Wing Vice Commander dropped in one night and exploded at the sight. This young colonel knew a serious situation like this demanded his special abilities as a leader and he proceeded to stand the young airman at attention and chew him out with words that would make Eddie Murphy cringe. The airman may have looked dumb, but he wasn't. Before coming to attention, he opened the microphone switch and fed the colonel's words to the entire Philippine audience who were amazed at his colorful approach to military discipline.

When last heard from, the colonel was assigned to a boy's military academy in Louisiana where he is helping prepare the next generation of leaders.

IN GERMANY AT AFN

Bob Cranston's military career is part of the fabric of the entire military broadcasting system; much of it at AFN. But if his military career is the warp of the fabric, Bob Harlan's is certainly the woof as the civilian counterpart. Eighteen commanders came and went during his thirty-five year tenure running from 1949 until retirement in 1986. During that period, he progressed from the lowliest of privates in the rear rank to the premiere civilian Program Director's position which he held longer than any of his predecessors which had included Roy Neal, Keith Jamison, Bruce Wendell, Emil Schweitzer, Hunton L. Downs, Donald J. Brewer and Francis T. McLaughlin.

Fiercely protective of the broadcast quality he demanded, he taught several generations of broadcasters the basics of their craft, and always in the politest way possible although they may well have strained to the limits his Southern-bred graciousness. He also taught — firmly but gently — a number of network commanders the difference between commanding a group of free-wheeling broadcasters and more disciplined types from combat-trained units. Serving with eighteen different

Bob Harlan, perhaps AFN's most distinguished alumnus, is shown here in his younger personna shortly after doffing his uniform and donning civies. Some may wonder why he is standing so far from the microphone. The fact is, Harlan's voice is so distinctive and so powerful it can pulverize granite tombstones three zip codes away. (Photo Courtesy R. J. Harlan)

One of AFN's most valuable allies for more than thirty years has been a German publication called **Your AFN TV Guide.** *Each month it publishes the television schedule for AFN (and a separate edition for SEB in Italy) and distributes it at no charge to American TV viewers in Europe. An invaluable relationship has developed and the magazine has rendered every bit of help requested by the network in keeping viewers informed about what is going on. It has sponsored promotional contests for the networks, printed reams of publicity and, during Desert Storm, even prepared a special Saudi Arabian edition for troops stationed there.*

Here Mary-Michaele Brooks, the magazine's advertising and merchandising manager, tries to convince Bob Harlan (R) to have another beer during a social gathering. The author (C) already has his. (Photo Courtesy AFN TV Guide)

AFRTS stations, before TV began swallowing much of the budget, often went to any length to make their radio broadcasts sound credible. When AFN decided to do a Halloween spook show in which much of the action took place inside Cheop's Pyramid, it seemed only logical to send a team to Egypt from Germany to gain a bit of realism.

This group returned with a show guaranteed to make your hair stand on end, as well as a number of special features for use in other programs. Reading left to right, and not counting the Sphinx, are John Keel, writer, Jim Melton, engineer, Leo Zales, actor, Bob Harlan, producer/director, and John Leffingwell, actor.

Harlan decided he liked this sort of work and stuck around for a mere 35 years. In broadcasting, that roughly equates to being as permanent as the pyramid. (Photo Courtesy Bob Harlan)

commanders, he was able to explain to many that you could order a service-person to build a bridge, climb a pole, dig a hole, repair a tank or construct a latrine but no way could you order them to be funny or clever. Being one himself, he understood broadcasters and, as he moved from the rear ranks to the front office, he was able to protect the broadcasters from the sometimes lamentable ignorance of line officers who found themselves assigned to an organization, the purpose of which often eluded them. Such training was not always easy. How do you politely tell a commander not to have a young announcer court-martialed because he did not have a return feed from the network and screwed up a remote broadcast? The commander insisted the man should have had the sense to get one from the supply room before he went out on the job.

Of course, even Harlan had to learn his trade although he had worked for a short time at a Gainesville, Florida, station during his college years. Still, he was fairly green when he debarked from the USS Patch in Bremerhaven in 1949, given a ticket to Hoechst, Germany and told to report to AFN. He soon found himself a member of an unforgettable cast of characters whom he describes today as "talented announcers, aggressive newsmen, inventive writers, enthusiastic director-producers, soldier and civilian, German and American, male and female, alcoholics and teetotalers, gays and straights and all super-talents. The level of original creative energy was wild and unreal. The scope of partying and sexuality was intense, equally wild and unreal. The senior, single non-commissioned officers even had their own private house with servants."

His first assignment was to keep the Sports Chief from passing out while tape inserts were running during his evening sports show. He was also appointed liaison with the *Goldener Rose* gasthaus across from the castle housing the network. When the military called an alert, which was frequent in those early days, his job was to dash to the gasthaus and get the staff enjoying their steins of good German beer back on duty. Handling tough jobs like these led to his promotion as Night Supervisor and soon he was also narrating documentaries, reading news, doing dance remotes, special events and celebrity interviews.

Although the hundreds of budding broadcasters who passed through his hands during his thirty-five years with AFN will find it hard to believe, even Harlan had his share of mental disconnects. His very first DJ show found him searching for a clever ad libbed introduction to a song and listeners heard, "Now let's all crawl on those Dinning Sisters as they take that Night Train to Memphis." Describing a military change of command ceremony from Heidelberg, his colorful description of a group of female soldiers marching past his microphone as "a group of crack soldiers" followed by a long pause as he realized what he had said probably was something less than generally appreciated in some quarters.

In spite of such occasional gaffes he continued to be assigned to the type of special events which made life at AFRTS stations a constant joy and challenge to the staff.

On one evening, he found himself assigned, along with partner Mel Riddle, to interview Frank Sinatra and new bride Ava Gardner, Tony Curtis and his new bride Janet Leigh, Rhonda Fleming and songwriters Jimmy Van Heusen and Johnny Burke. It must have been a memorable interview because twenty-five years later Miss Fleming and he met in Hollywood and she commented, "Oh, yes, we met in Wiesbaden a long time ago." It made his day. One memorable assignment was being sent, along with a portable tape recorder, to London and Paris to gather interviews to commemorate AFN's tenth anniversary in 1953. In Paris, he interviewed Bing Crosby, (whose son Gary was later to join the AFN staff,) and Art Linkletter. The Linkletter interview was postponed a day because Harlan made the mistake that every broadcaster makes — once — in his career. He arrived at the scene of the interview, in this case the Ritz Hotel, and discovered he had forgotten to bring tape. In London he managed to get interviews with Robert Taylor, Ava Gardner, Mel Ferrer, Gene Kelly, Lana Turner and Pier Angeli. Back stage at the Palladium, he talked to the then team of Dean Martin and Jerry Lewis. Douglas Fairbanks, Jr., hosted him at his London townhouse and he taped Frank Sinatra backstage.

Another trip was to Cairo to produce a Halloween spook special. It was actually recorded in Cheop's pyramid. These were the great days of AFRTS radio. The military draft provided a constant supply of well trained, imaginative, professional broadcasters. Television had not yet drained the talent pool. Money was available and interference from higher commands was generally minimal. AFRTS stations everywhere had their share of exceptional talent as did AFN which during the 50's and early 60's boasted such talent as Nick Clooney, Rosemary's brother, Art James, Jay Merkle, Frank Brooks, actor George Kennedy, Country-Western performer and Johnny's brother, Tommy Cash, Ted Daniel, Alan Landsburg, producer Steve Binder, Ed DeFontaine, Bill Ramsey, New York Times columnist Bill Safire, ABC Radio sports chief Shelby Whitfield and TV host Gary Collins.

During this period Harlan's term of military service expired and he was offered a civilian position which he accepted. As the years passed, he became network chief of production, program director at Kaiserslautern, Munich, AFN France and, following the untimely death of Fran McLaughlin, was appointed Program Director for the entire network in 1967. It was a position he filled with distinction and one from which he set the tone of the entire organization until his retirement in 1986.

IN WASHINGTON

While all the normal, and abnormal, broadcasting activities were going on around the world, AFRTS and its parent organization in Washington continued valiantly to support the affiliated stations and networks. Although the Congress was generally sympathetic and helpful, it wasn't stupid and realized that AFRTS had a higher mission than playing rock 'n' roll records and John Wayne movies. These were the sugar coating on the pill to get people to watch and listen. Inside the pill was the message which it was, and is, considered necessary to get to the audience. These messages fit in two general categories. Many are of only local interest and are produced by the stations. They include such subjects as reminders that the commissary will be closed on Tuesday, what is playing at the local movie, entertainment coming to the various local clubs, educational opportunities at the education center, frequent weather reports and anything else the audience needs to know about its "home town." They also keep local audiences advised during natural disasters such as typhoon warnings on Okinawa or volcano reports in the Philippines. At such times, the local command generally has only the AFRTS station as a conduit to the people for whom it is responsible. Keeping the local audience advised of what is happening it its own area is the primary function of any AFRTS station.

This is one of the ways AFRTS communicates with its affiliates around the world. Located adjacent to the building, satellite dishes receive signals from all the domestic satellites and the programs selected can either be recorded, edited, duplicated and sent out on videotape or can be immediately retransmitted to stations around the world.

So sophisticated are modern day techniques — such a far cry from the original shortwave signal and acetate disk — that both stereo music and voice channels can be transmitted along with the television picture and sound. (Photo Courtesy of AFRTS)

Batteries of videotape recorders, only a portion of which are seen here, receive programs from fiber optic circuits, co-axial cable or satellites and are recorded for editing and shipment overseas. Complex patching equipment allows for multiple recordings from the same or separate sources. (Photo Courtesy of AFRTS)

On a more cosmic level, the Department of Defense also has a number of subjects which it must pass on to the overseas audience. This, early on, was a fairly hit or miss proposition. Stations and networks were supplied with a subject list and produced to the best of their abilities, spot announcements on subjects as dull as Financial Planning, Safety and Military History. Larger organizations often turned out reasonably credible announcements. Smaller ones with a staff of two or three broadcasters were simply not able to perform their local function and produce acceptable material. The AFRTS program center in Los Angeles produced some spots but in general they were done inexpensively and looked it. Audiences were now used to commercials that used the latest production values and a spot with a voice reading an announcement over a series of tacky slides just didn't grab them. Besides, there were so few of these spots, which were inserted into the programming distributed by Los Angeles, that they would show up, it seemed, in every other program.

In 1981, AFIS formed a group originally called IMPO standing for the Internal Media Program Office. Later it was changed to RTPO for Radio Television Program Office. In a burst of brilliance on the part of someone, it was decided that commercial production houses were better prepared to turn out spots that looked like Stateside products. Money was obtained and contracts were let to producers. Because most of them were somewhat in the dark about the military audience, it was necessary to find someone who could educate and direct the process. Who better than...(drumroll and cymbal crash)...AL EDICK.

Back he was dragged to Washington from Italy, leaving a trail of tears and tomato sauce. He formed the office, wrote job descriptions for the staff and came up with a mission statement and operating plan. Here's the way he tells it:

"The first problem was, how do we pick the subjects? Our first idea was to go to the Services and ask them what THEY wanted to tell the troops. Back came the answer: the troops want to see real commercials about beer, cigarettes and Chevrolets. That being out of the question, we went back and said, 'Give us something we can use. What are your problems?' Back came the same tired old subjects: Financial Planning, Safety, Military History.

"We decided to pick the subjects ourselves. From my own experience, I had a good idea of the problems.

"Our first series dealing with a 'social issue' was on rape. We worked carefully with experts at the Rape Prevention Center in Washington and decided to take a different approach from past Public Service Announcements. We decided to talk to the men, not the women as previous announcements had done. We would talk about the myths men believe about rape. (When she says 'no" she really means 'yes.") When the spots were finished, we began inserting them in programs and the firestorm hit. We heard from Commanders, all of whom claimed they had no rape problems and spots were inappropriate. Members of the audience wrote in that they didn't want stuff like this in their living rooms where children could see them. Then letters started

The SEB outlet as Sigonella, Sicily, has a two-fold mission. As a station, it serves the large base there. It is also a duplication facility (DUPFAC) which records and duplicates timely programs for ships of the Sixth Fleet deployed in the Mediterranean and gets the programs aboard the ships by the fastest means possible — usually by aircraft. Crews aboard can then see the latest news and sports with minimum delay.

Shown here a portion of the master control facility at Sigonella. (Photo courtesy Miltrends Verlags GmbH)

For years, while all the bureaucratic infighting to grow into a major broadcast network was going on, SEB in Vicenza worked out of this building on the military base. They still do, but old timers would never recognize it now. Another story has been added and room provided for everyone. No longer does the commander's office echo because it was in a former bathroom with tiled walls. (Photo Courtesy Miltrends Verlags GmbH)

to come in that made us realize that there was indeed a real problem and it was one that Commanders simply did not want to recognize.

"Example: I made a trip to the Far East that year and met with a U. S. Commander who complained about the rape spots. He had no rape problem in his command, he said. Later, in a private lunch with the Public Affairs Officer, I learned that there had been eight rapes on that base in the past month.

"Some subjects got a poor reception. When **Stars and Stripes** began running a lot of letters complaining that the folks at home no longer supported the troops and families overseas, we produced a series called 'Thanks for Serving.' It featured people in all walks of life thanking the people overseas, used a lot of music and dancers and was a really glitzy spot series. We, and the overseas commanders we previewed it for, were highly enthusiastic. The whole thing screamed 'America is behind you.' The troops hated it. No wonder. The spots had barely begun to take effect when President Reagan announced that there would be no pay increase for the military that year."

Edick or his staff, including Sharleen Dunn, Jennifer Moore, Warren Lee and others, accompanied the production crews to the various locations to insure accuracy and, because most shoots were on military bases, to smooth the way. It didn't always work. Shooting an Air Force spot series at Andrews Air Force Base, everyone was warned that there were some highly restricted areas and not to wander around. In spite of this, an inquisitive actor decided to take a look at the Presidential aircraft and soon found himself spread-eagled on the tarmac with weapons pointed at him. After much talking, he was released but the crew was escorted to the main gate and cordially invited never to return. That night the script was rewritten and the Air Force spot became a Navy spot shot at Norfolk, Virginia.

A drug spot featured a real FirePERSON, a young service woman serving with a base fire department. It showed her doing her job and the tag line had her turning to the camera and saying about drugs, "For me it just doesn't make sense." Four months later she was bounced from the service for Drug Abuse, but on the sensible theory that not too many people knew her, the spot continued to run.

Another drug spot featured a group of service men playing touch football. Edick asked the base public affairs officer to furnish 15 men and he did. The shoot went well and the spot showed the men playing a rousing game, at the end of which one of them broke away and, looking into the camera, extolling the virtues of being "clean." It wasn't until the next day they discovered that the entire group of 15 were druggies who had just started a rehabilitation program. They kept the spot, though.

Working with a large number of production companies turned out to be a bad idea in that too much time had to be spent teaching the creative people the ins and outs of military life. Long haired males kept appearing in spots. Once the chevrons on a uniform were upside down. They were little things but ruined the credibility of the message with the audience who, if nothing else, were experts on haircuts and chevrons. Today the spot announcement contracts are with two companies, Northwest TeleProductions and Video Ventures. Both of them, after all their experience in turning out a total of more than 3,000 spot announcements about military subjects, are bona fide experts on what the military viewer needs and wants. Helping them stay that way are Dave Johnson, now a creative producer at Northwest TeleProductions and, Al Edick at Video Ventures.

❖ ❖ ❖ ❖ ❖

ANNOUNCER: TIME NOW FOR ME AND YOUSE, TO BUCKLE UP OUR COMBAT SHOES. OFF WE GO, ON A TROPICAL CRUISE.

❖ ❖ ❖ ❖ ❖

Originally named CFN, the Panama network was renamed SCN (for Southern Command Network) shortly before the arrival of Gerry Fry who was to be its program chief for twelve years. He later moved on to be Chief of Programming for Navy Broadcasting in Washington and later still Programming Chief for the entire AFRTS system, based in Los Angeles. This is essentially the way the headquarters in Panama looked when Gerry arrived to take over as head civilian.

The interior of the Panama station. Seen here, the reception area and news room. (Photos courtesy AFRTS)

CHAPTER TWENTY-THREE

In which is covered a number of fascinating topics from the tropics.

We're Beaming on a Slight Isthmus

Although forced to leave the air following the attack on Pearl Harbor so as not to act as a homing signal for possible attacking Japanese aircraft, as World War II began to concentrate action far across the Pacific, AFRTS again signed on from the Isthmus.

Panama was, and remains, a political time-bomb like many of its Central American neighbors. Some of its fiercely patriotic inhabitants saw the Panama Canal, operated by the U.S., as a ten-mile-wide insult to its national integrity. Although the great majority of the population recognized the political and economic benefits of the gigantic canal operation and the American troops stationed in their country, some minority groups, with Latin volatility, continued to stir up pots of trouble in the form of riots and demonstrations.

The AFRTS stations in Panama were known first as the Caribbean Forces Network, CFN. In 1957, the program director of the AFRTS stations in Panama, Jim Pattison, was covering the ceremony marking the first display of the U. S. and Panamanian flags being flown together as a symbol of unity. As his station cameramen filmed him, a Panamanian, apparently suffering from poor eyesight, leaped out of the crowd and proceeded to try to filet Pattison with a knife. Pattison required hospitalization and found out later that the attacker had mistaken him for "Fats" Fernandez, an anchorman on a Panamanian station.

Pattison, a contract civilian, finished out his contract but, for reasons best guessed at, declined to renew it. He departed in late 1963, thereby missing the bloody riots of January 1964. These, too, were caused by flagwavers — both American and Panamanian. Twenty-nine persons died following riots touched off after American High School students waged their own demonstration after Panamanian officials refused to permit them to fly the two flags simultaneously. In the aftermath, relations between the two countries broke down and led to continuing negotiations and the eventual signing of the Panama Canal treaty.

At this point, enter Gerald M. Fry. Then a young broadcaster with a boss he detested at KCRA in Sacramento. Three months married and eager to move onward and upward, he heard about an opening at what had now been renamed SCN, an acronym for the Southern Command Network. The "Southern Command" is the military element in charge of that portion of the globe and is headquartered in Panama. At that time, it was commanded by General Andrew P. O'Meara who had taken over control of the network as permitted under the emergency conditions caused by the riots.

Gerry Fry has had a distinguished career with AFRTS — and he looks the part. Well over six feet tall with a commanding presence, he is topped by a senatorial thatch of thick gray hair. Now worldwide chief of programming for AFRTS at the broadcast center in Los Angeles, his office is still decorated with mementos of his long stay in Panama. He tells here how he got his first job in Panama:

A KCRA-TV colleague also applied for the job, and when he got a telegram stating he should come to AFRTS-LA on McCadden Place for an interview and I didn't, I was furious. I called somebody named Bob de la Torre in LA and asked why I wasn't being asked to interview. He said it was an oversight and to "come on down!"

There I met an Air Force lieutenant colonel named Lou Churchville, public affairs officer of the U.S. Southern Command, and a young Air Force captain named Charles Crawford (later to join CBS News in New York and now an anchor on CNN), who was the officer in charge of SCN. The interview took place in a room that 18 years later would become my own office!

I'll always remember the interview because of Churchville's fidgeting while Crawford asked some intelligent questions. After about five minutes, Churchville said: "Chuck, you carry on. I've got to take a shit!" He left the room and did not return for the rest of the interview.

I heard nothing for a couple of months. I pestered Crawford on the phone and the Army's civilian personnel office in Washington, D.C., trying to get information about who they were going to hire. My colleague heard nothing either. One day Crawford told me they had selected me, but that I'd get confirmation from Washington. It finally came in August. I guess I should have been suspicious when I received my travel orders, cut by the Army Corps of Engineers personnel office in Sacramento, giving me the GS-13 salary plus 25% tropical differential, but sending me to Okinawa!

Fry suffered the same perceived indignities as most newly hired civilians when he got the orders straightened out and arrived in Panama. His assigned quarters were unlivable among other things. So vociferous were his complaints that exceptions were made and he was assigned quality bachelor quarters and his wife, Marty, was given special permission to join him until decent family quarters became available. It was a nice present. She arrived Christmas Eve 1964.

Not so nice was discovering that on the plane

ride down, the government had cut the tropical pay differential from 25% to 15%. He felt that a short plane ride should not have cost him $2,000 a year and appealed the decision. He lost, although the military kindly offered to pay his way back to Sacramento should he care to pursue it.

Within 15 minutes of reporting for duty, the members of the news department made him feel welcome by showing him footage of Jim Pattison, his predecessor, being stabbed while on the job.

He could see immediately that news was something of a problem for the network, working as it did in a highly volatile situation. General O'Meara had assumed control of the network following the riots, partially because of the public relations disaster caused by biased anti-American reporting from AP and UPI Panamanian stringers. Although this happened prior to Fry's arrival, he learned that the command's public affairs officer, Lieutenant Colonel Lou Churchville had vowed to get both news agencies to establish credible bureaus in Panama. He offered to get help from O'Meara's Southern Command. This primarily consisted of SCN paying for AP and UPI newspaper, radio and Spanish language news wires delivered to the command's Fort Clayton headquarters and, at government cost, running on to the bureau's offices downtown. UPI accepted. AP retained its Panamanian stringer, Luis Noli, who forever after maintained he had written the anti-American stories of the riots while a member of the *Panama Guardia Nacional* held a gun to his head.

O'Meara's decision was also influenced by the actions of the then network commander, a major famed for spending his evenings with a bottle and a bimbo in his office. The night of the riot he decided to play radio and, although officers are generally discouraged from going on the air by AFRTS, took the SCN microphone to the scene and began giving play-by-play accounts of the action. He described the flight of a military helicopter "over Panama City". As Panama had broken off relations with the U.S., they used a tape of this as proof that the U.S. was invading the republic. He was immediately relieved of duty and replaced by a man who knew when to talk and when to stay behind his desk.

FRY REMEMBERS GENERAL O'MEARA:

"My first and only experience with General O'Meara was in early December of 1964. I took our Auricon SOF camera to his office with a crew to film his annual Christmas and New Years message to the command. I have never met a meaner looking person in my life! He looked as if he'd bite your head off for breathing, and that's exactly how he looked as he glowered into the camera and said '...and Mrs. O'Meara and I want to wish you all the merriest of Christmases and the happiest of New Years.' I politely asked for several retakes, but he never was able to give even a hint of a smile. It was the first and only time I've aired a holiday greeting from someone who seemed to be sneering as he wished us all happy holidays!

"O'Meara was the guy who got hell from the State Department, I was told, over the CFN broadcasts of the riots. State told him to shut off the play-by-play coverage Kingsbury was giving them. When O'Meara asked what SCN should be airing, the reply was allegedly: "Tell them to play Christmas music!" And so the word was passed and SCN was told to play Christmas music in January —and they did! I never heard the end of that even though I wasn't there. Talk about taking your orders literally!!"

The broadcast equipment he found at SCN wasn't too bad. The radio control rooms were equipped with the old, reliable Gates boards and 16-inch turntables. The record library had every disc from day one, The television studio was equipped with RCA TK-14 vidicon studio cameras, a step above the original AFRTS-issue Dage industrial vidicons that were still there and which were used for set cover shots. The studio was flat lighted with scoops to provide as much light as possible for these vidicons. There was one RCA film chain with two 16mm RCA projectors and a slide unit.

Soon after arrival, he and Bob Botzenmayer, a 12 year veteran of the network, were sent to Guatemala City to check out some TV equipment that was surplus to USIS there. It had been obtained with funds from John F. Kennedy's Alliance for Progress program for USIS to produce kinescope recordings that would be used to teach Guatemala Indians how to speak Spanish. Obviously, the project had fallen flat on its face and the equipment was just sitting there. They scrounged two General Electric 4-inch image orthicon tube studio cameras with pneumatic pedestals, a GE film chain, audio control console, switcher, tons of studio lighting with control panel and a kinescope recorder. They returned to Panama and USIS had all the equipment packed and shipped. It was then installed and SCN got out of the studio vidicon era and into semi-modern television production.

Film was not only the basic medium for telecasting stateside programs, but was the only medium available to shoot local news stories and documentaries. Fry got the idea for documenting in a 30-minute show the work of a Panama Canal pilot. With the Auricon and two Bell and Howell 16mm cameras, he and his crew shot more than 2,000 feet of black and white film aboard a huge tanker transiting the canal from which they were to assemble the program. They got some terrific footage telling an exciting story. When the film came back from the Corozal Army photo lab, which processed all their motion picture film, the entire batch had a huge scratch running vertically through the center of every frame! It was all ruined. They later found out that someone at the Army lab had dropped a beer can opener into the processor, and it landed point up as the wet film, emulsion side down, rolled by it.

Here again is Gerry Fry...

"Although SCN when I arrived was run by SOUTHCOM, the Panama Canal Company, operators of the Canal, thought they owned it. They had been given a green light by my predecessor to stage an annual Combined Federal Campaign telethon, during which they auctioned off TV sets and all kinds of other prizes with blatant commercial plugs for the downtown merchants who had donated them. They also did several remote TV football games, during which SCN would announce things like "Today's game is brought to you by Ron Cortez — que bueno es!" I put a stop to all of this and immediately went on the Panama Canal hate list, which was fine with SOUTHCOM because there was no love lost between those two agencies. I also stopped the weekly Saturday night remote telecasts from the Balboa Bowling Alley. These involved enlisted broadcasters dismantling two studio cameras and all control equipment, lights, microphones and everything else, lugging them down a flight of stairs, loading the truck, setting up in Balboa, doing the two-hour live show, striking, and putting it all back together again in time for the next day's studio newscasts! And who cared about Balboa bowling anyway?

"The SCN staff were really doing some terribly creative things. They did a twice weekly cooking show featuring a military wife and using a set made of cardboard boxes painted to look like a kitchen! It didn't look bad through vidicon tubes!! They were doing a daily two-hour live kids show with a military host and a weekly pet show. The only show that didn't make any sense and was clearly a plaything for the two guys who spent all week preparing it was the Sunday hour- long news-in-review show called "Chronicle." These two guys, one of whom was very effeminate, spent the week re-writing Time Magazine and clipping pictures from it to illustrate their stories! That's all they did! While informative and innovative, it was hardly legal, and I put an end to it about the third week I was there. I proceeded to contract for daily delivery of UPITN Newsfilm from the U.S. so we could put some life into the nightly newscasts and compile a real TV news-in-review weekender for Sundays."

SCN Radio operated at 790 kHz and 5,000 watts on the Pacific side of the Isthmus with the transmitter at Corozal, and at 1490 kHz and 1,000 watts on the Atlantic side with the transmitter at Fort Davis. It served 11 military installations and covered all of Panama City and Colon. Television was on Channel 8 Pacific and Channel 10 Atlantic, operating at 5,000 watts each from Ancon Hill and Fort Davis respectively. Because of the hilly terrain, the TV signal needed boosting at several locations in the Canal Zone, but blanketed most of Panama City and Colon. In 1965, SCN moved its Pacific-side AM transmitter from Corozal (alongside the banks of the Canal) to the middle of Sykes Field, the Fort Clayton parade field that was seldom used. General Robert Porter, then CINCSOUTH, wanted to increase the AM power to 50,000 watts so it could be heard clearly in San Jose, Costa Rica, to counter the Castro propaganda coming out of Radio Havana. Fry had to delicately convince him that what he wanted was not the AFRTS mission, and that 50,000 AM watts radiating so close to the TV microwave equipment would cause RF problems in the TV picture. He wouldn't give up, but was talked down to 10,000 watts. Construction began on a 360-foot high tower and transmitter building in the middle of Sykes Field.

Fry could see the new construction about a mile away while driving to work. One morning he couldn't spot it. A sudden shudder went through him as he rounded the corner and there in the middle of Sykes field was a twisted mass of steel and guy wires! One of the installers had tightened a guy too tight causing it to snap and set the whole tower into vibration. Fortunately, it missed the transmitter building and nearby barracks

This might be the SCN staff orchestra. On the other hand, it might not be. To be truthful, we're not sure what it is except it is part of a local production being produced in the Canal Zone. (Photo Courtesy of AFRTS)

when it came crashing to the ground.

A replacement tower was rushed in by air and the signal was loud and clear way up into the interior some 80 miles away where U.S. personnel would often go to vacation or spend weekends.

Back once more to Gerry Fry:

Major General James D. Alger, the Army component commander, was a great guy who was a George Patton fan, however. He wore a scarf and carried pearl-handled revolvers just like his idol. Flying over Fort Clayton in a helicopter one day he spotted the new AM tower and transmitter building in the middle of Sykes Field. He stormed into the post commander's office voicing his outrage What have you done to my parade field? he shouted. Apparently they had forgotten to inform him of the construction project. Alger ordered it torn down. When convinced it was too late for that, he ordered it to be camouflaged so he wouldn't have to look at it. The post engineers painted the building green the next day and transplanted hundreds of jungle plants. Finally, all you could see was a tall red and white tower spearing the bright, blue sky. The jungle caused terrible problems with the antenna's ground grid, the roots cutting into the wires that formed it. The day after Alger left, the post engineers removed that jungle and the green transmitter building once more became visible.

Alger wasn't the only eccentric Army commander. His successor came from Korea where he had used a program of zero defects he called Circle U. The symbol was a capital letter U inside a circle the same symbol that to Jews means kosher! Major General Chester Johnson obtained SOUTHCOM permission to work directly with SCN to promote his campaign. Nobody was to know what Circle U meant Johnson actually had it classified Secret! Initially, SCN-TV just flashed the symbol on the screen several times each night with no audio in a teaser campaign. That was followed by our booth announcer saying Do you know what Circle U is? Later he would say Circle U is coming to USARSO! and Only a few more days to Circle U day! Johnson wanted the big announcement to be heard simultaneously by all his troops; since they were scattered in several locations across the Isthmus, he couldn't be in all places at the same time, so he recorded his speech at the Army photo lab and ordered the tapes to be played at all post theaters at the same time that he was giving it live in the Fort Amador Theater.

Two days before Circle U Day, I was told by the OIC that Johnson was furious because the tape he made was of inferior quality and that he was coming to SCN that afternoon to record it again. Since it was still a classified subject, I was to engineer the session and only I was to hear what General Johnson said and I was to tell no one. I went into our only production studio and set up a boom microphone over a small table. Since I would be in the same room with the general, I checked out the headphones I would wear to hear the signal. He arrived and we barricaded ourselves in the production studio. He spoke for about 20 minutes and made absolutely no sense whatsoever. He told the troops that they are the U in Circle U, and that from that day forward they would observe the highest standards of readiness, symbolized by Circle U logos that would be placed on all clocks, desks and doors in the command! Further, all Army officers would carry with them long tire pressure gauges (like swagger sticks) with Circle U and their military rank insignias on them. These officers would be required to check the air pressure on any Army vehicle they saw, no matter where it was! When he finished, he looked up at me and asked Do you think I made myself clear? I

Although the control room of SCN is about as basic as they come, the network managed to be the first overseas station in the AFRTS system to receive and retransmit live satellite pictures from the States. This happened to be the Army-Navy game in 1968. The local West Point Society agreed to pay the somewhat hefty satellite charges but finding they only had enough money for half a game, they asked Fry whether they should buy the first or second half.

Is it any wonder he now directs programming worldwide? He recommended they buy the second half. (Photo Courtesy AFRTS)

didn't see any point in telling him the truth, and asked if he wanted to hear it played back. He said he didn't have time, and left. To my utter horror, when I played it back on the monitor speaker the entire tape was full of 60-cycle ground hum! A bad mike cable. The headphones hadn't revealed the hum, and here I was with a defective tape that was to be duplicated into seven other copies and distributed to all Army post theaters! I can only guess that those listening were so intent upon trying to figure out what the hell this man was talking about that they didn't notice the hum. I never heard a word about it! Circle U logos permeated Army posts for the next three years, and when General Johnson left, Army officers had a symbolic burning of the tire gauges at the Fort Amador Officers Club!"

While Circle U was vexing, program restrictions ranked, then and now, right up there with bubonic plague on the list of problems a network doesn't need. Because SCN uses the U.S. standard NTSC television signal, as do the Panamanian stations, not only could the target American audience of about 50,000 see its programming, so could the Panamanians. Suppliers of programming, being of necessity profit driven, were constantly concerned that furnishing programs to SCN at a moderate price would prevent sales to the Panama networks. There were two commercial TV networks in Panama, *Televisora Nacional* (Channel 2), partly owned by ABC, and *RPC* (Channel 4), the CBS/NBC outlet owned by Carlos and Fernando Eleta (Carlos later owned boxer Roberto Duran). Chuck Crawford was bilingual and had made great strides in dealing with management at both networks in an effort to obtain release from the increasing restriction problem.

Programs they purchased or were thinking of purchasing could not be shown by SCN-TV even though we received them in the weekly program units from AFRTS-LA. Crawford introduced Fry to these people and he became good friends with Lindy Paredes of Channel 2 and Jaime de la Guardia of Channel 4. He visited them often to enlist their cooperation in telling program suppliers what programs they had no objection to SCN-TV airing. He obtained release of many top shows in this manner, and by letting them know in advance all the new shows SCN would be getting. They knew none would air that they had an interest in.

For the most part, this close coordination paid dividends for the SCN-TV audience. More than 1/3 of every week's program package was restricted, and always the best shows and movie titles (which continues to this day). Both the network and the audience found it extremely irritating when a series would begin and, after a couple of episodes, be forced to stop it because a Panama network had expressed interest.

Fry self-restricted SCN if he found out either Panama network had an interest in a series; conversely, if he found out they had no interest in a restricted show, he would work with Vince Harris and Bob Vinson in the Industry Liaison office at AFRTS Los Angeles to get the restriction lifted. Both de la Guardia and Paredes were good about sending letters to their network and syndicator contacts telling them they had no objection to SCN-TV using certain titles, and he frequently obtained release on shows that otherwise would have remained restricted.

Working closely with Vinson and Harris, he went to Los Angeles and searched the AFRTS archived film vaults for programs which had previous been restricted but were now approved for broadcast. Eventually he built one of the largest permanent libraries in the world. Even so, programs had to be repeated with monotonous regularity. Although the troops who rotated on a regular basis might see a program only once during their tour, the permanent American residents lived in a constant state of *deja vu*.

Numerous talented staff members served during this period. Among them Fry includes Robert Loggia and Chris Kelly, who later became a CBS news correspondent. Tom Kirby later became an executive with Gannet Broadcasting and president of that group's Oklahoma City station. Jack Harry returned to Kansas City and became the dean of sports anchors there. Al Loman of KFI, Los Angeles' popular Loman and Barkley comedy team was an early staff member. Navy Journalist Del Vaughn left the service to join CBS and was later killed when his helicopter crashed while he was covering floods in Harrisburg, PA. Fry personally picked Paul Hogan (no, not THAT Paul Hogan, but another one) and feels he was perhaps the most versatile announcer he worked with. Hogan is now a newsman with WMAQ, Chicago.

SCN has transmitters on both sides of the Panama Isthmus. This is the Ancon Hill transmitter serving the troops and their families on the Pacific side. (Photo Courtesy AFRTS)

Here, again, is Gerry Fry

But talented people could also cause trouble, and often did. A young airman named Bill Moss was about as versatile with radio production as one could be. He turned out spots that were nothing short of great, and his morning wakeup DJ show was full of fantastic humor and well-balanced music. In 1968, though, on the day of the anti-war march on the Lincoln Memorial, his show sounded odd to me after three or four records, it became clear he was playing nothing but music considered to be anti-war with almost no patter in between. When I called to ask him what in hell he was doing, he admitted he had carefully planned this as his personal protest against the war and that the show was on tape! We yanked him off the air and kept him off for the rest of his tour at SCN. From that point on he was only allowed to produce pre-recorded products, and turned out some great spots for us. He left the Air Force and under the name of Terry Moss established the L.A. Air Force, producer of humorous production aids and publisher of a widely-used radio newsletter.

The music that SCN Radio played caused lots of controversy during the Vietnam era. The gung-ho Army types, many of whom had either just returned or were on their way to Vietnam, complained about anything that even hinted at anti-war lyrics. I particularly remember one Army lieutenant colonel, then the Army component command comptroller, writing a two-page diatribe to his commanding general recommending all funding be withdrawn from that Commie, pinko SCN organization for playing Jose Feliciano s controversial version of The Star Spangled Banner.

The Southern Command Network was totally dependent, as were all AFRTS outlets at that time, on shortwave for stateside news and sports. The service came out of AFRTS-New York via various leased VOA transmitters. SCN was the only AFRTS station with 24-hour shortwave service beamed in its direction. The reception was frequently lousy, with typical shortwave fading and background noise. All major radio newscasts were read from wire copy and incorporated actualities from the shortwave service when they were listenable. Many of the hourly newscasts also had to be read locally when shortwave quality was just too bad.

When the Armed Forces News Bureau (AFNB) was established in 1967, there was talk of using the new satellite technology to send a dependable signal. The Office of Information for the Armed Forces (OIAF), under John Broger and Henry Valentino, arranged for leased satellite circuits to the major AFRTS networks, fed from AFNB. Although of disappointing quality (less than 3 kHz), the signal was at least dependable and SCN and the other networks around the world were able to consistently take radio news from AFNB. Line noise, particularly during pauses, remained a problem for years to come, however.

Satellite technology brought SCN-TV a first in 1968. Panama had just installed an earth station at Utibe, some 20 miles from SCN. The 95-foot antenna was absolutely stunning, and the facility was operated by a company called *Intercomsa*. They wanted to use the earth station, so the local West Point Society dreamed up a plan to present that year's Army-Navy football game live from Philadelphia. The society agreed to pay all costs and Intercomsa said they'd donate most of the satellite time which was very expensive in those early days.

Two days before the event, the West Point Society president, who had been trying to raise donations to pay for the event, announced he only had enough for one half of the football game and he wanted to know which half he should order. Needless to say, Fry told him to buy the second half and SCN would tell the audience to listen to the first half on radio. That half of the 1968 Army-Navy Game was the first live satellite event carried by any AFRTS outlet.

And a few more words from Gerry Fry:

"Color television sets were now available in the exchanges and the occasional satellite events whetted the SCN-TV audience s appetite for color. All priority sports from AFRTS-LA were received two weeks after the event on 16mm black-and-white film, a real downer after seeing live color satellite sports events. I persuaded the AFRTS-LA gang Bob Vinson, Vince Harris and John Cosgrove that the two-inch videotapes (recorded in color from networks) from which their black-and-white kinescope recordings were made should be mailed to me for playback in color from our newly-acquired RCA videotape recorders. They gave in, and with some persistent needling from me, our engineers figured out a way to get a usable color signal from those tapes through our transmitters and antennas and on to the air. This service was extended to include other timely programs, so we were able to expand our weekly color offerings. I didn t have the money to be able to ask for the entire AFRTS entertainment package to be shipped on two-inch videotape.

"The first weekly program package on color videotapes arrived in 1975 on IVC one-inch reels. The Navy had adopted this standard for it s SITE systems on board Navy ships and we were excited that our package programming was now going to be in color. We had a problem that most other AFRTS stations didn t have, though. Since a good part of our weekly schedule was pulled from the film library, we still had a large portion of our prime time airing from black and white films. I used a weekly column in The Southern Command News to explain this and our restriction problems to the audience.

"There were ten officers in charge during a 12-year tenure which doesn t speak too well for SCN leadership. When Chuck Crawford decided to leave the Air Force before the end of his tour, the young second lieutenant who was the first editor of The Southern Command News, Peter Jaensch, took over both duties for a few weeks. The Army then re-assigned Major Phil Christoffersen from a USARSO element to SCN. Phil tried his best, but liked to spend his time nights and weekends screening AFRTS restricted programs and movies in the film library. Air Force Major Paul Holter was a breath of fresh air after Phil, and ended up being our downstairs neighbor in Fort Clayton duplex quarters. He was highly intellectual and interested in broadcasting; it's likely that without a Sergeant Major named Sam Sumner he would have been one of the better OICs. He was followed by an Air Force captain named Newell Sawyer, but there was an interim period prior to Sawyer's arrival where the editor of the paper once more, this time Army Lt. Hank Benson, was in charge. After Sawyer, the Army sent us Lieutenant Colonel Dan Maguire. Dan had been in charge of the Army's Command Information Unit in Washington and knew and loved broadcasting. He was terrific and led the network to new heights. But Dan had a fierce temper that almost led to murder one day when a bulldozer operator severed the cable between Building 209 and the radio transmitter in the middle of Sykes Field not ten minutes after Maguire had warned him to be careful of it! We were off the air on radio for 36 hours. When Dan left, an Air Force captain named Charles McFatridge, who had been the newspaper editor, took over, but was relieved a few weeks later in a humiliating and totally uncalled-for public ceremony in the SCN TV studio by then SOUTHCOM PAO, COL Bob Bryant. Bryant placed in charge one of his staff, LtCol Major Bell, who remained until an Air Force major from Scott AFB, Red Viguerie arrived in 1975."

By now life was changing for the Americans in Panama and Fry and his wife decided to move on. Buzz Rizer, now head of Navy Broadcasting, offered him the job of Chief of Programming for the Navy. In August 1976 the Fry family, augmented since their arrival by two daughters and two Amazon parrots, departed the Isthmus for Washington D.C.

SCN, of course, continued to operate and continued to suffer serious programming restrictions which became a matter of concern to the Industry Relations department in Los Angeles. Providing a viable program package to Panama became almost a full time job.

In 1988, SCN once again came into the news. That story after this.

The SCN camera crew covers a remote from the roof of what would now be a valuable vintage Chevrolet. It's difficult to tell exactly what it is they are covering although it's possible it is a documentary on the need for road repairs in the Canal Zone. (Photo Courtesy AFRTS)

STATION BREAK NUMBER FOUR

A Brief Pause for Station Mortification

Time now to step out of the Central American jungle for a few moments and take a quick trip around the various AFRTS networks and stations to recall a few of the odd, unusual, singular and unique people and events which have become the stuff of legend throughout the system.

★

TOKYO. Actor Hans Conreid was the first morning DJ when the American troops landed in Japan following the surrender in September 1945 and as a spoil of war, moved in and took possession of Radio Tokyo. Conreid had one peculiarity. As an actor, he felt he had to maintain a certain amount of individuality. To this end, he took to wearing red socks with his corporal s uniform. A neat feat, all except the officers agreed; however, no one was ever able to make him change his sartorial habits, perhaps because court martial papers charging a man with wearing red socks would look pretty silly to higher headquarters.

★

STILL MORE TOKYO Bill Doty was Conreid s replacement and recalls that as a fifth anniversary salute to AFRTS they would do something different. Like destroy Tokyo. Orson Wells famed *War of the Worlds* broadcast which had terrified the country was still fresh in their minds. The staff decided to see if people were still as gullible as they had been in 1938 when Wells convinced everyone Martians were invading the United States.

The AFRTS Tokyo broadcast started out like a normal musical program but was soon interrupted with fake news bulletins about a mysterious creature seen in the waters of Tokyo bay. The news reports became progressively more excited as the creature began making his way across Tokyo. Could audiences have been so gullible? You bet your Atwater Kent they could. Battalions were put on alert and began hunting the creature. They couldn t find him for two reasons: the station was using fictitious locations of his sightings and, more important, he didn t exist. One battle group, having an organizational party, was called back to its barracks and put on their battle gear.

The broadcast was actually being done from a remote unit in the alley behind the station to give a credible sound. About midnight, one of the announcers broke in to say he had joined a group who had located the beast in the corner of the Emperor s palace and was going to try to work his way through the crowd so the audience could hear for themselves the horrible sounds coming from the monster. He pretended to work through the mob and got to the monster in time to hear it sing, in a faggy falsetto:

Happy Birthday to you,
Happy Birthday to you,
Happy Birthday, dear
Armed Forces Radio
Happy Birthday to you.

Suddenly everyone realized they had been had. One of them included Supreme Commander General Douglas MacArthur who was, as Queen Victoria used to say, not amused. Nor were the participants in the hoax part of the show. They were on their way to isolated assignments the next morning. In a way it seems unfair. Orson Wells made a gold mine out of his show. The AFRTS people only got the shaft.

★

For reasons known only to Absurdite, the goddess of Broadcasting, and the man-hating computer in the Pentagon which assigns, with electronic malice, lawyers to motor pools, mechanics to legal offices and egomaniacs to broadcasting organizations, AFN in Frankfurt one day found itself under the command of a gentleman whose acquaintance with the English language was tenuous at best.

A lieutenant colonel, he knew in his heart that he was automatically qualified by right of rank to take on any job assigned him by an infallible army. The entire staff knew he was wrong and such proved to be the case. He wandered the halls mumbling *non sequiturs* such as I wants you to detonate your thinking. Seeing bits of audio tape on the floor of an editing booth, he announced he would be back in ten minutes and expected to see the place looking despicable. Having been told by a general officer how much the general admired one of the staff, the commander called in the complimented gentleman and proudly told him that the general had had some very nice and disparaging things to say about him.

He developed a set speech, complete with chart, which after days of wiping perspiration from his brow, he managed to memorize. Down the chart was the word   I
　　　　　　　　　　　　　　　　M
　　　　　　　　　　　　　　　　A
　　　　　　　　　　　　　　　　G
　　　　　　　　　　　　　　　　E

Each letter was followed by a matching word. I was imagination.

M was morale. A for attention to detail. G , if it wasn t, should have been garbage. The speech took 19

Beautiful Bo Derek visited AFN Berlin for both a radio and television interview session, accompanied by husband/manager John Derek. For some reason, John insisted cameras get no closer than ten feet from his luscious bride. Later, being seen interviewed here by Sergeant Rik Delisle, it looks like he put the same restriction on microphones. Luckily, Delisle has long arms.

Sergeant Vicky Washington of AFN Berlin visits with and interviews Kirk Douglas at Berlin's famed Kempinski Hotel. He is in town promoting "The Final Countdown." From the looks of the drinks on the table, one is drinking water. The other one isn't. There is no way of knowing which is doing which.

Peggy March became a large star as a young teenager with her hit **I Will Follow Him**. *Moving to Germany, she became a superstar there and a sell-out performer in concerts and a regular on top TV variety shows. She also became a great favorite with AFNers who knew they could always count on her to do anything from sing a jingle to perform for the American audience. She was probably interviewed more times than any other performer — and here is Mark White doing it again. no wonder everyone at AFN loved her. (Photos courtesy of Mark White.)*

179

minutes and 11 seconds to deliver and was embossed on what passed for a brain. If interrupted for any reason, his memory banks automatically went in to fast rewind and he was back at "I" again.

One day the commander in chief of the European theater made a visit to the network headquarters. The AFN commander immediately whipped out his chart and went into action. Along about "G" the four star general interrupted him. Without pausing, the commander whipped into reverse and said, "'I' stands for Imagination."

The commander in chief muttered something about "I" standing for idiot and walked out, leaving the commander still talking at the easel. Shortly thereafter he found himself transferred to a new and less demanding job and, later, became a mail carrier in New Jersey.

★

Readers have already met Joe Ciokin, the Navy broadcaster who, throughout his long career with AFRTS, always seemed to be caught up in the center of whatever nasty action was taking place. When last seen, Ciokin was in Lebanon as the Marine barracks are blown up and he is thrown completely across the room by the blast.

This type of thing was to plague Joe from of his earliest assignment which was to Panama and the Caribbean Forces Network in 1959. Although new to the staff, he was soon put on to the news staff as News Editor and anchor for the three daily TV newscasts. One bright, sunny day the station was doing a radio remote broadcast from a public park. President Eisenhower had recently ordered that the Panamanian flag should fly alongside the Stars and Stripes inside the U.S. controlled Canal Zone to demonstrate shared responsibilities and interests of the two countries. The broadcast that day was to describe the first double flag-raising ceremony.

It was Ciokin's day off and by sheer coincidence he was strolling through the park where the ceremony was being broadcast. Network program director Jim Pattison was doing the vocal description of the events and Joe stopped nearby to watch. Suddenly a man rushed from the crowd and attacked Pattison with a knife. Two Canal Zone policemen standing next to Ciokin tackled the man with the knife and Ciokin, still not fully comprehending what was happening, ran to Pattison's side. Pattison was in shock but managed to say, "Joe, take over."

So great was Pattison's shock that his hand instinctively clutched the microphone and a tug of war began. Ciokin soon ended up holding the live microphone and uttered the understatement of the century: "Ladies and Gentlemen, we are experiencing technical difficulties." This perked up the ears of the staff on duty back at the station. His next sentence really woke them up. "Our announcer has just been stabbed," he said. At the station, there was pandemonium. "Did he say what I though he said?" "Take him off the air." "No, put him back on." "What the hell is going on out there."

At this point both flags were starting up the flagpole and he went on to describe the ceremony while the medics attended to Pattison's wounds. The next week Ciokin, who had only been out for a walk on his day off, found himself immortalized in **Time Magazine.**

Such disruptions of a quiet life continued to follow him. Later in his career he was anchor for the TV newscasts in Saigon. Shortly after the TET offensive, he was doing his thing when the station was hit by a 500-pound satchel charge. The UPI wire service immediately reported that the staff had been wiped out. In fact, the damage was moderate, no one was killed, but it took 15 hours to get UPI to correct the story.

His luck was holding. He reported the events of TET from the U.S. Embassy. His luck continued two days later when Army Sergeant Nick Palladino, also a member of the news section, decided that "we" were now in a position to go clear across town to find out what had happened to Palladino's Vietnamese wife and his house. Nick, later to become AFN Television news chief in Europe, was married to a cousin of General Lon Loc Loan who became famous for the widely published photograph of him shooting a captured Viet Cong in the head.

Nick and Joe set out on a motorcycle and headed across town. They were soon stopped by MPs at a roadblock who warned them that a bunch of bad guys were thought to be nearby. Understandably anxious, they pushed on and several miles further along came to a Vietnamese roadblock. The Vietnamese smiled and waved them on.

After about a hundred yards it hit Ciokin. "Hey, Nick," he yelled. "Did you notice those guys weren't wearing any shoes?" They were Viet Cong alright and Ciokin is certain to this day he is still alive only because the VC troops were trying to stay undetected. They didn't, though. As the two roared away, a furious fire fight broke out behind them as a Vietnamese patrol arrived on the scene.

Rounding the corner to Nick's house, they found the entire block razed. All that was left was the metal gate to his fence and his young puppy dog. They grabbed the dog and found one of Nick's wife's relatives who told them his wife was safe with other family members.

Then they took another — and faster — route out of there. Way out.

★ ★ ★ ★ ★

ANNOUNCER: YOU THINK JOE CIOKIN WAS JINXED. YOU AIN'T HEARD NOTHIN' YET. STAY TUNED AND FIND OUT WHAT HAPPENED TO ANOTHER GUY NAMED NORIEGA. HE HAD A WHOLE NATION, PLUS THE MIGHT OF THE AFRTS MUSIC LIBRARY, GUNNING FOR HIM.

★ ★ ★ ★ ★

CHAPTER TWENTY-FOUR

In which Manuel Noriega becomes a listener, like it or not, to AFRTS...and he likes it not.

Dropping in on Noriega

Until March, 1989, the most important problem faced by the Southern Command Network (SCN) was that of program restrictions placed on it by the owners of the material. A group from the station had been sent to the AFRTS Broadcast Center in Los Angeles in 1987 to comb the film vaults for programs cleared for broadcast in Panama and, with the help of Vince Harris and the Industry Liaison Office of AFRTS, a number of programs which had previously been restricted were released for the use of the network.

SCN didn't know what problems were until March. At about that time problems began flying at them like pies in a Mack Sennett comedy. By April the problems had become predicaments. In May, these turned to dilemmas. With June they had a crisis.

So did Panama, and it directly affected SCN and the American residents there — troops, their families, employees of the Panama Canal Commission and numerous civilian business people. Panama was in the midst of a political upheaval as it prepared for elections in May. Demonstrations and mini-riots were day-to-day occurrences. Panama strong-man Manuel Noriega, full-time dictator and indicted part-time drug smuggler, had orchestrated a continuing series of harassments toward the Americans. Each presented a different problem for SCN. As the primary means of communication between the military command and the Americans living in Panama, it fell to the network to keep its listeners fully informed of developments; to pass on information pertaining to their safety; to provide information on ways to avoid dangerous areas and to help in every way possible to avoid panic as the situation daily grew more tense.

SCN is somewhat different from other AFRTS outlets in several respects. Unlike others, it is the only outlet under the direct control of the local unified military command. Although it normally follows the AFRTS "rule book," the Panama command could, and did, break the rules when the situation required.

Following the elections, SCN found itself in the untenable position of having two host governments, both at odds with each other, while it tried to preserve its hard won objectivity as one faction, led by Noriega, became increasingly adversarial. Efforts to maintain objectivity reached ridiculous proportions when trying to report on local events and not to offend either political faction. Whatever was broadcast could be considered by one of the Panamanian groups as "anti-government propaganda."

Euphemisms became so commonplace that the American audience soon came to recognize that "traffic congestion" meant there was a street demonstration in progress and for "street blockage," to read "riot."

In April, the then Commander in Chief of the United States Southern Command (USSOUTHCOM), General Fred F. Woerner, told the Senate sub-committee on defense that "...we are dealing with problems...which are going to be vastly more expensive to fix later, if we do not remove them now." His words proved to be both accurate and prophetic as the U. S. found out months later.

Noriega and his band of trained thugs, the Panama Defense Forces (PDF), daily became more abrasive and authoritarian in their dealings with both the Panamanians and the Americans. The U. S. military command developed a system of alerts which were named Personnel Movement Limitations but, in the acronym-oriented military community, were called PMLs. The purpose was to provide Americans an instant notification of the situation in the community at that moment and insure the safety of U. S. citizens in a rapidly changing environment. SCN was the prime method used to notify the population.

At first there were four levels of PMLs. All military personnel, their families and other Americans had drilled into them that PML Alpha meant the situation was normal and movement was unrestricted although caution should be exercised.

PML Bravo, the next step up, meant to keep as low a profile as possible. Reduce time in public places, avoid large groups and stay near your home or place of work unless absolutely necessary to travel.

PML Charlie meant things were starting to get serious. It required Americans to stay out of public areas of Panama and travel directly from home to job site. Never leave unless you tell someone where you're going, what route you're taking and when you expect to arrive. Keep your radio or TV tuned to SCN.

PML Delta was really serious. Do not travel. If you live in quarters not under U. S. military control, be prepared to move immediately. Keep important papers such as passports, insurance, automobile registration and prescriptions with you at all times. If you live on a military installation, be prepared to share your residence with another family.

The PML announcements were used on TV in the form of a "crawl" across the screen preceded by six beeps. Other such announcements were broadcast in the same way, but without the beeps. The SCN audience was soon to be hearing the beeps with increasing regularity.

The first major incident occurred in early March when the Panama Defense Force, presumably under Noriega's orders, detained a convoy of 21 school busses taking approximately 100 American Junior High School students to class. The PDF claimed the busses had the wrong license plates although they were the normal U.S. Government plates authorized under the Panama Canal treaty. Panamanian troops ordered passing cars to wedge themselves around the busses to prevent movement and tried to tow one bus away before American military police vehicles blocked them.

From then on, military busses were used, but the American community was now beginning to feel the fear that can only come from acts of terrorism.

SCN began fine-tuning their emergency procedures in April when it became obvious that such acts would continue to escalate. Four teen-agers were detained all day on a nebulous charge. A U.S. military vehicle was rammed and pushed off the road by a PDF vehicle and a young American dependent, Kurt Muse, was arrested for "crimes against the security of the state" which were never specified.

Knowing that the Panamanian election day, scheduled for Sunday, May 7, could well be a day of violence, the network began planning a "Super Sunday" of blockbuster programming on both television and radio. It was hoped this would help keep Americans home and off the streets during what was sure to be an inflammatory situation. The Army Broadcasting Service in Washington, the American Forces Information Service and AFRTS all contributed long hours clearing programming for the event.

Realizing that Noriega viewed SCN as a threat to his somewhat biased approach to news coverage, special security measures went into effect and were strengthened as the May elections drew near. The station, located just yards from a main thoroughfare, was an easy target and, in fact, the wall facing the street was defaced with anti-American graffiti.

Prior to, and during, the election the Noriega faction of the government electronically jammed SCN during newscasts in an effort to keep eavesdropping Panamanian citizens from getting accurate reports. The staff worked around the clock for weeks preparing television graphics showing possible trouble spots to broadcast if needed. Telephone hotlines were set up and secondary "live" broadcast origination points were prepared at locations outside the permanent studios. All personnel were placed on a standby basis. Telefax and other special circuits were installed between the station at Fort Clayton and the public affairs office at SOUTHCOM headquarters at Quarry Heights, several miles away. It was over these circuits that requests for SCN to announce changes in the PMLs were received.

Radio went to an all-request format and it broadcast listener requests around the clock with great success. This meant that the four people assigned to radio worked twelve hour shifts, seven days a week. This backbreaking work schedule continued throughout the month of May.

The election was a fraud and a farce. Foreign observers were unanimous that the Noriega candidates had lost and that Guillermo Endara was the rightfully elected president. Noriega solved that problem very typically. He merely put his Panama Defense Force on the street and annulled the election. To make things absolutely clear, he had one of the opposition presidential candidates and two of the vice-presidential candidates picked up and beaten. In Washington, George Bush called the election a fraud and asked Noriega to resign; a request which, in hindsight, was as effective as King Alfred asking the tide to stop coming in. President Bush called on the Panamanians to overthrow Noriega and took the even more positive step of augmenting U. S. forces stationed in Panama.

With the unfathomable logic for which armies are famed, while additional troops were arriving in Panama, they were pouring out of SCN. Under the computer-chosen operation name of "Blade Jewel," a majority of family members were returned home. This in turn meant that the military members had their tour of duty in Panama shortened. Naturally, those that had been there longest and were the most experienced and valuable, were the first to leave. In June and July alone, ten highly experienced broadcasters packed and left for other assignments. The personnel hemorrhage continued throughout the remainder of the year and totaled 23 missing-from-action before it was over. Replacements, for the most part, consisted of eager and willing but semi-trained broadcasters fresh out of the Information School in Indianapolis although there were some transfers of more experienced broadcasters from other AFRTS outlets elsewhere in the world.

In such a taut situation, rumors are bound to abound and stay around until quelled. The network ran interviews with community leaders and produced spot announcements giving the straight word on major pieces of misinformation. There were so many questions that special programs were created to answer them on the very wise assumption that bad news doesn't get better with age.

Everyday life in Panama became increasingly difficult. There were increasing numbers of riots, barricades on major streets, demonstrations and increased security. Under terms of the treaty, Americans had unlimited right of access, but were often stopped and harassed. The U. S., to assert its rights, increased the number of armed convoys patrolling the streets. This naturally led to increased confrontations.

On September 1, President Bush, in a speech carried over SCN, declared the Noriega regime illegitimate, and pointed out his brutality, his disregard for democratic processes, renewed the charges of drug trafficking, and again told the people of Panama the U. S. would stand beside them in their fight for self-determination until a democratic government was restored.

Noriega's reply was to again begin jamming SCN newscasts. Incidents continued. A helicopter crashed while shepherding a ground convoy. An AK-47 bullet was blasted through the window of an American child's bedroom. PDF exercises were scheduled, night and day, near American compounds and the U. S. embassy.

General Maxwell R. Thurman replaced General Woerner as commander in chief and in his speech at the change of command ceremony, broadcast live by SCN and jammed by Noriega, he reiterated President Bush's statements. He promised to "confront tyranny in all its forms" and the confrontation was now intensified.

The Countdown Continues

Three days after Thurman assumed command, gunfire broke out at the *Commandancia* — the headquarters for the bully-boys of the PDF. Not knowing precisely what was happening, the command went to Personnel Movement Delta, the highest level of movement restriction then on the books. Every man, woman and child knew that Delta was to be used only in the most grave circumstances. Tensions were not eased when mothers, settling down for a cup of coffee and knowing their children were sitting in their school classrooms heard an announcement over SCN:

MILITARY POLICE ARE PROVIDING SECURITY AT ALL DEPARTMENT OF DEFENSE SCHOOLS. SCHOOL REMAINS IN SESSION. DO NOT PICK UP YOUR CHILDREN FROM SCHOOL. YOUR CHILDREN ARE SAFE. Memories of the March school bus incident came flooding back and concern among parents grew no matter how many times they heard SCN announce, "YOUR CHILDREN ARE SAFE." In fact, they were safe although busses returning them home ran late and children who lived in distant areas of Panama City were bunked down for the night with families closer to the school.

Movement restrictions were eased somewhat the following day although a curfew was put into effect running from 8:00 p.m. until 6:00 a.m. In a few days this was shortened by an hour at each end, but tensions continued and SCN went to a war-time posture.

The gunfire at the *Commandancia*, turned out to have been a coup attempt which the Noriega PDF forces quickly put down. Hopes had been raised and then dashed for Americans and Panamanians alike that an end would be put to the untenable situation. The atmosphere stayed at a near explosive level for weeks.

During the month of October, SCN ran a total of 637 crawls on the screen in order to keep its audience constantly in touch with the situation going on around them. More than 700 special radio announcements were also run. The network was operating with 25 assigned production personnel thanks to the personnel drawdown. The normal number of people assigned to this area had been 41.

By the end of October, the movement status was lowered to Charlie but it was obvious that the situation was reaching the flash point. General Thurman and his public affairs advisers, including SCN, developed a war-time status one step higher than Delta to be called, according to the military verbal alphabet, PML Echo. The network developed a seven minute videotape explaining this new category and used interviews with officials and announcements to make certain everyone understood what Echo was. To put it briefly, it was instructions on how to react if a war started. Stay home. Stay down. Stay out of the way of the fighting.

As November rolled around, tensions — known in the military as "the pucker factor" — increased still further. Bomb threats were received almost daily. Cars were forbidden to park closer than 75 yards from government buildings. Obstacles were placed at the entrance to military installations and many streets were barricaded completely. Because Operation "Blade Jewel" was still in effect, personnel continued to disappear from SCN as the work load increased and they were partially replaced by bright-eyed, eager young entry-level broadcasters in search of adventure.

One thing the network could not complain about was something every station in existance would like to have — a true captive audience. The commanding general issued an order that everyone should keep their radios and television sets tuned to SCN for the latest directives and information.

The political situation continued to deteriorate in November. Convoys were blocked by armed PDF groups and Americans were forced to push their way through. To further intensify the hardships, General Thurman issued an order placing the Holiday Inn in Panama City and the Colonel Sanders Kentucky Fried Chicken outlet in Colon off limits. While he was at it, he included "all bordellos" in the off limits order. Chicks, fried or otherwise, were now off the menu.

SCN spent a large part of November preparing for the worst. Programming material was collected and scattered in various locations so it could be recovered and used in the event the station was immobilized. The TV van was moved to a new location, ready to act as an emergency transmission point. Radio equipment was gathered for use from other locations.

December arrived with no peace on earth, no goodwill toward men. Worse, Murphy's Law kicked in with a vengeance.

The SCN news director, obviously a key member of the team, was forced to return to the States on emergency leave. Quickly thereafter, the Program Director and Chief of Broadcast Operations had to do the same thing. Meanwhile, the exodus of staff members continued under the imposition of "Blade Jewel."

As this was happening, Noriega decided to make a speech. He announced that a "state of war" now existed between the Republic of Panama and the United States. He encouraged Panamanians to gather at the banks of the canal "to watch the bodies of Americans float by." The line between bluster and muster had become paper thin. On the 15th, he named himself Head of Government and his hand-picked 510-member People's Assembly unanimously voted him unlimited power.

To the world, the U. S. intervention in Panama began on December 20. To the Americans living there, it had begun even earlier and many mark December 16 as the day the shooting started. On that day, the PDF attempted to stop an automobile containing four Americans at a PDF roadblock. Knowing they had the right under the various treaties to right of passage, they tried to push through and were fired upon. U. S. Navy Lieutenant

Robert Paz was hit in the back and died before reaching the hospital. The driver was hit in the foot. Witnessing the scene was a Marine captain and his wife who were dragged from their car. The captain was taken to an unknown location, beaten and tortured. His wife was also roughed up and repeatedly threatened with sexual abuse.

SCN broadcast a taped interview with the three survivors of the shooting incident and, later, interviewed the captain and his wife. Copies were also sent to Washington.

The command went to PML Delta. On the afternoon of December 19, Lieutenant Colonel Bob Gaylord, the SCN commander, called the supervisors together and directed a "test" of the unit's operational readiness. A mobile van team was sent to Quarry Heights, the military headquarters for Panama, to set up a microwave link to the studios at Fort Clayton. An engineer was sent to Ancon Hill; the mission being to be able to deploy and set up a return microwave link from the van to SCN from any location on the Pacific side of the Isthmus. A TV news team was attached to the command's public affairs office and two more teams were assigned to prepare equipment for standby use. All personnel were instructed to remain on duty until released and operation and contingency plans were reviewed by all.

This required a total restructuring of duties within the network in order to make the plans work. People were moved from less critical jobs and placed where they could be better utilized. Teams were set up to work 12-hour shifts. Quarters on-post at Fort Clayton were identified so the now sleep-deprived staff could have areas in which to bunk down. Those living off-post were moved in with others living on Fort Clayton.

Normally the 11 meter satellite dish was programmable to bring in signals from a number of satellites but the decision was made to lock it on to Galaxy I to guarantee constant coverage of CNN and CNN Headline News. The giant dish had a history of having its motor burn out and the network wanted to take no chances that it would become immovable and unable to pick up continuing incoming news in time of crisis.

This meant that a number of entertainment programs could not be received from the satellite and, crisis or not, a number of vociferous complaints were received from soap opera fans who would rather watch General Hospital than the war going on outside their windows.

Although only Gaylord had been notified that the abilities of his group would soon be put to the test, it soon became obvious to everyone that this was something more than a routine test.

Shortly before midnight, a camera crew was sent out with only a destination known. They were told only that they would cover a news conference with three people participating "inside a house." It turned out to be the swearing-in ceremonies of the legally elected Panamanian President, Guillermo Endara, and his two vice-presidents. This footage was used both by SCN and supplied to all U.S. networks.

Accompanying the television team was a member of the SCN radio staff who made audio recordings of the event and was then asked to stay on for the next three days to provide radio assistance in support of the new Endara democratic government.

Word came down from headquarters at Quarry Heights to be prepared to go on television live at about 25 minutes after midnight. The crew and news team was standing by and Gaylord handed the anchor a telefax. A minute later the anchor announced to the audience that an armed conflict was in progress in Panama.

He also announced that PML Echo was in effect and stated the action was directed only against Noriega and his rogue Panamanian Defense Forces. Within minutes the local CNN stringer, Berta Thayer, had picked up on the story from SCN and was reporting it to the world over CNN.

Operation Just Cause, another Washington-selected name for an operation, had begun.

As the paratroopers began dropping in and aircraft began bombing strategic targets all about them, a conscious decision was made by SCN not to go to continual live news coverage in the interests of operational security. The networks were so extremely timely in their reports that concern for the safety of the fighting troops dictated that some items be held up for not more than thirty minutes. In this electronic age, watching military actions in progress can be of immense help to the enemy. Live coverage that night of December 20 was provided of White House spokesman Marlin Fitzwater's announcement of the invasion and, later, of the President's address to the nation.

SCN radio went to a format of thirty minutes of music and thirty minutes of back-to-back newscasts. The musical portions were call-in request shows which were designed to satisfy those who were housebound under the rules of PML Echo.

By 4:30 a.m. SCN was prepared for the arrival of the media pool and they arrived at the station by helicopter at 6:05 a.m. Their greeting will be remembered by many for some time. As they were unloading their gear and moving it into the station a mortar shell landed on the field nearby. Luckily there were no casualties. A briefing was conducted and tapes of the Endara swearing-in were passed out.

Some of the media were better prepared technically than others and SCN engineers were kept busy for days modifying equipment, or loaning SCN equipment, so the various U.S. news organizations could get their stories out. SCN also assisted in the construction of the satellite up-link facilities at Quarry Heights which carried the bulk of the pool feeds back to the U.S. — and which were received back in Panama for broadcast there. Early in the fighting, more than one correspondent reported the war while trapped in downtown hotel rooms, giving graphic descriptions of the action as they watched it on SCN.

All major U.S. networks made good use of the SCN television van which was used to review their video footage and make necessary copies for editing into live reports. One SCN broadcaster was appointed part of the media pool and he and an NBC cameraman totally supported the pool with video for the first several days. They covered looting downtown, the destruction of the *Commandancia* and President Endara's address from the Presidential Palace. Later the SCN man became the cameraman at Quarry Heights for all network live standups from Panama back to the States.

People in Panama had lived with, and trusted, SCN for years. As the fighting spread throughout the city and became an urban guerilla situation, they began phoning SCN with information of sightings and movements of Noriega and other wanted individuals, arms caches, sniper locations, PDF hideouts and other types of information. Almost every call was ended with tearful expressions of gratitude to the U. S. for their intervention.

Noriega successfully evaded capture and, after several days, it was learned he had asked for sanctuary inside the Papal Nunciature, the Vatican's headquarters in Panama. This was soon surrounded by troops and fighting throughout the city began to ease up.

Increased, however, were requests for songs to be played over the SCN radio request shows. Some of them struck the staff as somewhat strange. Strange, that is, until it was learned after several days that the Psychological Operations group had set up giant speakers outside the Nunciature and was subjecting Noriega to cruel and unusual punishment by bombarding him with music selected by Americans and dedicated to the barricaded dictator.

Following are some of the songs which will probably never make Noriega's list of favorites but which were dedicated to him at peak volume from outside his window:

All I want is You, U2; **Born to Run**, Bruce Springsteen; **Cleaning Up the Town**, The Bus Boys; **Dead Man's Party**, Oingo Boingo; **Don't Close Your Eyes**, Kix; **Don't Look Back**, Boston; **Guilty**, Bonham; **Hang 'em High**, Van Halen; **The Pusher**, Steppenwolf.

I Fought the Law and the Law Won, Bobby Fuller; **Renegade**, Styx; **It Keeps you Running**, Doobie Bros.; **Judgement Day**, Whitesnake; **Screaming for Vengence**, Judas Priest; **Run Like Hell**, Pink Floyd; **The Party's Over**, Journey; **Wanted Man**, Ratt; **Voodoo Child**, Hendrix.

Your Time is Gonna Come, Led Zeppelin; **Nowhere to Run**, Martha & Vandellas.

Perhaps, after a few days of this, a cozy jail cell began to look good to Noriega. Which is what he got.

He got a real earful as well. The music came from AFRTS disks, and was played over some very special speakers used by the Psychological Warfare people from Fort Bragg. Even the loud speakers and loud is the operative word have a definite AFRTS connection.

The system was designed and built by Video Masters, Incorporated, of Kansas City, the CEO of which is a former AFRTS staffer who has served at stations in a number of locations. He is Bob Cleveland and he and his staff have been designing esoteric equipment for the Department of Defense for a number of years. He learned about speakers with enough power to turn your ears wrong side out while helping to set up Bob Hope Shows, and others, in Viet Nam. Speakers had to be loud enough to be heard in a crowd of 10,000 people. Later, for the Psy War people, he developed a speaker backpack for use in El Salvador. It worked fine but the poor guy carrying it wasn't all that happy. It weighed 68 pounds which the man had to carry along with his regular equipment. Later, for the Philippine government, Video Masters was able to bring the system down to 50 pounds but this was still pretty heavy for the Filipinos who generally are considerably smaller in stature than Americans. The system used in Panama weighed only 30 pounds and included a wireless remote control allowing the operator to manipulate it, if necessary, from several miles away. This is a handy feature which is much appreciated by the PsyOps people who realize that going anywhere near those speakers in a wartime situation is liable to get you shot by a disgusted enemy.

By January 5, things were nearly normal. At last the SCN staff could begin, slowly at first, to return to a normal work schedule and even plan on a full night's sleep once in awhile.

When Army Broadcasting headquarters in Washington phoned and asked SCN's Sergeant Major, MSG Jerry Elliot, for a report, he told them: "It's just another day in Paradise."

Army General Norman Schwartzkopf, commander of U. S. Central Command, visits Lieutenant Colonel Randy Morger, network director, to thank him for AFRTS support to the U. S. forces during Operations Desert Shield and Storm. (Photo courtesy Jeff Whitted)

LTC Bob Gaylord, commander, and AF SMSGT Rafael Alcantara, the NCO in charge of network maintenance get ready to depart Riyadh to install an audio receiving dish at Thumama. (Photo courtesy Jeff Whitted)

Gaylord and Alcantara make it to Thumama and get the first of the field deployed vans in operation. The antenna is shown fully extended and the AFSTRS (Armed Forces Satellite Transmitted Radio Service) dish is up and operating, receiving its signal from the States. As for Thumama, it looks like an okay place to visit but we wouldn't want to live there. (Photo courtesy Jeff Whitted)

CHAPTER TWENTY-FIVE

"Keep Your Head Down and Your Volume Up!!"
—Desert Storm Radio Network
Motto during hostilities

"Pumpin' the tunes across the dunes..."
—Desert Shield Radio Network
Motto after hostilities

IT'S SADDAM SHAME, BUT IT'S BACK TO THE DESERT

So there they were. Combat broadcasting had come full circle. After almost fifty years, AFRTS was broadcasting once more from the sand dunes; this time in Saudi Arabia and the Persian Gulf rather than North Africa. The equipment is a little more modern, transistors have replaced vacuum tubes and CDs sound better than 16 inch transcriptions but broadcasting from a dinky little van is still not a broadcaster's idea of luxury. Chief Petty Officer Kevin Clarke, running the station at Dhahran, put the best face possible on the situation and said, "This is the Corvair of radio. It's ugly, but it works."

Once a year, the management and key staffs of AFIS, AFRTS, the Army, Navy and Air Force Broadcast Services and station personnel from around the world meet in Los Angeles for a series of seminars, discussions, and exchanges of ideas. It's an invaluable forum in which ideas to improve the entire system can be tossed around. Or out. Policy makers can get a good idea what the people in the trenches are thinking, needing, demanding or complaining about. And the people from those far away places can learn how others have solved problems and what "those guys in Washington and Hollywood" are going to do next.

At the April 1990 meeting, the newest broadcast van was on display in the parking lot of the conference hotel. It was described as a "contingency" van and most of the conferees walked through it and said, "My, ain't that cute." Built by T-ASA at the Sacramento Army Depot to specifications supplied by AFIS, it was indeed cute. Several people suggested that AFRTS recruit articulate dwarfs to operate it. The broadcast portion, to which was hooked a transmitter and generator van, measured a giant 8 by 12 feet and contained both a radio and television station. The radio side contained two turntables, two CD players, three tape cartridge machines and an audio console feeding a 50-watt transmitter. The television side had three videotape playback units, a 16mm film and slide projector and all the associated rack mounted equipment necessary to feed the transmitter. A separate receiver dish could be attached to receive and decrypt satellite feeds. True, it was small—perhaps a size 6—and while there certainly wasn't room to swing a cat, no one would want to anyway. AFRTS has rules about that sort of thing.

Little did the visitors realize that within months, many of them would be operating out of one of the four vans then completed.

Navy Chief Journalist Rich Yanku opened his microphone the morning of October 9, 1990 and bellowed "Goooooood Morrrning, Saudi Arabia!!!" as he spun the recording of *Rock the Casbah* by the group Clash. By doing so, he also opened up the 132d country in which AFRTS operates — sandy Saudi. Yanku happened to be in Dhahran and there were other stations also operating throughout the Kingdom. Here is how it all came about:

The broadcast vans are designed to be used by any of the three broadcast services depending on whose turf they are to be used. Until they are in place and operating, the Navy is charged with their care and feeding. On August 2, 1990, Saddam Hussein sent the Iraqi army marching into neighboring Kuwait and while the politicians began tub-thumping and harumphing, the Navy began preparing the vans for possible shipment.

As troops began to be dispatched to the Middle East, AFRTS sent a message on August 15 to all major commands advising that radio stations were immediately available and that television would become available as soon as the necessary host country permission could be obtained. AFRTS also asked for volunteers to operate the stations and the response was overwhelming. Initially seventeen soldiers, sailors, airmen and Marines from stations around the world were chosen from the lengthy list. In the meantime, the Navy Broadcasting Service sent CWO4 Tom Jones and ICC Cliff Baker to Saudi to survey possible sites and begin installing small FM transmitters in anticipation of approval from Central Command.

Jones and Baker soon found that low power transmitters in that flat and sandy area were never going to be able to cover the territory expected to soon be occupied by what was then estimated to be 150- to 200,000 troops. Early on no one expected almost a half-million Americans in the desert. They would cover a lot of square miles and if broadcast signals were to do the same thing, a little innovation was required.

Obviously, higher powered transmitters were one solution and the Army furnished these from their operations at AFN and SEB in Germany and Italy. SEB sent Sergeant First Class Roland Martel, an expert on radio propagation and frequencies, and he joined Jones and Baker in the installation of transmitters and antennas. AFRTS sent Lieutenant Colonel Ray Shephard as the senior representative to coordinate all activities. He was later replaced by Lieutenant Colonel Dave MacNamee.

The Northern Arabian desert area is not exactly laced with available telephone lines so another technique had to be used to feed programs from a headquarters station out into the operational areas where the majority of the troops were in place; mostly in small, extremely isolated locations. The technique, called "cascading," was used. The theory is simple. A headquarters studio station broadcasts a signal. Installers take off across the desert and drive as far as they can until the signal is at its outer usable limits. An unattended transmitter is then installed and picks up, amplifies and rebroadcasts the signal from station number 1. The team then takes off again and when the signal from station 2 starts to fade, another transmitter is installed picking up station 2's signal.

...and awaaaaay we go, hopping across the desert installing transmitters. Technicians found that a string of five to six transmitters could be effectively linked in this way. Before they were through, they had a network of stations along the northern perimeter of Saudi Arabia. There were problems. Maps were not always accurate and one sand dune looks very much like another. Units were often not where the maps said they were. One AFRTS group installing a transmitter came across a tank concentration which supposedly had technologically perfect navigational equipment. Asking the tankers to mark their position on the map, they got five different locations from five different tank crews. One put them near downtown Cleveland.

Two of the Navy contingency vans arrived by September 5 and were positioned at King Fahd International Airport in Dhahran where initial tests were conducted. The remaining two vans were sent to the Navy's support area in Bahrain. Low powered stations were set up at Al Jubayl and King Khalid Military City. Local live programming was approved on September 19 and broadcasts began on a limited scale from the four manned stations on that day. Within a week or two, it was a full-blown network — DSN, the Desert Shield Network.

A lot of lessons had been learned in past experiences of this type but one important one had been forgotten. The first waves of troops to be sent to the gulf were told not to bring radios. When this word got around at home, radios started flooding in. Local radio stations collected them. Manufacturers donated them. The Army and Air Force Exchange Service began shipping them in. Sony contributed 5,000. The USO sent another 5,000. The American Electricians Association delivered 11,000 and the National Association of Broadcasters pledged 25,000 more. Montgomery Ward donated 350 videocassette recorders and television sets. New troops being sent were finally advised to bring radios.

Both AFN in Germany and SEB in Italy continued to supply technical and programming help on a rotating basis. One was engineer Sergeant Chris Hopwood of AFN who arrived in Dhahran at midnight and found himself out in the desert installing stations at 7 a.m. the next morning. Broadcast equipment from a lot of sources was pouring into the Gulf states and each service had its own break-down areas where equipment was unloaded, sorted and shipped to the end user. Hopwood recalls that he spent weeks hunting through the mountains of material being unloaded which, because of the peculiarity of the AFRTS name, was never at the right place. Unloading crews would see the "AF" in AFRTS stenciled on the box and ship it a hundred miles away to an Air Force supply center. Much of the arriving material was marked "Navy Broadcasting System" and this often ended up in the

As the hostilities come to a halt, a mobile van is set up in Kuwait City. Originally working out of the embassy, the van was moved to this site near the airport to provide wider coverage. (Photo courtesy Jeff Whitted)

This is a close as any AFRTS station came to getting blown off the air. That piece of junk in the foreground is a Scud missile which has been knocked out by a Patriot and has landed 100 yards from the station at KKMC — King Khalid Military City. The staff hacked off hunks to frame or wear on their dog tags. (Photo courtesy Mark Conner)

Navy supply yard. Hopwood and his compatriots often found it easier to fabricate things like antenna towers from "found" pipe than to dig through mountains of boxes. Aramco, the giant oil conglomerate, had a series of towers in place for their internal communications system and their offer to share tower space with AFRTS was gratefully accepted. Aramco also volunteered to let AFRTS technicians use their shop facilities to fabricate antennas.

Transportation was another problem. AFRTS was a little orphan compared to the giant units such as divisions, wings and armored cavalry regiments. Perhaps for the first time since Paris taxis took French troops off to World War I, AFRTS opted for commercial transportation. They went to the nearby Hertz and Budget rent-a-car offices and rented vehicles, most of which will probably never be the same after a few months of off-road use in the sands of Arabia.

Network headquarters was set up in Riyadh with Lieutenant Colonel Bob Gaylord in charge, Captain Jeff Whitted as deputy, and long-time AFRTS broadcaster, Command Sergeant Major Bob Nelson as chief of broadcasting. This group was liaison between AFIS, AFRTS and the men in the field but there were times when the men in the field were so far afield that it was difficult to stay in touch as they roamed the desert in rented cars looking for transmission sites. One member of such a lost patrol explained that, for the most part, these groups were self-dependent for such things as billeting and food. It tended to make them ingenious, though. In perhaps the financial coup of the century, one group found a Japanese party with a two and a half ton truck. They managed to trade two small transistor radios and a portable TV set for the truck. They felt they came out quite well considering that the two things in the world the Japanese are not short of are radios and TV sets.

Most of the Navy broadcasters were on temporary duty in Saudi from their regular assignments and the turnover was almost 100 per cent every 90 days. By early January, the Army had extended the duty to 179 days and the total group had grown from the original 17 to more than 50. Joining the regulars was an Army Reserve unit from Rome, Georgia, the 209th Broadcast Public Affairs Detachment.

The programming on the early radio stations and the television stations to go on line shortly thereafter was somewhat different than the troops were used to hearing and seeing. Because many of them were originally stationed in Europe, such as the 2nd Armored Cavalry and the 3rd Infantry Division, the AFRTS stations there sent recorded greetings from their families remaining at home base. And the process also worked in reverse. U.S. stations sent personalized greetings as well. One big difference was in the broadcast announcements. "Be sure to shake the scorpions out of your shoes before putting them on in the morning" is guaranteed to get the listener's attention. Instructions on what to do if captured by the Iraqis was an attention grabber, too. So were frequent warnings to stay by your vehicle when it broke down in the desert. "Don't forget to make a will," was a spot guaranteed to cheer up the listeners.

Listeners were able to crank up their field phones, call the stations, and dedicate songs to buddies in other outfits. The only rule was, don't call the unit by its real name or mention its location. No point in telling Saddam Hussein where the 82d Airborne was on that particular day.

Because the audience was so highly and emotionally involved in the Desert Shield operation, news was a key ingredient of the programming mix. The stations collaborated on producing a daily feature called "Desert Update" which each day was sent by phone line to AFRTS Los Angeles and immediately retransmitted back to Saudi Arabia. Strangely, it was the most efficient way of getting it to all stations quickly.

Television presented its own set of programming problems. The Saudis were grateful for the allied presence but they were not about to change their centuries old religious beliefs in order that Americans could watch T&A. Although the rules had relaxed since their own Saudi national television had come into being in the 1960s and, among other things, would not then permit an unveiled woman to appear on TV (thus ruining the plot line of a Bette Davis movie,) they were not about to let some of the raunchier music videos or jiggle shows pollute their airwaves. U. S. Television thus consisted primarily of news and sports, excluding female jello wrestling. Quite obviously, the items which Saudis would find offensive were an inherent part of their culture and the Americans could not be expected to understand all the nuances of what was and was not considered acceptable.

A clear day in Kuwait City as several hundred burning oil wells pump gunk into the air. The AFRTS broadcast van is parked back-end-to in the line at the right. The tent behind it houses the engineers. Katty-corner across the street is a warehouse where the program staff set up billets. Way in the background can be seen a burning oil well. It soon became necessary to move the van under cover, so heavy was the fall-out from the flaming wells. (Photo courtesy Mark Conner)

The Joint Services Command to operate the AFRTS stations in the Gulf was formed December 2, 1990. This is the team that arrived to set up the headquarters in Riyadh and established the network. Left to Right, Command Sergeant Major Bob Nelson, the network commander, Army Lieutenant Colonel Bob Gaylord and the deputy commander, Air Force Captain Jeff Whitted. (Photo courtesy Jeff Whitted)

The first week of the war, Scud missiles began dropping into Riyadh although the staff had been told they didn't have the range to reach there. Wrong! The first week they came in like clockwork at 11:30 p.m. and 3:30 a.m. and each time the staff would put on their protective clothing, as they are doing here. Someone finally figured out that these times were chosen in order to reach the best news times on CNN back in the States. After a couple of weeks the attacks lessened. (Photo courtesy of Mark Conner)

The satellite downlink dish at Riyadh's Eskan Village went into service in December and was used to receive the radio and TV signals from the States. (Photo Courtesy Jeff Whitted)

Naturally everybody wants to meet Bob Hope. If you would like to, join the service. It's unimaginable that the U. S. would give a war and not invite Bob. He got to Saudi in December and as usual, AFRTS was there to interview him as it is doing here at the Riyadh station. The interviewer is Staff Sergeant Gregg Hearns. (Photo courtesy Jeff Whitted)

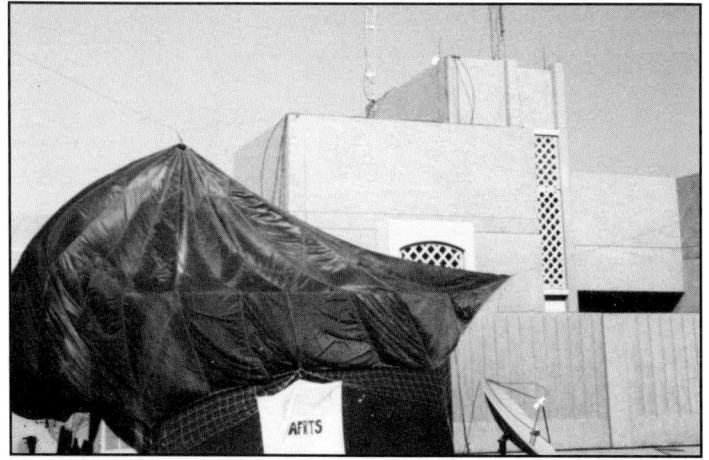

The Saudi Government built the villas in the background hoping the roving bedouins would settle down and move in. They never did, but the AFRTS staff in Riyadh was happy to. The Navy contingency van outside is covered with a parachute and the small receiving antenna dish is picking up the audio signal from the States. (Photo courtesy Jeff Whitted)

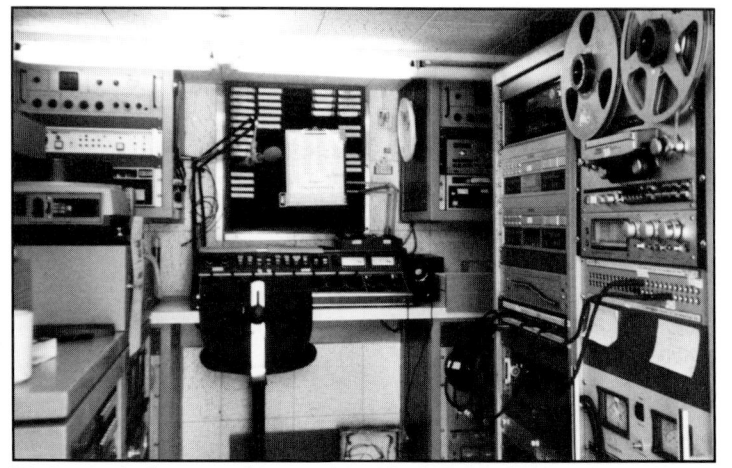

Underneath that parachute and inside the van, this is what it looks like. There were three other vans just like it — in Dhahran, Al Jubayl and King Khalid Military City. Each was configured identically. (Photo courtesy Jeff Whitted)

Down in Dhahran, one of the most popular DJs called herself "The Desert Fox." Her fellow staffers knew her as Specialist Brooke Perkins. Here she is in the "office" of the Dhahran station, preparing for her next show. (Photo courtesy Jeff Whitted)

Who says there are no Brooks in the desert. Here's another one: Brooke Shields, visiting the troops under the auspices of the USO. The troop she is visiting here in Bahrain is the deputy network commander, Captain Jeff Whitted. (Photo courtesy Jeff Whitted)

Modeling the latest in desert resort wear, Army Specialist Brigette Horau, DJ at the Riyadh station (L) and Air Force Chief master Sergeant Linda Arnold, Riyadh station manager, getting all dressed up for the nightly Scud party. Chief Arnold has always been well known during her years in the Air Force as an impeccable dresser. Here, however, even the most courteous would have to admit she's looking slightly peccable. (Photo courtesy Jeff Whitted)

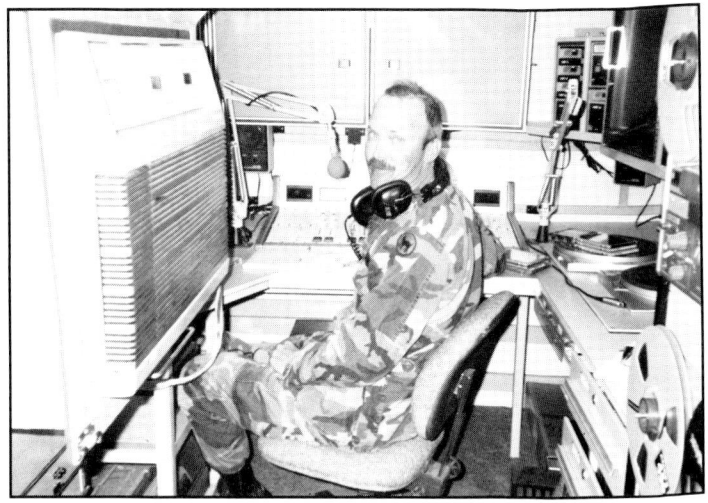

Returning from the bright lights and excitement of lovely downtown Thumama, Army Sergeant Russ McClanahan of the 209th BPAD, a reserve unit deployed to serve with AFRTS in the Gulf, takes his seat at the microphone in the van and gets ready to go to work. (Photo courtesy Jeff Whitted)

Meanwhile, back in Riyadh, Bob Gaylord dresses for the possible Scud attack. Here he tunes in CNN, which AFRTS also broadcast, to find out what in hell is happening. What kind of weird war is this, anyway, when the commander of his very own network has to have some news person in Atlanta, Georgia, tell him what's going on outside the window? (Photo courtesy Jeff Whitted)

You can't have a unit without having a meeting, so they had a meeting in Riyadh. In attendance: (Rear, L to R) SSGT Ed Whitener, Kuwait City station manager; Marine SSGT Terry Ruggles, KKMC station manager; AF MSGT Steve Napolillo, Riyadh station manager; Army SSGT Randy Wilkins, Dhahran station Manager. (Center, L to R) AF MSGT Rafael Alcantara, NCOIC of maintenance; Army Captain Frank Barron, Deputy network director; AF LTC Randy Morger, network director, AF CMSGT Mark Conner, chief of broadcasting; Army CWO Ronnie Motes, chief of logistics and engineering. (Front, L to R) Army Captain Bob Close, officer in charge of the Eastern region; Army First Lieutenant Mike Warren, OIC of Northern region and AF Captain Jeff Whitted, OIC Central region and chief of operations. (Photo courtesy Jeff Whitted)

191

An extensive list of no-nos was published but could never cover everything. The operative rule-of-thumb became, "If you have to ask whether something is offensive, you already have the answer." Entertainment programs were sent in separate packages to individual units who played them on VCR units for the private edification of the troops. There were also strictures against playing Christmas music during the 1990-91 season. *Jingle Bells, Winter Wonderland* and songs of that type were approved as were instrumental versions of some Christian songs. Not vocals, though.

On January 17, the multi-national forces began air strikes on enemy positions in Iraq and Kuwait. Programming was switched from a music to a news format on radio and television expanded to 24-hour news coverage from the States via SATNET. It was a weird situation for the troops sitting in hardened areas or bunkers watching on television the Scud attacks being directed at them. After a week or so, the schedule was revised to provide 30 minutes of music and 30 minutes of news each hour.

Because the middle East is Air Force territory according to the atlas in the AFIS office, this service took control of the network and, following fifty years of unwritten tradition, changed the name from the Desert Shield Network to AFDN, the Armed Forces Desert Network. Ten days after assuming operational control, the Iraqis welcomed them with a Scud missile which landed 150 meters from the station at King Khalid Military City. There were no injuries but the staff learned that you can get into protective clothing in a big hurry, given the right motivation. Two of the Air Force's outstanding broadcasters were in charge at various times and credit must be given to both. Chief Master Sergeant Linda Arnold who, during her career to date, has handled just about every tough job the service could dream up. She has done it all, from writing news copy and doing news reports in Berlin to planning and managing the duplication facility in Japan to running stations in Italy and Turkey. After returning from the desert, she returned to Air Force Broadcasting headquarters in San Antonio. So did Chief Master Sergeant Mark Conner, who, like Arnold, took every job handed him and did it well wherever in the world the service decided he could be best utilized.

Transmitters were installed in Northeastern Saudi near the Kuwait border in anticipation of forthcoming action in that sector but were not turned on in order not to give away allied combat positions. They became an on-air part of AFDN after February 22 as the ground offensive began. Once more programming reverted to a heavy dose of news on both radio and TV. By the 27th, Kuwait was liberated and on the 28th, President Bush called a halt to offensive operations. The next day the network returned to a 50/50 mixture of entertainment and news and sports. By the 4th of March, a two man AFRTS team was arranging for a station in Kuwait City and the next day nine more arrived with an Army broadcast van. Two days later it was on the air from the U. S. Embassy; two days after that it moved to the Kuwait International Airport and in two more days had hooked up a more powerful 5,000 watt transmitter which effectively covered the entire country.

All the activity was not on the ground. In both the Persian Gulf and the Red Sea, Navy and Marine personnel were prepared for action. AFRTS set up duplication facilities on both coasts of Arabia. A facility in Bahrain recorded material from satellite and flew it out to ships in the Gulf while a sister activity in Jiddah on the Red Sea did the same for ships in that area. While ships could easily receive and put the audio signal on their shipboard systems, television technology has still not reached the point where a consistent signal can be picked up from a receiving dish on a rolling deck, although a breakthrough in this desirable effort is expected one day. In the meantime, crews in Dupfacs continue to work weird hours because of the time differences, recording, duplicating and shipping out time-sensitive programs. Crews aboard ships in the Gulf and the Red Sea watched the NFL playoff games with as little as a five hour delay.

As peace broke out, the stations began to neaten up and settle down into more practical, more efficient and certainly more comfortable quarters. Network headquarters moved from downtown Riyadh to nearby Eskan Village, and later, moved to Dhahran. A 5,000 watt unmanned repeater station replaced the manned station at Al Jubayl and the staff moved to Dhahran which began feeding the stations in Saudi Arabia. An independent radio outlet in Kuwait City served the forces in that country.

Host country sensitivities continued to plague the network and television broadcasts were discontinued at King Khalid Military City. All available mini-TV equipment was shipped there and duplication facilities there and elsewhere went into overdrive to provide approximately 6 hours of programming daily to troops to watch on closed circuit. Later on-air television will be resumed when a microwave distribution system is completed.

Finally, the story of the network in the Gulf States closes with a bang as an ammunition truck near AFDN-Kuwait catches fire and the station and nearby personnel are evacuated as exploding ordnance begins peppering the area. As usual, the station is back on within an hour with half the staff volunteering to search for unexploded ordnance and the other half manning their microphones, explaining what happened and telling the audience everything they need to know to avoid getting their fannies blown off.

Information like that could be considered the Bottom Line of AFRTS. That's what it's all about — going where the troops are overseas in war and peace; giving them large doses of home; keeping them informed and entertained; giving them the information they need to know to function in a strange environment and doing it all with short money, short staffs and long hours.

As AFRTS completed its fifth decade, the world was changing. Some long-time stations will be closing. Others will be consolidating or moving. Perhaps new ones will be needed in places not yet dreamed of. But whatever the future brings, AFRTS and the dedicated men and women who make it the most unusual broadcast entity on Earth, will be there.

❖ ❖ ❖ ❖ ❖

BOOTH DIRECTOR: "FADE TO BLACK. AND GO TO ONE. UP ON THE CHRONOLOGY WHICH IS GOING TO MAKE THIS WHOLE THING CLEAR."

❖ ❖ ❖ ❖ ❖

Navy Petty Officer Rick McGraw (L) and Army Sergeant Phil Tracy check out an infra-red transmitter link used to provide AFRTS television service to the Joint Information Bureaus at the Hyatt Regency Hotel in Riyadh. (Photo courtesy Jeff Whitted)

Peace broke out when the station set up in Kuwait. So did sweat on the forehead as this cache of high explosive ordnance was discovered about 100 yards from the 5,000 Watt transmitter just installed. Kneeling among it is Chief Warrent Officer Ronnie Motes, Chief of network engineering. (Photo courtesy Jeff Whitted)

Early troops to arrive in the Gulf didn't have radios but soon they began to arrive from numerous sources. When AFRTS people travelled, they tried to carry as many as possible to pass out. Here their convoy headed for Kuwait to set up the station there is stopped at a checkpoint and Army SSGT Patty Cunningham starts passing them out to the troops in the truck in front. (Photo courtesy Mark Conner)

One of the two navy contingency vans made up the station at Dhahran. That's an AFSTRS audio receiving dish in the foreground. (Photo courtesy Mark Conner)

...and this is the office and administrative headquarters of AFRTS Dhahran. Visiting the station (L to R) LTC Bob Gaylord, Randy Kaafka and Command Sergeant Major Bob Nelson. (Photo courtesy Mark Conner)

The Kuwait station set up this table with the listing of all the songs in their library. Troops could drop by, pick a record, and have it dedicated to a friend or, if they had a recorder, dedicate it to a loved one back home, record it and send the tape back. (Photo courtesy Mark Conner)

1) KEFLAVIK, ICELAND. Visitors to the station at Keflavik entered the building into this reception area. Stations everywhere used their imaginations to try and make their facility look as much like a stateside station as possible. They had to use their imaginations because there seldom was any money for such fripperies. (Photo courtesy of AFRTS) 2) GUANTANAMO BAY, CUBA. In 1961 the station looked like this. The Navy has always been proud of this facility and for good reason. Because the base is surrounded by a fence and the Americans restricted in their movements, the radio and TV station has become critically important to their morale. (Photo courtesy of U.S. Navy) 3) KODIAK, ALASKA. This exciting photo shows how exciting things can get in the Aleutians. One exciting thing is that there is a tree in the picture and they are rare. One suspects that the station mascot, shown guarding the entrance, really stays around because of the tree. (Photo courtesy of AFRTS) 4) DHAHRAN, SAUDI ARABIA. The 7122d once ran a small station for the U.S. Air Base here and this picture shows the kind of imagination that goes into operating in such an environment. Someone wanted, for totally inexplicable reasons, a fireplace. A fireplace requires logs. The nearest tree was approximately six area codes away. So the logs are made of cork. Very clever, these Air Force guys. (Photo courtesy U.S. Air Force) 5) ADAK, ALASKA. It gets pretty lonely and isolated out there in the Aleutians, but back in 1958 the Navy television station there did all they could to entertain the families and, as here, the kids. This is "Kiddie Kapers." the hostess is Mrs. L. Kehoe and the host is P.R. Mumford. (Photo courtesy U.S. Navy)

1) WHEELUS FIELD, LIBYA. As Vice President under President Eisenhower, Richard Nixon and Mrs. Nixon made a swing through the Mediterranean area. He is shown here talking to the troops through the 7122d Support Squadron station in Libya. Don't ask us to explain why the microphones are bandaged. A person could wonder, though, why some over-zealous Public Affairs Officer gave Mrs. Nixon enough calla lilies to stock a mafia funeral. (Photo courtesy U. S. Air Force. 2) KEFLAVIK, ICELAND is another Navy-run station and just about every Navy broadcaster who has served there has nothing but nice things to say about the assignment. And why not? The island seems to be populated by gorgeous blonds. The fishing is good, too, but most of the sailors seem to prefer the women. (Photo courtesy of AFRTS) 3) SIDI SLAMANE, MOROCCO is now long gone. The 7122d Support Squadron (AFRS-TV), as its was then known, ran a number of stations on air bases in Morocco including Nouasseur near Casablanca. A French dependency, the French returned it to the Moroccans who returned the Americans to whence they came, but keeping the majority of the American equipment. Sorry to say, but they didn't get much when they got this. (Photo courtesy U.S. Air Force) 4) WHEELUS FIELD, TRIPOLI, LIBYA. From the earliest beginnings after WWII, AFRTS stations almost invariably had teens participate. Schools formed radio clubs. So did the Scouts or church groups. Wherever you went in AFRTS country, you could count on hearing or seeing programs prepared by teens. One teenager named Sonny Craven, in Poitiers, France, was a member of the high school radio club, and returned to AFN years later as a Lieutenant Colonel and commander of the network. (Photo courtesy U.S. Air Force) 5) KANUA, ASMARA, ERITREA. Another totally isolated posting for the Air Force where the station did its best to keep the kids happy. the caption on this photo reads, "Wild Wes talks to guests on AFTV." We haven't a clue who or which Wild Wes is although if Wes is truly wild, it is possibly the bird who seems to be enjoying itself more than anyone else. (Photo courtesy U.S. Air Force)

CLOSING CREDITS

My special thanks and gratitude to the more than three-hundred present and former members of the AFRTS family who so generously responded to my pleas for the use of their memories, anecdotes, memorabilia and advice. This is their book.

I am grateful beyond measure for the cooperation extended by the American Forces Information Service (AFIS) who provided a wealth of data including written material from various previously unpublished sources, as well as photographs and videotapes.

The support of the Armed Forces Broadcasters Association was invaluable in the preparation of this book. This group, made up of those with any connection, past or present, with AFRTS has been unflaggingly helpful in involving the membership in this project.

Present members of AFRTS Washington, AFRTS Broadcast Center in Los Angeles and AFRTS staff broadcasters in every corner of the world have made invaluable contributions and prove once again that the word "service" in their name is more than just an idle word.

Finally, a very special "thank you" to a number of individuals who worked particularly closely with the author to check facts, provide specialized information, dig for pictures, supply clearances and generally make life easier. They include -- but are certainly just the iceberg's tip -- John Bradley, Buzz Rizer, Mel Russell, Gerry Fry, Dorothy McAdam, Normand Lucier, Keith Anderson, Bob Cranston, Bob Harlan, Al Edick, Mary Carnes, Bob Cleveland of Video Masters, Inc., Robert Beltz, Mary-Michaele Brooks and Paul Ullrich of AFN TV Guide, Howard Medici, Walt Cleary, Linda Arnold, Mark Conner, Jeff Whitted, Roger Maynard, Bill Hart, Jack Brown and the Pacific Pioneer Broadcasters, Bob Stoffel and Bob Matthes.

The author and Announcer Paul Macko luxuriate in their private stretch jeep as they prepare to cover a training exercise in the snowy forests of Germany. The author is smiling because he's a civilian and knows how foolish he looks in a steel helmet. Macko is not smiling because he's in the army and would rather be someplace else.

Later, he got out of the army and joined AFN as a civilian newsman, and still later as Station Manager in Stuttgart. (Photo: John Pilger)

The author tries in vain to find out where Phil Donahue gets that beautiful grey color in his hair so that he, too, can become a successful broadcaster. Donahue, obviously frightened of the competition, refuses to reveal his secret formula. (Photo: Trent Christman)

CHRONOLOGY

Telling the story of hundreds of people working at dozens of networks and stations on every continent over five decades requires a certain amount of literary leaping around through time and space. The reader can easily be excused for becoming confused about where events described fit into the total picture. It is hoped that the following Chronology will, like a Richard Nixon speech, "Make everything perfectly clear."

Pre-1941

February 1939. KGEI, a General Electric Company shortwave transmitter at the San Francisco World's Fair begins operation furnishing news and entertainment to the Philippines.

February 1940 The British Broadcasting Corporation begins "The Forces Programme" directed to British forces on the continent.

July 1940 The War Department forms the Morale Division of the Adjutant General's Office. Sections include Army Motion Picture Service, Recreation and Welfare, Exhibits, Decorations and Morale Publicity.

1941

January Troops in Panama are dispersed to isolated anti-aircraft and artillery sites in jungle. Begin broadcasting music and information over the tactical circuits. Broadcasts cease on December 7.

February General Marshall calls a meeting of Morale Officers of all major units and finds most major units don't, and never did, have one.

March. Marshall forms a Morale Branch reporting directly to the Chief of Staff of the Army.

October KODK, begun by soldiers in Kodiak, Alaska, and becomes the first station to continue without interruption through World War II.

December 7 Pearl Harbor attacked by Japanese forces and America goes to war.

1942

January American troops in the Philippines fall back to the Bataan Peninsula and Corregidor. Use a small transmitter to rebroadcast KGEI as only source of news.

January 15 The "Morale Branch" has its name changed to **Special Services** as the demands of War increase the scope of its responsibilities.

February Movie Director Frank Capra commissioned a Colonel in the Signal Corps and is later transferred into Special Services.

February Privates Charles Gilliam and Robert Nelson begin Station GAB (for Gil and Bob) at Sitka, Alaska. Call letters later changed to KRAY (for Fort Ray.) Station, originally unlicensed, receives FCC approval in November and assigned call letters WVCX.

March 1 First *Command Performance* produced at CBS studios in New York, starring Eddie Cantor, Harry Von Zell, Cookie Fairchild's Orchestra and Bert Gordon, "the Mad Russian."

April 12 *Command Performance*, after six shows from New York, moves to Hollywood in order to take advantage of the larger talent pool. First Hollywood production stars Gene Tierney, Betty Hutton, Gary Cooper, the Andrews Sisters, Ray Noble Orchestra, Edgar Bergen and Charlie McCarthy, Ginny Simms, Bob Burns and announcer Paul Douglas.

May 26 The Armed Forces Radio Service is formed as a part of the Morale Services Division of the War Department with the charter to provide service personnel world-wide with similar radio programming as they would hear at home. On the same day, Tom Lewis's commission as a major is approved and he goes to Washington to present his initial planning document on the formation of AFRS.

June Tom Lewis and Murray Brophey, Chief of the Office of the Coordinator of Information, go to Alaska to learn first hand the needs of the troops in the field. There they learn that the initial needs are already being filled by the stations in Kodiak and Sitka. These operations provide the genesis for future expansion.

July 19 True Boardman joins the staff of AFRS direct from the Harem movie set where he is acting as dialogue director. He becomes a top assistant to Tom Lewis.

August Proposed AFRS program schedule is developed and includes programs such as *Mail Call* which becomes one of the most popular with the troops and survives the war intact.

August 11 First *Mail Call* recorded from the CBS studios. True Boardman produced, Bob Lee and Jerry Lawrence wrote and Loretta Young, Bob Hope, Frances

	Langford and Jerry Colonna starred.
November 1	Generals Eisenhower and Marshall approve the concept of an American radio network in the United Kingdom.
November 8	The U. S. invades North Africa through Morocco. André Baruch tries to describe the action from the deck of the battlewagon USS Texas with singular lack of success.
December 15	Baruch and his group, with General Patton's permission, get the first station broadcasting from the middle of a ground war on the air.
December 15	Control of *Command Performance*, previously under the direction of the Bureau of Public Relations, is transferred to newly formed AFRTS.

1943

March	Neither knowing nor caring that AFRS has responsibility for troop broadcasting, Charles Vanda and a group put a second station on the air in Morocco. Major Martin Work from AFRS in Hollywood arrives to see what in the devil is going on here.
April	Major Work arranges to have Baruch transferred to AFRS and combines the two stations. Baruch is appointed Chief of American Stations in the North African Theater.
May	True Boardman, at the request of the Army, makes a tour of the Pacific area to determine if, and where, stations should be located, and to work out operational details with the Navy.
May	Captain John Hayes in England begins assembling a staff for AFN and now has four officers and 13 enlisted men.
July 4	AFN signs on the air in England using BBC studios at 11 Carlos Place.. Syl Binken, announcer, makes the opening announcement but to this day no one knows what it was.
August 13	Fighting having ended August 7, Baruch moves a station into Sicily.
September 1	The Office of War Information (OWI) ceases production of programming directed at troops overseas and turns over the entire project to AFRS. In addition, it makes shortwave equipment available to AFRS for overseas transmissions.
October 23	AFRS station opens in Naples as American Fifth Army troops continue to move North.
December	Dresser Dahlstead, popular radio announcer and later to be a producer with the Ralph Edwards organization for many years, joins AFRS and becomes Tom Lewis's assistant. He and Elliot Lewis develop the technique to fill the program time taken out of programs when the commercials are removed.
Year's End	140 stations around the world are now broadcasting AFRS programs. This early in the life of the system, many stations are foreign national broadcasters and share time with the Americans who must wait to begin English language programming until the locals decide to end their broadcast day.

1944

January	There are now eight stations operating in the Mediterranean area, reaching from Casablanca to Naples.
January	Having completed a training program for personnel destined for overseas assignments, AFRS also begins shipping self contained stations to the Pacific Theater, each complete with transmitter, studio gear and record library.
February	The first self-contained station arrives and is placed on the air in Noumea, New Caledonia. This ultimately becomes the key station of the Mosquito Network of seven stations in the Southwest Pacific.
February	By now the troops have moved so far North in Italy, the Naples station is out of their listening range. Back to the mobile units.
February 26	Following Boardman's exploratory journey in 1943, the first Jungle Network station signs on from Port Moresby, New Guinea. This is a joint U.S.-Australian effort.
Early Spring	A station is opened in Auckland, New Zealand, and continues to operate through December when troops have moved far to the North.
May	AFRS in Hollywood is now producing 106 different programs each week totalling 40 hours in length. An additional sixty programs are from commercial sources and are edited to remove commercial announcements. According to *Daily Variety*, it would cost $10 million dollars to produce the AFRS output if normal commercial rates were applied.
May	Captain Edgar Tidwell begins opening purely U.S. stations to form the Jungle Network. First on: Nadzab, New Guinea. Finschaafen, which becomes network headquarters, signs on in June. Ted Sherdeman becomes network commander.
May	Tom Lewis arrives in England to supervise preparations for radio operations during and after the Allied Invasion of the European Continent -- D-Day.
June 6	**D-Day.** Allied troops land successfully in France. Broadcasting is now a joint British/American/Canadian operation called the Allied Expeditionary Forces

July	Program, AEFP. AFN continues to broadcast to Americans remaining in England and prepares mobile units to follow the American armies. In England there are 60 transmitters connected to new AFN studios at 80 Portland Place in London.
July	Fifth Army AFRS station reaches Rome and begins broadcasting counter-programming to "Axis Sally" a week after the city is captured.

1945

February 15	The "Command Performance" *Wedding of Dick Tracy* is produced for AFRS under the direction of Pat Weaver who modestly calls it, "The greatest radio show in history."
February	The Philippines have fallen and new Jungle Network commander Graf Boepple moves the headquarters from New Guinea to Manila. He also changes the name to "The Far Eastern Network." It now has 14 affiliates.
March	AFRS presses its one-millionth disk for shipment overseas. This one is presented to Colonel Tom Lewis. For trivia buffs: It was *GI Journal* featuring Bing Crosby, Linda Darnell, Betty Grable and Abbott & Costello.
March	There are now 154 stations worldwide including the Middle-East, Europe, the China-Burma-India theater, Alaska, the Caribbean and the Pacific. There are also 143 AFRS operated public address systems and hospital bedside networks. AFRS supplies each with 126 separate programs each week on transcriptions.
April 12	AFRS receives the news that President Franklin D. Roosevelt has died at 5:49 EWT. The first bulletin is broadcast worldwide at 5:52 EWT.
June 10	Bob Light opens the first AFN station on German soil in Munich at 15 Kaulbachstrasse, the mansion of famed artist Kaulbach and used during the war as the home of the German *gauleiter*. Broadcast circuits from there feed 100,000 watt transmitters in Munich and Stuttgart.
July 15	AFN opens headquarters and studios on Kaiser Sigmund Strasse in Frankfurt near the headquarters of Supreme Allied Commander Dwight Eisenhower.
July 28	The Combined Allied broadcasts over the AEFP end and AFN is able to supply around-the-clock broadcasts in the American language and using American performers.
August 6	The Atomic Age begins as the B-29 *Enola Gay* drops the first atom bomb on Hiroshima.
August 12	Ben Hoberman at AFN calls the Japanese military attache in Switzerland and scoops the world on announcing the Japanese surrender.
August 15	AFN moves into the Von Bruening castle, started in the 12th century, in Hoechst on the outskirts of Frankfurt. This will remain its home for the next 21 years
September 2	AFRS covers the Japanese surrender from the deck of the USS Missouri anchored in Tokyo Bay. Staff members quickly begin looking for permanent sites for AFRS stations on the Japanese mainland.
September 7	General MacArthur's headquarters advises AFRS that there are four mobile stations operating in Japan and one in Korea, five days after the surrender. One, in Northern Kyushu, signed on September 1, one day before the surrender.
September 15	Using Japanese studios, AFRS goes on the air in Osaka.
September 23	WVTR, the headquarters station, begins broadcasting from Tokyo and the name officially changes from "Far Eastern Network" to "Far East Network." It must have been a good name because it's still in use.
October 25	His job done, Colonel Tom Lewis returns to civilian life and Martin Work is appointed AFRS commander.
Autumn	Chief engineer Walt Cleary of AFN discovers a German invention which miraculously records sound on a band of peculiar looking tape. Finally getting it to work, he puts it in the system and AFN becomes the first American station in the world to utilize audio tape.
October	AFN opens studios in Paris.
November	Colonel John Hayes assigns Corporal Harold Burson to cover the Nuernberg war trials. Coverage continues until all defendants are hung, imprisoned, or commit suicide.
December 31	The London headquarters of AFN signs off for the last time and Frankfurt becomes the new headquarters.

1946

January 23	**Army Information School** established at Carlisle Barracks, PA.
June	With most Americans now out of Paris, AFN attempts to close its station. So many French civilians protest, the American ambassador orders it to remain open which it does until the end of 1947. The then Chief Engineer Walt Cleary insists it finally closed because of orders from a Colonel whose sleep was disturbed by his neighbor playing his radio too loudly.
Fall	Bing Crosby, after seeing from his shows on AFRS that the audience really doesn't

mind recorded programs, leaves NBC and joins newly formed ABC which permits recorded programming. This breaks the dam and other stars demand the same privilege. Live network programming starts to become a thing of the past.

1947

Year's start With the war over and post-war economies in effect, AFRS production of original programs drops to fourteen hours a week. An additional 41 hours of off-network programming is also supplied to AFRS stations.

Post-war agreements with AFTRA, the actors union, and AFM, the musician's union, are mutually arrived at and continue to this day although in slightly altered form.

Mid-year The *Far East Network* now has 16 stations, down from a high of 39, as troops go home. Now there are seven in Japan and others in Manila, Okinawa, Guam, Iwo Jima, Saipan, the Admiralty Islands and three in Korea.

1948

April 9 Secretary of Defense James Forrestal directs the establishment of the Armed Forces Information School at Fort Slocum, N.Y.

June 26 The Russians blockade all land access to Berlin and the allies begin supplying the isolated city by air. For more than a year, AFN covers every aspect of the Berlin Airlift, keeping troops informed and signifying to the beleaguered Berliners that the West was not deserting them. The Berlin station acts as a friendly homing beacon for the incoming aircraft which fly one minute apart around the clock, loaded with fuel, medicine and food.

1949

May 12 Berlin Blockade ends and AFN Berlin broadcasts the festivities as crowds go wild.

Bob Harlan joins the AFN staff and will remain a key player in the network for more than 35 years.

1950

All Year AFRS gradually gets out of the creative end of radio production and, with the exception of a few informational and educational programs, ends the type of original entertainment programming which has been its hallmark. The number of "canned" syndicated and network programs supplied to stations is increased to 60 hours weekly.

June 25 North Korean troops invade South Korea and attack Seoul. AFRS station there assists in the evacuation of troops and civilians and goes on the road as "Radio Vagabond." After a short stay at Taegu, it moves to Pusan in the far South of the peninsula. Surrounded, it retreats to Japan.

September 27 General Order by General MacArthur creates AFKN, the American Forces Korea Network. They are back on the air by October 4 in Seoul, which has been recaptured. Soon they are again driven out and join their sister stations as "stations on wheels." By May 1951 they are once again back in Seoul. This time for good.

1951

April 5 It took a little while, but the Armed Forces Information School opens at Fort Slocum.

April 11 President Truman fires General Douglas MacArthur as commander in Korea. The General learns this interesting fact while listening to AFKN.

1952

General Curtis LeMay, Strategic Air Force Commander tells his buddy Arthur Godfrey that SAC has morale problems at its isolated bases and Godfrey suggests installing television might be a help.

1953

Early The Korean War delayed things a bit, but General Curtis LeMay again starts trying to get television into SAC Bases. Limestone AirBase in Maine is selected as the first site although it is within the borders of the U.S. The rationale apparently is that it is so far out in the boonies that it might as well be on Mars.

February AFN opens radio station in Kaiserslautern, Germany.

June The House Appropriations Committee closes Fort Slocum, and the Information School along with it. Courses continue at various service schools.

Mid-Year The Japanese radio industry issues a report that estimates about 10-million Japanese listen regularly to FEN. This is demonstrated when the network runs a "Diskathon" soliciting musical requests and the entire Japanese telephone system, including emergency circuits, is overloaded and shuts down. The government decrees that FEN never again honor phone requests. A Japanese newspaper revises the figure of local

October 28	national listeners to 14-million. (In honesty, it must be noted that then the Japanese stations all stopped broadcasting at midnight while FEN continued all night.) The Secretary of Defense signs a memorandum officially permitting military television.
November	Headquarters, Far East Network moves from Tokyo to military installation at South Camp Drake, Japan. It controls 20 outlets stretching from the Philippines to Guam.
December 23	Under direction to go on the air Christmas Day, the equipment for the Limestone AFB TV station arrives. In spite of the fact the wrong antenna has been sent, the staff gets on the air -- barely. Film for the opening day arrives late Christmas eve. Some sadist who apparently hated TV located the station in the uninsulated metal shack housing the elevator machinery on the roof of the base hospital.

1954

April 24	Facing reality squarely, the Armed Forces Radio Service (AFRS) becomes the Armed Forces Radio **and Television** Service (AFRTS.)
(Later it changes to the **American** Forces Radio and Television Service. Still later it was changed back to "**Armed.**" Let's not confuse things and just stick to **AFRTS.**)	
October 17	Not about to let the Strategic Air Command get all the glory, MATS, the Military Air Transport Service, opens its own television station at Lajes Field in the Azores.
December 22	MATS does it again, and opens a station at Wheelus Field in Libya.

1955

March 1	MATS opens a TV station at Keflavik, Iceland.
July 4	MATS opens a station at Kindley AFB in Bermuda and becomes the first Air Force command to have television at all its remote bases.
July	The 7122 Broadcasting Squadron (Air Force) is formed at Ramstein AFB in Germany to supervise five radio stations and two TV stations. It is the first Broadcasting Squadron in the armed forces and plans are soon underway to expand to eleven additional bases.
November	Clark Air Force Base becomes the first Far East Network station to add television.
December	FEN Okinawa becomes second FEN station to add television.

1956

September 15	AFKN Seoul, Korea, becomes the first Army TV outlet and operates a film-only outlet with no capability for live broadcasting.
The Air Force	begins television broadcasting from Ramstein Air Base near Kaiserslautern, Germany. Coverage is limited to a small area but the word soon gets around and EVERYBODY wants TV. The Air Force expands coverage to other air bases in Spangdahlem and Bitburg. This doesn't satisfy the Army troops who wipe out entire forests supplying newsprint for the **Stars and Stripes** to print their gripe letters.

1959

January 4	AFKN Seoul finally gets the capability to televise live programming.
Early	After long negotiations, AFN returns to France with a network of about 60 small 50-watt FM transmitters. The French Network is connected to the AFN headquarters in Frankfurt, Germany, but exercises considerable autonomy in programming. French headquarters are in Orleans with other studio locations in Poitiers and Verdun.
March	Control of AFRS Okinawa moves from FEN to the Air Force island commander.

1960

December 24	FEN Misawa becomes the first FEN television station on the Japanese mainland. The Japanese government insists it be UHF so as not to interfere with national VHF signals.

1961

Secretary of Defense Robert McNamara and Assistant Secretary Arthur Sylvester begin action to reopen the Information School.

1962

The 6120th Broadcasting Squadron is formed and FEN is placed under its control.

Congress begins pressing a long and futile fight for consolidation of the broadcasting activities of the three services. Hearings are heard, studies are studied, recommendations are recommended but the services continue to want to control their own fiefdoms and little is done.

The Von Bruening castle, home of AFN, is purchased by the *Farbewerke Hoechst* chemical combine and the Americans and the Bonn government begin looking for a new home

	for the network. A site is found near the headquarters of the German First Network in Frankfurt.
July 6	Although there are only about 6,000 Americans in Vietnam, Radio Hanoi begins propaganda-filled broadcasts directed to them from North Vietnam, following in the tradition of Axis Sally and Tokyo Rose of earlier days.
August 15	AFRTS signs on with radio from Saigon using a vintage World War II transmitter which has been stored in the Philippines. Studios are in the Rex Hotel. Later in the month they move to the Brink BOQ.
December 24	The Viet Cong blow up the Brink BOQ, killing two, injuring scores and destroying the studios. Radio service is restored within twenty minutes using auxiliary equipment.

1963

February 11	AFRTS stations worldwide are ordered to broadcast a program prepared by the United States Information Service titled "Today's Analysis of Events from Washington." Most stations balk at the obvious propaganda and the threat of news management. WTOP, Washington, Jack Anderson and others make it a crusade. AFN Commander Robert Cranston leads fight to keep it off the air and succeeds in France but not Germany. The fight drags on, most stations playing it during late night hours. Finally, on March 13, 1967, USIA gives up and cancels the program to the relief of all.
November 22	President Kennedy assassinated. AFRTS stations around the world carry practically uninterrupted coverage from the U.S. by either cable or shortwave through the funeral and remain in a solemn mode for days thereafter.

1964

February 21	Secretary McNamara issues the charter for the Information School to open.
	AFKN headquarters station in Seoul moves both radio and television studios to Yongsan compound, where it remains today. In addition, it now begins to use up-to-date Image Orthicon cameras.
	Ground is broken for the new AFN Headquarters building and studios. The German government assumes $2.3 million in building costs.

1965

May	Work begins on building the "Blue Eagles", flying radio and television stations for use in Vietnam using C-121 Super Constellations.
July	Local commander assumes control of FEN station at Clark Air Force Base. It becomes the headquarters of a three station Philippines network which includes Subic Bay and San Miguel.
September	Department of Defense Information School opens at Ft. Benjamin Harrison.
December 2	Deputy Secretary of Defense Cyrus Vance approves plans for the establishment of ground television facilities in Saigon and on December 24, the U. S. and Vietnamese governments officially authorize AFRTS to begin operations.
Year's End	AFKN is now a network in fact with all affiliates linked to the Seoul headquarters.

1966

February 5	Cyrus Vance approves the installation of four high powered radio ground stations in Vietnam.
February 7	The **Blue Eagles**. which arrived in Vietnam in January, begin daily operations, transmitting to the U. S. troops on Channel 11 and to Vietnamese on Channel 9.
March 24	AFRTS Saigon is connected to AFRTS Los Angeles by means of an underseas cable, allowing it to receive broadcast quality news and sports from the West Coast instead of by unreliable shortwave. Cable is two-way, allowing Saigon to transmit material back to the States.
April 13	Viet Cong mortar fire damages three **Blue Eagle** aircraft parked at Ton Son Nhut air base. Only one aircraft is flyable and hours of operations are cut. A full schedule is resumed on May 11, 28 days after the attack.
Mid-Year	AFN Building is completed in Frankfurt and the operation begins a phased move into the new facility which features a central core containing the studios mounted on gigantic springs to absorb noise and vibrations.
September 26	A number of broadcast vans, radio and TV, arrive in Vietnam and the first one goes into operation at Qui Nhon. The local PX sells 1,000 television sets in anticipation.
October 21	The second van, at Da Nang, goes on the air.
October 25	Saigon, the flagship station of the AFVN network, signs on although the **Blue Eagles** keep flying in order to cover areas over the delta farther South.
December 9	The Armed Forces News Bureau, a unit of AFRTS which supplies around the clock shortwave news and special events broadcasts to overseas affiliates, including sports, moves from New York to Arlington, Virginia, in a move to consoli-

December 23	date operations. There is an immediate outcry from numerous quarters that this is an attempt to manage the news. Saigon station takes small arms fire during the midst of its Christmas party. No casualties to people but the party dies.

1967

Spring	MACV, the Military Assistance Command Vietnam, orders AFRTS to submit for approval all quotes by certain politicians who are against the war. Among them: Wayne Morse, Robert Kennedy, Lyndon Johnson. At same time Secretary of Defense McNamara reaffirms his position that DoD policy is that there will be maximum disclosure with minimum delay. He insists on a free flow of information. Things improve for about a year but complaints to congress by the staff result in a number of high level investigations and low level morale.
Mid-year	Charles DeGaulle withdraws from the military arm of NATO and AFN is obligated to close its network of 60 stations throughout France. Headquarters of the Supreme Allied Commander of NATO forces moves to SHAPE, Belgium, near Mons, and AFN installs a station there which receives its network feed from Frankfurt.
August	The Armed Forces News Bureau renamed AFRTS-W, the "W" standing for Washington, and charges of possible news management again begin to fly. Charges soon die out when it becomes apparent that material distributed originates from commercial network and independent sources.
October 27	William Slatter, deputy news director of AFN Europe, in a speech claims news reports are censored by the military. Chief of Public Affairs in Europe denies charges. Investigative team, led by Bob Cranston, former AFN commander and now head of AFRTS under John Broger in Washington, finds no cases of censorship since the McNamara dictum.
December 11	The station at Nha Trang comes under enemy mortar attack. By good fortune, the broadcast facilities suffer no damage, but the mess hall and lounge are destroyed. No casualties.

1968

January 7	Da Nang's Red Beach radio transmitter is knocked out by enemy fire but operations are restored within two hours.
January 31	The AFVN station in Hue is attacked during the Tet Offensive. The staff holds off the North Vietnamese Regulars, who control most of the city, through the night and for the next five days. With one killed and every staff member wounded to some degree and with only 100 rounds of ammunition left, they break for safety. Sergeant John Anderson, station manager, and others are captured and spend next five years in North Vietnam prison camps.
May 3	AFVN headquarters building in Saigon suffers heavy damage from a car bomb containing an estimated 110 pounds of explosive.

1969

December 21-22	AFKN covers release of the Pueblo internees from North Korea, getting the only available video, and shares it with world-wide stations.

1970

April 10	An F-4 Phantom fighter, returning from a mission over Vietnam, crashes directly into the AFRTS station in Udorn, Thailand. Both crewmen are able to eject but nine AFRTS staff members die in the inferno as the station goes up in flames.
September	The **Blue Eagles,** no longer required, get both off the air and out of the air as they return to their base at Patuxent River.

1971

Secretary of the Army Robert Froehlike visits Europe and	hears little from the troops except that they all want television. He agrees and sets the wheels in motion to expand the Air Force limited system to all areas.
The Navy hires	Jordan "Buzz" Rizer to run its AFRTS operations. He is charged with operating and maintaining all outlets under Navy jurisdiction and establishing policy and guidance for its affiliates. This provides the template for the other broadcast services later to be formed by Army and Air Force. Rizer's first mission is to place closed circuit TV aboard all ships with crews of 350 or more. To do so he and his staff develop "SITE" (Shipboard Information, Training and Entertainment) systems which can broadcast from the typically small spaces aboard ship.

1972

All Year	U. S. Forces begin their drawdown in Vietnam, and so does AFVN. A number of stations close as troops move back to rear areas. Hue and Qui Nhon go off the air in February; Cam Ranh Bay in April; Nha Trang in June.

All year	Action begins to expand television in Europe to reach maximum numbers of troops and their families. First phase of the expansion is an Air Force project, in conjunction with European Command, Signal Corps, AFN and others.		**1976**
		May	Following the Mini-SITE systems, Navy develops the Sub-SITE, small enough to find space on a submarine. Testing starts on the USS **Lewis and Clark.**
	1973	July 1	The Secretary of the Navy establishes NBS, the Navy Broadcasting Service, under the direction of Buzz Rizer. Completing actions begun as early as 1972, Rizer consolidates complete control of personnel, funding and equipment under NBS. In theory at least, the Navy is now independent of control by local commanders.
Early year	In Vietnam, equipment from deactivated stations is turned over to the Vietnamese Government or returned to the U. S. Saigon is the last station to close.		
Early year	Television expansion continues in Europe and General Michael Davison announces he has decided that troops in the most isolated areas should get it first. Although a morale booster to these troops, it slows up the carefully planned expansion program which followed a logical geographic path for the microwave system. Meanwhile, the Air Force continues to telecast from studios at Ramstein Air Base to the slowly growing network.	October 28	AFN is finally scheduled to sign on at noon in full color from the completed Frankfurt studios. At 9 a.m. the German government asks that more time be given for testing. Ramstein has been gutted during the night and equipment moved to Frankfurt. Major Larry Pollack works a miracle and sign-on comes off as planned.
			1977
March 22	Message from AFVN Commander Col. Felix Casipit to AFRTS: "AFVN ceased to be as of 2400 hours 22 March 73."	All Year	Work continues on expansion of the AFN television system. Eventually a total of 300 communities or bases are served.
July 4	With Army television audiences now outnumbering Air Force viewers, control of the television system in Europe is transferred to AFN. System is still black and white and operates out of the same studios the Air Force had used in Ramstein, 90 miles from AFN's Frankfurt headquarters.		**1978**
			FEN moves from long-time headquarters at Camp Drake to new and modern facilities at Yokota Air Base from which it serves the entire Kanto Plain area.
		October	SATNET, the AFRTS satellite system, receives official sanction to begin planning for a 24-hour a day feed of news, sports, information and time-sensitive programming from the Broadcast Center in Los Angeles using real-time programming from network and commercial sources.
	1974		
May 7	In a rare honor, Bob Cranston is awarded the second highest military award, the Distinguised Service Medal, by President Nixon. He becomes the first broadcast officer to ever receive this award.		
December	The Navy has now installed SITE systems on larger ships and developed "Mini-SITES" for use on ships with fewer than 350 crewmen. The test is done on the USS **Miller.**	November	FEN at Yokota adds television. Japanese regulations require signals at all bases be closed circuit which are fed from Yokota to other affiliates by microwave.
		December	Television satellite service begins on a test basis to four sites in Alaska. Although not part of the test, Panama begins construction of an earth station, expedited to be in place before possible restrictions are placed on it under the terms of the Panama Canal Zone Treaty.
	1975		
All Year	AFN, while continuing to operate radio from Frankfurt and television from Ramstein, spends a miserable year rebuilding the Frankfurt headquarters to accept a full-color television operation. Staff works through jack hammer noises and constant dust as walls disappear, workmen tramp through studios and conditions deteriorate.		
			1979
		April	Panama receiving site completed. Dish can read a domestic satellite and provides some news and sports programs.

1980

Early	The Congress directs the Air Force and the Army to centralize the management of their AFRTS assets along the lines of the Navy in order to avoid unnecessary duplication of management functions. This is the result of a number of studies at various levels which express dissatisfaction with unstructured management. The services are not particularly happy with restructuring or placing control under central management.
March 4	The decision is made for the services. The Deputy Secretary of Defense issues a directive to reorganize. Army Broadcasting Service and Air Force Broadcasting Service are authorized and, with NBS, are given responsibility for operating their own stations while reporting upward to the Assistant Secretary of Defense for Public Affairs through the Armed Forces Information Service. The ASD(PA), acting through AFIS, now can approve or disapprove actions, direct the opening and closing of stations and the configuration of networks.
October 1	Both the Army Broadcasting Service and the Air Force Broadcasting Service are activated. ABS, like the Navy, headquarters in Washington. AFBS choses San Antonio, Texas, as home.

1982

Early	Los Angeles begins feeding material up to a U. S. domestic satellite for use by Panama, Guantanamo Bay, Cuba and Roosevelt Roads, Puerto Rico.
October 12	The Marines, having just gone ashore to help preserve the fragile peace in Lebanon, requests Navy Broadcasting Service provide radio and television.
December 15	A mobile broadcast unit is dispatched from Virginia to Lebanon by air.
December	Los Angeles Broadcast Center links up to the Atlantic Satellite with a signal footprint reaching from Diego Garcia in the Indian Ocean to Iceland. Diego Garcia becomes operational immediately.

1983

April	Ground facilities completed in Keflavik, Iceland, and station joins SATNET.
July 5	The station in Lebanon is attacked by mortar fire.
October 3	Three mobile stations are flown into Honduras in support of U.S. troops there.
October 22	Shortly after the AFRTS station in Lebanon is moved to the Marine compound for greater safety, a suicide bomber explodes a car bomb, killing 241 marines and seriously damaging the station.
November	Lajes, Azores, begins receiving live satellite broadcasts from Los Angeles.
Fiscal Year	AFRTS can now claim, on the basis of its reach, to be the world's largest network. With a budget of only $27.9 million, it serves 1.2 million active duty military in more than 100 countries plus the civilian component and military family members. The AFRTS programming center staff numbers only 96 civilians and 53 military.
Fall	Tests begin using the Pacific satellite to feed stations in the far East.
October	AFKN, Korea, goes on line with satellite reception.

1984

January	FEN, Clark, in the Philippines, joins SATNET.
February	AFN distribution system essentially completed and 11 meter receiving dish is installed. Network now capable of receiving and re-transmitting live programming from the States.
All Year	A new building has been obtained in Sun Valley, California, a suburb of Los Angeles, and work begins on a new home for the AFRTS Broadcast Center.

1985

February	Bermuda joins the SATNET family.
Fall	Satellite service to Okinawa and Japan begins on a test basis.

1986

Mid-Year	Almost all full-service land based television stations are now linked to Los Angeles by satellite. Programs are almost all transmitted "live" directly from the source and AFRTS personnel must cut out the commercials and insert AFRTS public service spots in real time.
Intermittent	In phases, various AFRTS departments move from the former drop forge factory in Hollywood, home since World War II, to the sparkling new broadcast facility in Sun Valley. No airtime is lost and all equipment has been installed and tested prior to the move.

1987

Fiscal Year	**Variety** reports the AFRTS total operating budget to be $27,400,000 this year and estimates that the programming received from all sources to be worth $117,000,000 in the commercial marketplace.

1988

May 20 — Colonel Tom Lewis, the man who shaped and molded AFRTS into the worldwide voice of home for millions of servicemen and women who served overseas, passes away at age 87.

1989

All satellite transmissions from AFRTS are now totally encrypted to prevent piracy of material by unauthorized sources. Industry cooperation remains outstanding and programming now costs the taxpayer approximately $3.50 for every $100 worth of programs (figured at commercial rates.)

1990

All Year — Government hiring freeze causes considerable difficulty in replacing personnel and most AFRTS operations work under severe personnel restraints. "Doing More With Less" becomes a key phrase, more honored in theory than practice.

Spring — Using a government-issued M-l Crystal Ball, AFIS orders T-ASA to build four completely equipped broadcast vans for contingency use.

August 2 — Iraq invades Kuwait

August 15 — AFRTS advises U. S. Central Command of the broadcast assets available and selects 17 broadcasters for possible deployment from a lengthy list of volunteers.

September 5 — First two mobile stations arrive in Dhahran.

September 12 — Broadcasts begin in Dhahran and Riyadh.

December 2 — Joint Services Command for the Armed Forces Desert Network (AFDN) formed, headquartered in Riyadh. Installation of additional stations continues.

1991

January 17 — Allied air strikes against Iraqi troops and positions begin. Program schedules are adjusted to give listeners a minimum of 30 minutes of news each hour. Some newscasts advise listeners that Scud missiles are falling on them, a fact readily apparent to them.

February 11 — Air Force Broadcasting Service assumes operational control of what is now named AFDN, the Armed Forces Desert Network.

Early — Mount Pinatubo in the Philippines erupts. Volcanic ash covers Clark Air Base rendering it unusable. Weight of the ash collapses the roof of the AFRTS station and destroys much of the equipment. Operations move to Subic Bay but the future of the stations remains questionable as the Philippine senate debates whether to renew the base leases.

February 27 — Kuwait City liberated.

February 28 — President George Bush calls for a halt to military offensive operations.

March 6 — AFDN signs on the air at noon from Kuwait. Installation of low powered stations throughout the area where U.S. troops remain continues. Because of requirement to observe strict Moslem laws concerning content of programming, network programs are heavy on news, sports and non-offensive musical selections.

September 16 — André Baruch, who brought radio into a combat zone for the first time during North African invasion of 1943, dies in Los Angeles.

1992

AFRTS celebrates its Golden Anniversary. From tiny, tentative beginnings, it has served its overseas audience wherever fate brought them. The 1942 baby has now earned the respect of its audience and its peers as, fifty years after its birth, it has developed into the largest -- at least geographically -- and most unique broadcast entity on earth.

★Each week it supplies some 300 outlets on every continent with:
- 80 hours of top rated videotaped television entertainment programs including series and feature films.
- 8-12 hours of videotaped family programming to areas which include families.
- sends 50-60 hours of television programs to about 180 totally isolated locations for playback in day rooms or mess halls.
- Using satellites in orbit around the world, sends out 24 hours daily of timely programming including live sports and newscasts from all major network sources. Viewers overseas can see events as they happen just like their friends and family at home. Missing, although not necessarily missed: hemorrhoid, diaper and deodorant commercials. In their place, audiences see and hear AFRTS contract-produced announcements from a constantly growing library of 3,500 television and 3,200 radio spots.
- On the satellite audio channels, AFRTS piggy-backs a voice channel and several music channels, including stereo, for use by overseas outlets.
- Operates "Superstations" in Germany, Italy and Turkey. These stations uplink their local program schedule to another satellite

and it is decrypted and rebroadcast at more than 400 isolated locations which heretofore could not receive a signal.
- Operates duplication facilities which can receive the satellite signal, duplicate it and get it to ships operating in their area. "Dupfacs" are located in New Zealand, Sicily, Bahrain, Spain, Diego Garcia, Subic Bay, and Puerto Rico. One operated in Jiddah, Saudi Arabia, during Desert Storm.
- Radio continues to be of prime importance. By its very nature it can react to breaking events more quickly. Currently AFRTS operates 400 radio outlets. Each manned station is sent a weekly package of approximately 80 hours of recorded programs, the latest albums and singles, a priority package of the hottest hits, a compact disk library. In addition, radio outlets have 24 hours of programming available to them on both International Maritime Satellite channels and audio channels on the television satellites giving them a wide choice of music, talk, news, sports and special events.

Fifty years! The technology has changed, but the AFRTS mission remains the same -- to bring the friendly voice of home to Americans overseas, to tell them what they need to know and to bring them the news as it happens without censorship or manipulation.

In his long career in the AFRTS system, during which he served with Army, Air Force and Navy — and is now boss of the whole shootin' match as head of American Forces Information Service — Buzz Rizer (L) received a whole chest full of decorations. He is seen here at a celebration at which the Hoechst Schlossgaard dropped by in full regalia to honor AFN. (Photo Courtesy of AFRTS)

At one of his last public appearances before his untimely death, the daddy of AFRTS — Colonel Tom Lewis — proudly shows off the award named after him. The Colonel Tom Lewis Award goes to the AFRTS staff member who has contributed the most to the worldwide system during the preceding year. (Photo Courtesy AFRTS)

❖ ❖ ❖ ❖

...t-t-t-t-that's all, folks!

❖ ❖ ❖ ❖